# New Paradigms in ENVIRONMENTAL BIOMONITORING USING PLANTS

# New Paradigms in ENVIRONMENTAL BIOMONITORING USING PLANTS

Edited by

**SUPRIYA TIWARI**

**SHASHI BHUSHAN AGRAWAL**

Department of Botany, Banaras Hindu University, India

ELSEVIER

Elsevier
Radarweg 29, PO Box 211, 1000 AE Amsterdam, Netherlands
The Boulevard, Langford Lane, Kidlington, Oxford OX5 1GB, United Kingdom
50 Hampshire Street, 5th Floor, Cambridge, MA 02139, United States



ISBN: 978-0-12-824351-0

For Information on all Elsevier publications visit our website at https://www.elsevier.com/books-and-journals

*Publisher:* Candice Janco
*Editorial Project Manager:* Aleksandra Packowska
*Production Project Manager:* R.Vijay Bharath
*Cover Designer:* Matthew Limbert



Typeset by Aptara, New Delhi, India

# Contents

# Contributors

**Madhoolika Agrawal**
Laboratory of Air Pollution and Global Climate Change, Department of Botany, Institute of Science, Banaras Hindu University, Varanasi, Uttar Pradesh, India

**Shashi Bhushan Agrawal**
Laboratory of Air Pollution and Climate Change, Department of Botany, Banaras Hindu University, Varanasi, Uttar Pradesh, India

**Timothy Ashibel Anake**
Department of Mathematics, College of Science & Technology, Covenant University, Ota, Nigeria

**Winifred Uduak Anake**
Department of Chemistry, College of Science & Technology, Covenant University, Ota, Nigeria

**Hamza Badamasi**
Department of Chemistry, Federal University Dutse, Dutse, Jigawa State, Nigeria

**Debasis Chakrabarty**
Council of Scientific and Industrial Research-National Botanical Research Institute (CSIR-NBRI), Lucknow, Uttar Pradesh, India; Academy of Scientific and Innovative Research (AcSIR), Ghaziabad, Uttar Pradesh, India

**Nivedita Chaudhary**
Department of Environmental Science, School of Earth Sciences, Central University of Rajasthan, Ajmer, Rajasthan, India

**Hethesh Chellapandian**
Department of Biotechnology and Microbiology, National College, Tiruchirappalli, Tamil Nadu, India

**Pulak Das**
School of Human Ecology, Ambedkar University Delhi, Delhi, India

**Sonali Dubey**
Institute of Biotechnology, Amity University, Noida, Uttar Pradesh, India

**Archana Dwivedi**
Ecotoxicology, Bioremediation and Nanotoxicology Laboratory, Department of Botany University of Lucknow, Lucknow, U.P., India

**Sanjay Dwivedi**
Council of Scientific and Industrial Research-National Botanical Research Institute (CSIR-NBRI), Lucknow, Uttar Pradesh, India; Academy of Scientific and Innovative Research (AcSIR), Ghaziabad, Uttar Pradesh, India

**Meenu Gautam**
Laboratory of Air Pollution and Global Climate Change, Department of Botany, Institute of Science, Banaras Hindu University, Varanasi, Uttar Pradesh, India

**Gereraj Sen Gupta**
Department of Botany, Institute of Science, Banaras Hindu University, Varanasi, Uttar Pradesh, India

**Didem Gökçe**
Department of Biology, İnönü University, Malatya, Turkey

**Subodh Kumar**
Ecotoxicology, Bioremediation and Nanotoxicology Laboratory, Department of Botany University of Lucknow, Lucknow, U.P., India

**Parvati Madheshiya**
Department of Botany, Institute of Science, Banaras Hindu University, Varanasi, Uttar Pradesh, India

**Srishti Mishra**
Laboratory of Air Pollution and Global Climate Change, Department of Botany, Institute of Science, Banaras Hindu University, Varanasi, Uttar Pradesh, India

**Oluwole Akinwumi Odetunmibi**
Department of Mathematics, College of Science & Technology, Covenant University, Ota, Nigeria

**Banita Kumari Paswan**
Ecotoxicology, Bioremediation and Nanotoxicology Laboratory, Department of Botany University of Lucknow, Lucknow, U.P., India

**Shiv Prasad**
Centre for Environment Science and Climate Resilient Agriculture, Indian Institute of Agricultural Research, New Delhi, India

**Dheeraj Rathore**
School of Environment and Sustainable Development, Central University of Gujarat, Gandhinagar, Gujarat, India

**Ansuman Sahoo**
Department of Botany, Institute of Science, Banaras Hindu University, Varanasi, Uttar Pradesh, India

**Manju Shri**
Gujarat Biotechnology Research Centre, Gandhi Nagar, Gujarat, India

**Anoop Singh**
Department of Scientific and Industrial Research (DSIR), Ministry of Science and Technology, Government of India, Technology Bhawan, New Delhi, India

**Aditya Abha Singh**
University Department of Botany, Babasaheb Bhimrao Ambedkar Bihar University, Muzaffarpur, Bihar, India

**Pradyumna Kumar Singh**
Council of Scientific and Industrial Research-National Botanical Research Institute (CSIR-NBRI), Lucknow, Uttar Pradesh, India

**Priyanka Singh**
Laboratory of Air Pollution and Climate Change, Department of Botany, Banaras Hindu University, Varanasi, Uttar Pradesh, India

**Jeyachandran Sivakamavalli**
Department of Biotechnology and Microbiology, National College, Tiruchirappalli, Tamil Nadu, India

**Supriya Tiwari**
Laboratory of Ecotoxicology, Department of Botany, Banaras Hindu University, Varanasi, Uttar Pradesh, India; Department of Botany, Institute of Science, Banaras Hindu University, Varanasi, Uttar Pradesh, India

**Rudra Deo Tripathi**
Council of Scientific and Industrial Research-National Botanical Research Institute (CSIR-NBRI), Lucknow, Uttar Pradesh, India; Academy of Scientific and Innovative Research (AcSIR), Ghaziabad, Uttar Pradesh, India

**Poornima Vajpayee**
Ecotoxicology, Bioremediation and Nanotoxicology Laboratory, Department of Botany University of Lucknow, Lucknow, U.P., India

# About the editors

Dr Tiwari is presently an Assistant Professor at the Department of Botany, Banaras Hindu University, India. Her groundbreaking research is focused on the antioxidant response of plants growing under the stress of ground level ozone and involves evaluating the function and adaptation of plants antioxidant machinery to cope with this stress. As well as producing 1 authored and 1 edited book, and over 30 publications, which have appeared in various well-reputed National and International journals, Dr Tiwari's research findings have gained recognition through National and International Awards, including the Young Scientist Medal of the Indian National Science Academy and the esteemed Green Talent Award, from the Ministry of Education and Research, Germany.

Prof Agrawal is Professor of the Department of Botany, Banaras Hindu University, India. He is an eminent and experienced researcher, actively involved in research related to stress biology and conducted field experiments to ascertain the effects of Ozone and UV-B on plants. He has worked as a Visiting Scientist at USDA, Maryland, USA and is the Fellow of National Academy of Science, a prestigious science academy of India. He has published more than 200 research papers, edited four books, completed several research projects and has guided 18 PhD students.

# Preface

Environmental biomonitoring using plants have been commonly practiced for many decades to assess the alterations occurring in the different ecosystems. However, in the last few years, the studies based on the concept of biomonitoring have gained special momentum, as they are cost-effective, sustainable, and environmental friendly compared to the traditional physiochemical methods. Although morphological parameters like foliar injury and biomass have been used as important indicators, recent years have witnessed the use of biomarkers and genomic parameters in biomonitoring studies. The present book not only discusses the traditional biomonitoring concepts but also highlights the newly developed biomonitoring techniques.

Chapter 1 emphasizes the use of the lichen-based index of atmospheric purity (IAP) to monitor the gaseous pollutants and heavy metals in the environment. Chapter 2 accentuates the perspectives and challenges faced while phytomonitoring of heavy metals and discusses the different morphological, anatomical, and physiological features of the plants that can be used in biomonitoring. Chapter 3 talks about biomonitoring in aquatic ecosystems by studying the changes in the primary producer community in response to the alteration of water quality and the significance of aquatic biomonitoring in providing valuable information related to environmental degradation. Chapter 4 discusses the use of tree species for biomonitoring of atmospheric trace element deposition at the urban site. In Chapter 5, the concept of chemoinformatics is used in establishing a relationship between environmental DNA (eDNA) and marine biodiversity. Chapter 6 brings new insights into the applications of DNA metabarcoding in ecological investigations and biomonitoring programs. Chapters 7 and 8 explicate the role of biomarkers as a biomonitoring tool in assessing environmental pollution impacts. Chapter 8 further highlights the significance of foliar injury as biological indicators to monitor ambient ozone levels. Chapter 9 highlights the importance of integrating different research tools along with common statistical models in selecting suitable plant species for biomonitoring of environmental pollution. Use of different indices like air pollution tolerance index (APTI), anticipated performance index (API), and dust capturing potential (DCP) was made in selecting sensitive plant species, which can be used as bioindicators. Chapter 10 has examined the genomic, proteomic, and metabolomic responses of the plants to arsenic stress and has extrapolated their use as biomarkers for biomonitoring of arsenic in the environment. Chapter 11 exemplifies the use of living organisms like algae, lichens, moss, and higher plants as bioindicators in the analysis of soil contaminated with organic and inorganic pollutants. In Chapter 12, metallic nanoparticles produced in response to metals and metalloids are hypothesized as promising tools for biomonitoring of heavy metals in the environment. Chapter 13 interprets the role of stress-induced chromosomal aberrations and DNA damage as biomonitoring tools for evaluating environmental pollution. Chapter 14 presents a comprehensive review of different biomonitoring tools and biomonitoring programs carried out around the globe.

The diligent and sincere efforts of the contributing authors, and their timely submission of the manuscripts are sincerely appreciated. Our profound thanks extend to all the learned reviewers for giving their critical suggestions on different chapters. We are also grateful to the Elsevier team, especially Aleksandra Packowska, for their pragmatic help and cooperation during compilation and submission of this book.

Last, our heartfelt thanks encompass to all our well-wishers and family members for their moral support, blessings, and inspiration during the publication of this book.

Supriya Tiwari
Shashi Bhushan Agrawal

CHAPTER 1

# Lichen-based index of atmospheric purity (IAP) for biomonitoring of air

**Pulak Das**
School of Human Ecology, Ambedkar University Delhi, Delhi, India

## 1.1 Introduction

Lichens are symbiotic association between fungus and algae. They have special characteristics of absorption of gases, water, and nutrients directly from the air, which makes them an ideal organism for air *biomonitoring*; a term used for application of biological organisms to monitor air quality, and changes in surrounding environment. Lichens absorb metals from the environment which uses modes such as intracellular absorption through an exchange process, intracellular accumulation, and entrapments of particles that contain metals (Richardson, 1995). During the late 1960s and early 1970s, a lichen-based mathematical formula (an Index) was introduced as "index of atmospheric purity" (IAP) in Canada. IAP, denoted as a number, uses an algorithm consisting lichen community parameters. The index is useful in characterizing the air quality in different settings such as urban areas, peri-urban areas, rural areas, industrial areas, forested areas, geothermal areas, etc. This index is regularly applied since then, sometimes with modifications, in different parts of the world. IAP (all versions) has been primarily used in biomonitoring of air pollutants such as $SO_x$, $NO_x$, and heavy metals in the atmosphere. In most of the studies, zonation of land (lichen zonation) is also done around a pollution source using IAP values. These zones indicate gradient of air quality from central zone (sometimes called as lichen desert) which are highly polluted, to the outer most zone indicating least polluted area. Initially there were two IAP methods proposed by LeBlanc and De Sloover (1970) (shown below as equations Eqs. (1.1) and (1.2)); using parameters such as number of species, frequency-abundance, cover-frequency, along with toxiphoby, and companion factor or sociability index or ecological index

*New Paradigms in Environmental Biomonitoring Using Plants*
DOI: https://doi.org/10.1016/B978-0-12-824351-0.00003-1

(LeBlanc and De Sloover, 1970; Gombert et al., 2005; Ghennam, 2021). Both toxiphoby and companion factor are denoted by Q. Toxiphoby can be understood as the degree of sensitivity to a given environmental stressor. Companion factor considers the number of associated lichens with a particular lichen in any tree or substrate. Frequency indicates the abundance of lichen patches in a quadrat and cover indicates the size of a patch in terms of its radius, length, breadth, etc. Biomonitoring using IAP assumes that frequency, cover, and companionship of lichens are affected to a large extent by the surrounding environment such as air quality, substrate quality, and climatic condition. IAP, proposed initially by LeBlanc and De Sloover (1970) ($IAP_b$ in Eq. (1.2)) used cover and frequency (F) where it ranged between 1 and 5 (1 being very rare and very low degree of cover, and 5 being abundant species with high degree of cover on most trees). Q is calculated using Eq. (1.3). All lichen species up to a height of 2 m from the ground level for 10–12 trees is considered for the calculation.

$$IAP_a = \frac{n}{100} \times \left( \sum_{i=1}^{n} \frac{Q \times f}{n} \right) \tag{1.1}$$

Q is the index of toxiphoby which is empirically derived from different mapping studies in Europe, f is frequency abundance of each species, n is number of species.

$$IAP_b = \sum_{i=1}^{n} \frac{Q \times f}{10} \tag{1.2}$$

$$Q = \frac{\sum_{j=1}^{m} \sum_{i=1}^{n} S_{ij}}{m} \tag{1.3}$$

Q is companion factor, n is the number of species; m is the number of trees where the species of interest are present and $S_{ij}$ equals 1 if species i is present at station j.

$IAP_b$ based on companion factor and frequency-cover, is used to study temporal air quality changes between 1968 and 1990 (LeBlanc and De Sloover, 1970; Beckett, 1995) in Sudbury, Canada. The method thenonwards is used by many authors with modifications also (e.g., Das et al., 2013). $IAP_b$ method is reported to be capable of indicating lichen desert, an area devoid of lichens. Originally, the idea of dividing any region into lichen desert, struggle zone, and normal zone was of Sernander (1926). The 'desert' mentioned in the study was observed to shrunk by around 20% in 6 years. It is important to note that during these 6 years there were reduction in major emissions and systematic plant closure in and around the study area. Lichen species in each zone (total 5 zones were demarcated) also experienced jumps. The next study (after 16 years) showed that lichen desert completely disappeared. Zones 2 and 3 experienced reduction in sizes and limited to pollution sources such as smelters. As suggested by Conti and Cecchetti (2001), IAP is one of the ways of bioindication, "*method-A*," used worldwide for mapping air quality of any area. There were around twenty different formulae (some are mentioned below) for IAP calculation reported worldwide (Kricke and Loppi, 2002), of which the one using frequency (also termed by some as $IAP_{18}$) ($IAP_i$ in Eq. 1.10) and the other using frequency-cover and companion factor (by LeBlanc and De Sloover, 1970) ($IAP_b$ in (Eq. 1.2)) is most widely used in different literature. Some researchers reported certainty of up to 97% of predicting pollution using the former one (e.g., Lo Porto et al., 1992; Gottardini et al., 1999; Gombert et al., 2004). The present paper reports a few more formulae used by different authors (as mentioned in Table 1.1).

$$IAP_c = \sum_{i=1}^{n} (Q \times C) \tag{1.4}$$

$$IAP_d = \sum_{i=1}^{n} (Q \times C \times F) \tag{1.5}$$

$$IAP_e = \sum_{i=1}^{n} \left( \frac{Q \times C \times F}{V \times S} \right) \tag{1.6}$$

$$IAP_f = \sum_{i=1}^{n} C \tag{1.7}$$

$$IAP_g = \sum_{i=1}^{n} \left( \frac{C \times F}{V \times S} \right) \tag{1.8}$$

$$IAP_h = \sum_{i=1}^{n} \left( \frac{Q \times C}{V \times S} \right) \tag{1.9}$$

$$IAP_i = \sum_{i=1}^{n} F \tag{1.10}$$

$$IAP_j = \sum_{i=1}^{n} (Q \times F) \tag{1.11}$$

$$IAP_k = \sum_{i=1}^{n} Q \tag{1.12}$$

Where Q is companion factor, C is cover, F is frequency, V is vitality, S is damage. The numbers on the right are equation numbers.

**Table 1.1** Classification criteria for IAP zones in some selected studies.

| Criteria 1 | | Criteria 2 | | Criteria 3 | | Criteria 4 | | Criteria 5 | | Criteria 6 | |
|---|---|---|---|---|---|---|---|---|---|---|---|
| IAP values | IAP Zones | IAP values | IAP Zones | IAP values | IAP Zones | IAP values | IAP Zones | IAP values | IAP Zones | IAP values | IAP Zones |
| <0 | Zone I | 0–12.5 | Level A | ≤6 | Zone I | 0–2 | Zone I | 0–1 | Poor | 0–7 | Class 1 |
| 1–24 | Zone II | 12.5–25 | Level B | 6–16 | Zone II | 2–6 | Zone II | 1–3 | Fair | 7–13 | Class 2 |
| 25–49 | Zone III | 25–37.5 | Level C | 16–24 | Zone III | 6–10 | Zone III | 3–5 | Good | 13–17 | Class 3 |
| 50–74 | Zone IV | 37.5–50 | Level D | 24–31 | Zone IV | 10–14 | Zone IV | 5–7 | Very Good | 17–21 | Class 4 |
| 75–99 | Zone V | >50 | Level E | 31–38 | Zone V | >14 | Zone V | | | 21–29 | Class 5 |
| >99 | Zone VI | | | >38 | Zone VI | | | | | 29–32 | Class 6 |
| | | | | | | | | | | 32–54 | Class 7 |

IAP examines the effects of a pollutant source on lichen communities, and the utility of IAP is proper when pollution gradient is high (Will-Wolf, 1988). It is a quantitative phytosociological approach requiring the collection of data such as frequency and/or percentage cover and a factor of tolerance to toxicity (Mulgrew and Williams, 2000). Gombert et al., (2004) compared the IAP with another index (index of human impact—IHI). The IHI takes into consideration of traffic, parking sites, housing sites, green areas, isolated trees, etc. They observed that such anthropogenic activities are not impacting the air quality to a large extent. Ghennam et al., (2021) also studied the relation between IAP with IHI, and respiratory disease such as asthma. Along with the "Q value," cover "r" is used in the formula of IAP. The study found a negative correlation of IAP with pollution, and asthma which was statistically significant.

## 1.2 Approaches to IAP based on frequency (only), and companion factor

IAP method using frequency (only) has advantage over other methods, as Q values need not have to be assigned for each species (Monaci et al., 1997). Calculation of Q is a cumbersome method. Talking about Q, Herben and Liška (1986) observed that it depends on frequency and distribution of species, total number of species, and study sites number and their distribution. Low Q indicates low number of companion species, i.e., it only occurs in species poor vegetation in a polluted area (Loppi, 1996). Q may be associated with certain errors especially when lichen species diversity is less (Wirth, 1988). In a project in Switzerland in the 1980s, a total of 10 different IAP formulas (which includes Eqs. 1.4–1.12) were applied using parameters mentioned for Eqs. (1.1) and (1.2) above, and additionally, vitality and damage of lichens (Kricke and Loppi, 2002). In the late 1980s and early 90s, IAP using frequency ($IAP_i$ in Eq. (1.10)) was used as a national guideline in Italy for monitoring gases such as $SO_2$ and $NO_x$. Originally, Kunze (1971) proposed the use of lichen frequency on single tree species which formed the basis of $IAP_{18}$ used in Swiss project, as national guideline for monitoring atmospheric pollution using lichens. $IAP_{18}$ divides the region into five lichen zones (which corresponds with respective air emission zones); lichen desert (critical air pollution), inner struggle zone (high air pollution), outer struggle zone (medium air pollution), transition zone (low air pollution), and normal zone (very low air pollution) (Herzig et al., 1989). The project proposed significant correlation between $IAP_{18}$ and pollution. The $IAP_{18}$ is also encouraged due to its simplification, rapidity, and objectivity in assessment of using lichen formula as IAP. Frequency is also used by Ammann et al., (1987) with slight modifications where one to four trees were selected at each location and a 30 × 50 cm grid was placed on the tree trunk between 120 and 200 cm. The method is found to be effective for pollutants such as $NO_x$, $SO_2$, and CO, and is used in areas of Italy by many authors during the 1990s. IAP—with—frequency is also used in areas with geothermal activity using a 30 × 50 cm grid at a height between 120 and 200 cm of the bole having highest density (Loppi, 1996). IAP calculated using frequency (only) method (with maximum ten sites) may be used to divide the region into five levels (Kommission Reinhaltung der Luft im VDI und DIN, 1995; Conti and Cecchetti, 2001; Conti, 2008), based on IAP values; Level A (IAP: 0–12.5; very high level of pollution), level B (12.5–25; high level of pollution), level C (25–37.5; moderate level of pollution), level D (37.5–50; low level of pollution), and level E (> 50; very low level of pollution). The grid used for this study is 30 × 50 cm.

## 1.3 IAP-like other methods

IAP-like methods with similar formula (using frequencies) but with different names are also used in many studies. For example, The German VDI (Verein Deutcher Ingenieure) guideline 3799 (VDI, 1995) was similar to the $IAP_i$, and is based on establishing a grid over the area to be surveyed, whose units would vary according to the availability of suitable tree species and the size of the area to be surveyed. Three to six trees conforming to established criteria are selected in each grid square for recording lichens. Another method, lichen biodiversity index (LBI) was used in Italy (ANPA, 2001), where each sampling unit is represented by a square of 250 m × 250 m and three to six trees per sampling unit is selected. A sampling grid of 50 cm × 30 cm is positioned on the part of the stem at a height of 100 cm, and frequency of lichens is noted. European guidelines (Asta et al., 2002) as lichen diversity values (LDV) is a standardized method to assess frequency of occurrence of lichen species on a defined portion of tree bark as an estimate of diversity, and as a parameter to estimate the degree of environmental stress.

Fatima et al., (2019) observed significant correlation between $IAP_b$ and LDV, along with species diversity and abundance. Here the IAP ranged between 16.19 and 79.82 while LDV ranged between 12.5 and 52.16. Trass (1973) calculated index of poleotolerance using lichens in Estonia.

Poleotolerance indicates tolerance to human disturbance, where lichen features such as frequency, bark cover, and their vitality are taken into consideration. Moore (1974) suggested that IAP method is restrictive because area to be mapped must be homogeneous, and choice of stations should be ecologically homogeneous, and therefore he suggested index of lichen abundance, to eliminate the influence of different substrates on IAP (Estrabou et al., 2011). It is calculated using the number of lichen species, a cover scale chosen by the investigator, degree of resistivity of each species (somewhat related to Q in IAP), and degree of substrate tolerance of each species or average number of companion species escorting a particular species on a particular substrate type.

## 1.4 IAP is evolving

Jeran et al., (2002) used IAP in an entirely different way in Slovenia ($IAP_l$ in Eq. 1.13). The method used cover and abundance of three morphological forms; crustose, foliose, and fruticose. The cover (c) is assigned values such as 0 (devoid of lichens), 1 (thalli up to 10% of tree surface), 2 (between 10% and 50% of surface), and 3 (between 50% and 100% of surface). The abundance (a) is also assigned values as 0 (devoid of lichens), 1 (1–5 lichens), 2 (6–10 lichens), and 3 (>10 lichens). IAP is calculated for three different heights from the ground (<0.5 m; 0.5–2.5 m; and > 2.5 m) as C (a + c) + F (a + c) + R (a + c) (elaborated in Eq. 1.13 below). The IAP results were then ranked in five classes (0; 1.3–13.5; 13.6–27.0; 27.1–40.5; 40.6–54.0) and correlated with trace elements measured separately with instrument. The authors observed that for Cr, Zn, Sr, Hf, As, and Mo, there was a distinct tendency to lower values with increasing IAP values. Although they also noted that lower loadings for IAP in factor analysis indicate that trace elements (such as As, Cr, Cd, Mo, and Zn) only partly explains the cause of poor lichen vegetation.

$$IAP_l = IAP_1 + IAP_2 + IAP_3 \tag{1.13}$$

1, 2, and 3 are different heights at <0.5, 0.5–2.5, and >2.5 m respectively from the ground and IAP at each height is calculated as C (a + c) + F (a + c) + R (a + c) where C is for crustose, F for foliose, and R for fruticose lichens; a is abundance and c is cover.

Gombert et al., (2004) modified $IAP_b$, and instead used only cover (r) (with companion species), ($IAP_m$ in Eq. 1.14) following Braun-Blanquet (1964) for cover estimate, which ranged as follows: 0.5 (for less than

1% cover), 1 (1%–5% cover), 2 (6%–25% cover), 3 (26%–50% cover), 4 (51%–75% cover), 5 (for more than 75% cover).

$$\mathrm{IAP_m} = \frac{1}{10}\sum_{i=1}^{n}(Q_i \times r_i) \tag{1.14}$$

where Q is companion species and r is cover.

Käffer et al., (2011) suggested that use of environmental classification factor (ECF) is to be complemented with IAP, specifically in areas with high lichen species richness. The ECF considers different forms such as crustose, foliose, and fruticose as a scale, considering their respective sensitivities towards pollution; crustose being more tolerant and fruticose being more sensitive towards pollution. ECF is calculated as addition of individual cover scales for all three morphological groups and multiplying it with IAP/100. Based on ECF values the region is divided into five zones; lichen absent (IAP: 1.0–5.5), lichen poor (5.6–15.5), transition area (15.6–35.5), normal (35.6–75.5), and excellent (>75.6). It is reported that incorporation of ECF has led to reclassify some of the "excellent" areas into "poor" or "transition zones," which according to them is the case in real, as evidenced by presence of built-up, vehicular pollution, thermal inversion, and/or decrease in vegetation.

Dymytrova (2009) used a modified IAP (details provided in table 1.2) based on Kondratyuk (1994). This modified IAP, interestingly termed as $IAP_m$, uses companion species and cover, where cover is given values such as 1 (<1%), 2 (1%–20%), 3 (21%–50%), and 4 (51%–100%). It was observed that $IAP_m$ showed higher correlation with air pollution than the IAP. In his formula, $a_{ij}$ indicates the coverage class of «i» species, $b_{ij}$ is the occurrence of species with the coverage of «j» class and "m" is number of the coverage classes of «i» species. Tanona and Czarnota (2020) allotted frequency-coverage score "f" for calculating IAP (Eq. 1.15 below). "f" for each species is calculated by compiling the constancy class (cc) of occurrence and the patch cover (pc) in the analyzed plot of sampled trunks (I: 20%; II: 21%–40%; III: 41%–60%; IV: 61%–80%; V: 81%–100%). "f" coefficient values are calculated with index of patch cover and constancy class as follows—1: (pc < 50 & cc = 1); 1: (pc < 2.5 & cc = 2); 2 (pc > 2.5 & cc = 2); 2 (pc < 5 & cc = 3); 3 (pc > 5 & cc = 3); 2 (pc < 2.5 & cc = 4); 3 (pc: 2.5–25 & cc = 4); 4 (pc > 25 & cc = 4); 3 (pc < 2.5 & cc = 5); 4 (pc: 2.5–25 & cc = 5); 5 (pc > 25 & cc = 5).

$$\mathrm{IAP} = \frac{1}{10}\sum_{i=1}^{n}(Q \times f) \tag{1.15}$$

where Q is ecological sensitivity or companion species and f is frequency-coverage.

The ecological monitoring and assessment network (EMAN), a Canadian national monitoring network mission developed an IAP method based on 18 epiphytic macrolichen. In this frequency-based method, lichens are collected from trees between 30 and 150 cm from around four to five trees in each location of a grid system and the values are used to develop contours using GIS (MacDonald and Coxson, 2013).

## 1.5 IAP, air pollutants, and heavy metals

IAP is used for monitoring number of air pollutants including heavy metals. Herzig et al., (1987) observed that IAP calculated using summation of frequencies are highly correlated with Zn, Cl, Cd, Pb, Cu, $NO_x$, and $SO_2$. Neidoni et al., (2018) observed in a study in Romania that low IAP values indicated areas with high atmospheric pollution whereas high IAP values indicated areas with improved air quality. They divided the study area into different qualitative grades or evaluation scale (low, moderate, high, and very high pollution) based on both IAP and heavy metals. Total amount of heavy metals such as Ni, Zn, Mn, Cu, Pb, Cd, and Cr, was calculated and categorized based on metal quantity for zonation; from low pollution zone (100–200 mg/kg of dry matter) to very high pollution zone (≥401 mg/kg of dry matter). IAP was also categorized from low pollution zone (IAP: 21–30) to very high pollution zone (IAP: 0–5). Although out of seven study sites three sites matched with each other while four sites did not match, overall it was concluded that lichens were good bioindicators of heavy metals in the air, and they can be used for identification of heavy metals in the air samples. Beckett (1995) observed in a study that IAP zones of air quality is directly related with sulfur or copper content of *Parmelia sulcata* lichen collected from trees in each zone in Sudbury, Canada. Zocchi et al., (1997) in a study in Italy observed high correlation between IAP values with $SO_2$ and $NO_x$ concentration measured by devices. Jayalal et al., (2017) in a study in Sri Lanka observed that high IAP values coincided with low $SO_2$ concentration in the ambient air; the formula involved both companion factor and frequency-coverage. Yatawara and Dayananda (2019) observed that IAP is an indication of air pollutants, where they found that as one goes from rural to urban areas in Sir Lanka, the IAP decreases whereas $SO_2$ and $NO_2$ increases. Agnan et al., (2017) in a study in France although observed that IAP which is based on frequency is a better indicator of metal pollution as opposed to IAP calculated using companion factor and frequency-coverage both. Tanona and Czarnota (2020) in a study in Poland also observed similar results applying IAP calculated through companion factor and frequency-coverage. They found that there are weak relations between IAP and pollutants but they also indicated that IAP certainly reflects overall ecological functioning in an ecosystem such as forest. Out of many elements, IAP shows a significant (negative) correlation with copper because it reflects environmental levels of phytotoxic gaseous pollutants such as $SO_2$ and $NO_x$ rather than those of trace elements, which are less phytotoxic and are mainly associated with airborne particulates (Monaci et al., 1997). Herzig et al., (1989) in a study in Switzerland observed that $IAP_i$ is positively correlated with calcium, and negatively correlated with group of elements including lead, iron, copper, chromium, total sulfur, zinc, and phosphorus. Elements such as aluminum, boron, potassium, magnesium, molybdenum, sodium, nickel, tin, and chlorine did not show any clear gradient with increasing $IAP_i$. Rhzaoui et al., (2015) observed a positive correlation between most of the heavy metals with IAP in a study in Morocco. Cislaghi and Nimis (1997) reported high correlation of sum of frequencies of lichen with certain pollutants such as $SO_2$, $NO_3$, dust, and $SO_4^{-2}$. The "r" value was 0.93, 0.87, 0.86, and 0.85, respectively; $P < 0.01$ in all cases. IAP is more related to phytotoxic gaseous pollutants such as $SO_2$ and $NO_x$, as compared with trace elements (Monaci et al., 1997). Attanayaka and Wijeyaratne (2013) in a study in Sri Lankan city observed that IAP is negatively correlated with $SO_2$ and $NO_2$, and IAP gradually increased with distance from the city. In another study in Sri Lanka, Jayalal et al., (2017) observed that low IAP corresponded with higher level of $SO_2$ in continuous forest as opposed to forest islands. Overall, they observed a high value of IAP (54.22) in Horton plains national park. The index is also related with metals, and other lichens also exhibited significant correlation with instances of lung cancer in Veneto region of Italy. There are findings in some studies which may not necessarily conform with above observations, for example Granger (1970), and Case and Krouse (1980) observed that the IAP values calculated by using $IAP_b$ did not relate clearly with $SO_2$. This is supported by Tulumello (2010) also who mentioned that since the IAP is a weighted measure of the relative abundance of lichen species in a sample set, the link between IAP and any one stressor is entirely unknown. The same author is of the opinion that inclusion of nitrophytes may increase the IAP values giving a misleading impression of good air quality. Agnan et al., (2017) is of the view that indices such as IAP do not point to the exact pollutants caused by disturbance, and therefore both qualitative and quantitative characterization is important together. While comparing LDV which they considered a measure of lichen abundance, and IAP (a measure of lichen richness), it was observed that IAP is not a good indicator of metal pollution, eutrophication, and pH. Some authors also view that IAP method may indicate about the presence of air pollution but not about origin and type of pollutant (Jeran et al., 1996). There are some studies which are indicating that IAP is not suitable for cities as many cities are experiencing decline in $SO_2$ (Gombert et al., 2004). Conversely, Badin and Nimis (1996) observed that IAP map correlates with land use and natural vegetation but not with climatic condition, and therefore is another reason for using IAP for pollution monitoring only. They also observed that high spatial density of recording instruments is a constraint to air quality mapping. Altitude is also an important component need to be taken into consideration before site selection. Regarding altitude, Monaci et al., (1997) observed that the heterogeneous altimetry of a town in Italy does not explain air quality as observed from IAP.

## 1.6 IAP, zones, and mapping

Mapping pollution using IAP method is an attractive approach facilitating easy identification of areas with disturbance gradients (Yatawara and Dayananda, 2019), and may be used in different situation (Conti, 2008); as a baseline study to know the existing condition, to monitor the ongoing compromised condition, and in restoration by controlling the quality. IAP is being used to delineate zones of air quality, for example, Estrabou et al., (2011) divided air quality based on IAP values as poor (0–1), fair (1–3), good (3–5), and very good (5–7), and suggested that IAP results are difficult to compare between sites. Results may vary (Table 1.1) between different areas and different studies; for example, it ranged from 5.9 to 71.7 by Gombert et al., (2004) in France, whereas it was in the range of 1–3 in a study by Stamenkovic et al., (2010) in Serbia.

Categorization of zones is not always based on same values, as either approach used is different or formula may have slight variations. Käffer and Martins (2014) in a study in Brazil delineated five zones; zone 1 (IAP: 1–5.5), zone 2 (5.6–15.5), zone 3 (15.6–35.5), zone 4 (35.6–75.5), and zone 5 (>75.6), as lichen absent, lichen poor, transition area, normal, and excellent. Zocchi et al., (1997) observed in a study that IAP values lower than 10 coincided with population density, industrial activity and vehicular movement while higher IAP coincided with mountainous area, and this is also confirmed by correlation between IAP, $SO_2$, and $NO_x$ concentration recorded by devices. Loppi (1996) used IAP to create zones (following Herzig and Urech, 1991) around geothermal plants. The steps involved are; calculating the mean IAP and the standard deviation of all IAP values; adding half the standard deviation to and subtracting half the standard deviation from the mean in order to obtain a pair of values defining the first zone; using the standard deviation as the interval between zones, with zero as the lower limit and the upper limit open. The IAP values were divided into four zones A) 0–25, B) 26–45, C) 46–65, D) >65.

Around a paper mill in India, Das et al., (2013) observed five zones based on IAP; zone 2 (IAP: 1–24), zone 3 (25–49), zone 4 (50–74), zone 5 (75–99), and zone 6 (IAP > 99). Zone 1 (or lichen desert) was absent around the paper mill. The IAP values ranged between 17 and 113. The study indicated a relation between zone shapes with land use, other pollution sources, and complex affinities of lichen assemblages with any particular pollutant. Stamenković et al., (2010, 2013, 2016) mentioned about air pollution zones as categorized using IAP such as normal zone, struggle zone, and lichen desert zone. Within a span of a decade the normal zone got converted into struggle zone, while struggle zone changed into lichen desert zone. IAP ranged between 1 and 31 (Dimitrovgrad); 3 and 32 (Blace); and 0 and 47 (Leskovac, Vlasotince, Lebane) in studies in Serbia. Djekić et al., (2020), also in a study in Serbia, observed that the IAP values ranged between 40 and 56 which as per the criteria of Conti and Cecchetti (2001) indicates that level of air pollution is low or very low. Gombert et al., (2004) defined five IAP categories of equal intervals (15); category 1 (IAP: 0–<15, very poor air quality), category 2 (15–<30, poor air quality), category 3 (30–<45, medium air quality), category 4 (45–<60, good air quality), and category 5 (≥60, very good air quality).

## 1.7 Standardization, application, and prospects of IAP

The present paper lists around 35 usages (Table 1.2) of IAP across the globe in last over 25 years from 1995 to 2021. There are different ways and formulae for calculating IAP. Different criteria are adopted to collect lichen species from trees and different ranges of IAP are observed.

Two primary formulae used to calculate IAP include either summation of frequency or multiplication of companion factor and frequency and/or coverage values (and their summation). Along with pollution, IAP may help in identifying species richness within forest, continuity of ecological processes, stand dynamicity (Tanona and Czarnota, 2020), and rural–urban ecosystems (Yatawara and Dayananda, 2019). It may be used

**Table 1.2** IAP used as tool for biomonitoring by different authors in over last two decades.

| Author | Formula of IAP | Study area | Phorophyte species | No. of lichen species | Total area of study ($km^2$) | Selection criteria for lichens | Height of collection (m) | Locations | Number of phorophytes in each location | Q | IAP value |
|---|---|---|---|---|---|---|---|---|---|---|---|
| Loppi and Corsini (1995) | $\sum_{1}^{n} F$ | Central northern Italy | One species (*Tilia platyphyllos*) | 34 | ns | 30 × 50 cm grid, divided into 10 units of 10 × 15 cm, part of the bole with highest lichen density | 1.2–2 | 26 | 1–4 | ns | 6–67 |
| Badin and Nimis (1996) | $\sum_{1}^{n} F$ | North east Italy | Multiple species | 60 | 473 | 30 × 50 cm grid, divided into 10 units of 10 × 15 cm, part of the bole with highest lichen density | >1.5 | 104 | ns | ns | <5–> 61 |
| Loppi (1996) | $\sum_{1}^{n} F$ | Central Italy | One species (*Quercus pubescens*) | 52 | 15 | 30 × 50 cm grid, divided into 10 units of 10 × 15 cm | 1.2–2 | 51 | 1–5 | ns | 0–82 |
| Monaci et al. (1997) | $\sum_{1}^{n} F$ | Central Italy | One species (*Tilia cordata*) | 46 | 6.8 | ns | ns | 25 | ns | ns | 9–54 |

*(Continued)*

**Table 1.2** Cont'd

| Author | Formula of IAP | Study area | Phorophyte species | No. of lichen species | Total area of study ($km^2$) | Selection criteria for lichens | Height of collection (m) | Locations | Number of phorophytes in each location | Q | IAP value |
|---|---|---|---|---|---|---|---|---|---|---|---|
| Zocchi et al. (1997) | ns | North Italy; South Switzerland | Two species (Tilia and *Quercus*) | 99 | 6500 | ns | ns | 425 | ns | ns | 10–70, <6–>38, >0–> 14 |
| Loppi (2000) | $\Sigma Fi$ | North Italy | Multiple species | ns | ns | Sampling grid of 50 × 30 cm, subdivided into 10 units of 10 × 15 cm | 1.5–2 | 2 | 1–5 | ns | <5–<35 |
| Jeran et al. (2002) | [C(a+c) + F(a+c) + R(a+c)] | Slovenia | Multiple species | ns | 20000 | ns | 0–0.5, 0.5–2.5, >2.5 | 546 | 6 | ns | 0–54 |
| Gombert et al. (2004) | $\frac{1}{10}\sum_{i=1}^{n}(Q_i \times r_i)$ | South east France | Multiple species | 83 | ns | All patches | >0.8 | 198 | 5 | 9–23 | 5.9–71.7 |
| Dymytrova (2009) | $\frac{1}{10}\sum_{i=1}^{n}(Q_i \times f_i)$ $\sum_{i}^{n}\frac{Q_i}{10}\sum_{j}^{m}\frac{a_{ij} \times b_{ij}}{m}$ | North Ukraine | Two species (*Tilia cordata* Mill. and *Acer platanoides* L.) | 65 + 20 bryophytes | 839 | All species recorded | <1.5 | 272 | >5 | ns | 0–173.8, 1.6–82 |

| Author | Formula of IAP | Study area | Phorophyte species | No. of lichen species | Total area of study ($km^2$) | Selection criteria for lichens | Height of collection (m) | Locations | Number of phorophytes in each location | Q | IAP value |
|---|---|---|---|---|---|---|---|---|---|---|---|
| Stamenković et al. (2010) | $\sum_{1}^{n} F$ | South east Serbia | Four species (*Acer negundo, Juglans regia, Prunus domestica, Tilia platyphyllos*) | 19 | ns | ns | ns | 13 | ns | ns | 1–31 |
| Tulumello (2010) | $\frac{1}{10}\sum_{i=1}^{n}(Q_i \times f_i)$ | Sarnia, Canada | Four species (*Acer platanoides, Acer rubrum, Acer saccharinum, Acer saccharum*) | 11 | 63 | Line intercept method | 0.5–1.5 | 63 | 8 | | 2.61–19.68 |
| Käffer et al. (2011) | $\sum n \frac{(Q \times f)}{10}$ | South Brazil | Multiple species | 131 | 497 | All patches | 0.5–1.5 | 30 | 10 | ns | 19.9–434.2 |
| Estrabou et al. (2011) | $\sum_{1}^{n} F$ | Central Argentina | Three species | 9 | 576 | Grid of 20 × 20 cm placed on the south-western side of the trunk | 1.5 | 341 | 3 | ns | 0–7 |

*(Continued)*

**Table 1.2** Cont'd

| Author | Formula of IAP | Study area | Phorophyte species | No. of lichen species | Total area of study (km$^2$) | Selection criteria for lichens | Height of collection (m) | Locations | Number of phorophytes in each location | Q | IAP value |
|---|---|---|---|---|---|---|---|---|---|---|---|
| Das et al. (2013) | $\frac{1}{10}\sum_{i=1}^{n}(Q_i \times f_i)$ | North east India | One species (*Artocarpus heterophyllus*) | 53 | 1800 | All patches | 0–2 | 17 | 5 | 8–38 | 17–113 |
| Attanayaka and Wijeyaratne (2013) | $\sum_{i}^{n}\frac{Q \times f}{10}$ | West Sri Lanka | Three species (*Cocos nucifera*, *Mangifera indica* and *Artocarpus heterophyllus*) | 47 genera | ns | Four 25 × 10 cm quadrat randomly placed/patches > 3 mm diameter | 0.5–1.5 | 31 | 8 | ns | 24–86.3 |
| Stamenković et al. (2013) | $\Sigma Fi$ | South eastern Serbia | Multiple species | 25 | ns | ns | 1.5–2 | 24 | 5 | ns | 3–22 |
| Vitarana (2013) | $\sum_{1}^{n}F$ | South west Australia | Live trees not used | 20 | 1685 | Ten fallen logs or old tree stumps/A cumulative area of 20 × 20 cm was sampled on each log/stump by using either a 10 × 10 cm grid or/and a 20 × 5 cm grid | >0.1 (in case of stumps) | 36 | 10 | ns | 21–58 |

| Author | Formula of IAP | Study area | Phorophyte species | No. of lichen species | Total area of study ($km^2$) | Selection criteria for lichens | Height of collection (m) | Locations | Number of phorophytes in each location | Q | IAP value |
|---|---|---|---|---|---|---|---|---|---|---|---|
| Käffer and Martin (2014) | $\frac{1}{10}\sum_{i=1}^{n}(Q_i \times f_i)$ | South Brazil | Multiple species | 193 | ns | Acetate method (20 × 20 cm transparent plastic) on two sides of trunk (N & S) | 0.5–1.5 | 12 | 5 | ns | 8.1–64.2 |
| Rhzaoui et al. (2015) | $\frac{1}{10}\sum_{i=1}^{n}(Q_i \times f_i)$ | North east Morocco | Two species, (*Quercus* and *Pinus*) | 44 | ns | ns | ns | 11 | 5 | ns | 2–24 |
| Stamenković et al. (2016) | $\frac{1}{10}\sum_{i=1}^{n}(Q_i \times f_i)$ | South Serbia | ns | 53 | ns | ns | 1.5–2 | 142 | ns | ns | 0–47 |
| Jayalal et al. (2016) | $\sum_{i}^{n}\frac{Q \times f}{10}$ | South Korea | Two species (*Quercus* sp. and *Pinus japonica*) | 65 | ns | Transparency lichen grid (50 × 50 cm) | 1.5 ± 0.3 | 3 | 20 | ns | 30.9–53.9 |
| Jayalal et al. (2017) | $\sum_{i}^{n}\frac{Q \times f}{10}$ | South Sri Lanka | Multiple species | 379 | ns | Using 20 × 20 cm grid | 0–2 | 12 | 1/3rd of all trees in a plot of 1000 $m^2$ | ns | 21.7–63.7 |

*(Continued)*

**Table 1.2** Cont'd

| Author | Formula of IAP | Study area | Phorophyte species | No. of lichen species | Total area of study ($km^2$) | Selection criteria for lichens | Height of collection (m) | Locations | Number of phorophytes in each location | Q | IAP value |
|---|---|---|---|---|---|---|---|---|---|---|---|
| Agnan et al. (2017) | $\frac{1}{10}\sum_{i=1}^{n}(Q_i \times f_i)$ | France | Multiple species | 92 | Entire France (7 sites) + 1 site | 4 cardinal points using ladder grid of 5 vertical square (10 × 10 cm) | >1 | 8 | 12 | ns | 94–263 |
| Khairuddin et al. (2017) | $\Sigma Fi$ | South Malaysia | One species (*Roystonea regia*) | 7 | ns | Grid of 30 × 50 cm subdivided into 10 rectangles/ Lichens facing vehicle transport | 1–1.3 | 4 | 8 | ns | ns |
| Firdous et al. (2017) | $\frac{1}{10}\sum_{i=1}^{n}(Q_i \times f_i)$ | Pakistan | Multiple species | 42 | ns | 25 × 10 cm quadrat on the trunk of each tree | 0.5–1.5 | 25 | 10–12 | 2–42 | 38–145 |
| Neidoni et al. (2018) | $\sum \frac{Q \times f}{10}$ | West Romania | Multiple species | ns | Entire city | All patches | 1–1.5 | 7 | 2–4 | ns | <5–<30 |
| Muhammad et al. (2018) | $\Sigma Fi$ | South Malaysia | One species (*Roystonea regia*) | ns | ns | Using 30 × 50 cm grid | 1–1.3 | ns | 8 | ns | ns |

| Author | Formula of IAP | Study area | Phorophyte species | No. of lichen species | Total area of study ($km^2$) | Selection criteria for lichens | Height of collection (m) | Locations | Number of phorophytes in each location | Q | IAP value |
|---|---|---|---|---|---|---|---|---|---|---|---|
| Yatawara and Dayananda (2019) | $\frac{1}{10}\sum_{i=1}^{n}(Q_i \times f_i)$ | Southwest Sri Lanka | Two species (*Mangifera indica*, *Cocos nucifera*) | 89 | ns | 20 × 20 cm transparent quadrat was placed randomly at north and south directions | 1.5 | 27 | 6 | ns | 17–90.26 |
| Alecsić et al. (2019) | $\Sigma Fi$ | North Kosovo | Two Species (*Aesculus hippocastanum* and *Picea abies*) | 16 | ns | Using 30 × 50 cm grid | 1 | 1 | 6 | ns | 34.67, 38.16 |
| Lopes et al. (2019) | $\sum\left(\frac{Qi \times fi}{10} \times C_i\right)$ | South east Brazil | Multiple species | 21 | 289 | Reticulated square (20 × 20 cm), subdivided into 100 reticles/east face of tree | > 1.2 | 4 | 5 | ns | 7.91–31.45 |
| Fatima et al. (2019) | $\frac{1}{10}\sum_{i=1}^{n}(Q_i \times f_i)$ | North Algeria | Multiple species | 62 | Ns | Transparent grid (20 × 50 cm) consisting of 10 squares of 10 × 10 cm each on the most species-rich trunk faces | 1–1.5 | 34 | ns | ns | 16.19–79.82 |

(Continued)

Table 1.2 Cont'd

| Author | Formula of IAP | Study area | Phorophyte species | No. of lichen species | Total area of study (km$^2$) | Selection criteria for lichens | Height of collection (m) | Locations | Number of phorophytes in each location | Q | IAP value |
|---|---|---|---|---|---|---|---|---|---|---|---|
| Tanona and Czarnota (2020) | $\frac{1}{10}\sum_{i=1}^{n}(Q \times f)$ | South Poland | Spruce | 61 | 70 | Around the trunk | < 2 | 39 | 10 | ns | 14.5–56.2 (1993), 6.6–50.7 (2013), 9.2–62.2 (2018) |
| Correa-Ochoa et al. (2020) | $\frac{1}{n}\sum_{i=1}^{n}(Q_i \times fi \times Ci)$ | North Columbia | Multiple species | 148 | ns | 100 × 50 cm acetate sheet | 0.5 | 5 | 28–30 | <20–<120 | >0–>500 |
| Djekić et al. (2020) | $\sum_{1}^{n} F$ | South Serbia | One species | 72 | 18.1 | A sampling ladder measuring 50 × 10 cm | 1–1.5 | 15 | 5 | ns | 40–56 |
| Ghennam et al. 2021 | $\frac{1}{10}\sum_{i=1}^{n}(Q_i \times r_i)$ | North Algeria | One species (*Olea europaea*) | 35 | 200 | ns | ns | 111 | 10 | ns | 0–27.93 |

ns, not specified.

to assess ecological conditions in natural areas such as forests (but not in the forests with poor air quality (Tanona and Czarnota, 2020)), as other conditions such as humidity, competition for space, species of trees, etc., also play important roles (Gibson et al., 2013). In a study in South Poland, it is observed that the increase in IAP in 5 years in a forest is attributed to change in dynamicity in tree stand due to bark beetle outbreaks and windstorms (Tanona and Czarnota, 2020). IAP method is also used for forest decline inventories (Jeran et al., 2002). Monaci et al., (1997) in a study in Italy observed that four lichen species *Physcia adscendens, Physconia grisea, Lecidella elaeochroma,* and *Hyperphyscia adglutinata* determined to a large extent the IAP values for the entire study area (r = 0.80). This means that if limited number of lichens can be identified, it may help in quick and reliable estimate of air quality next time. Some studies have also observed low IAP along with presence of nitrophilous lichens (e.g., *Physcia undulata* and *P. endochryscea*), and suggested that presence of such species relates with IAP and bad air quality (due to $NO_2$) (Estrabou et al., 2011). Tulumello (2010) mentions about line intercept method for IAP calculation. The method sums the widths of all lichen thalli that made total contact with a painter's tape that was wrapped around the tree at 1.5 m in height. Percent lichen cover is calculated using (Circumference – Total Line Intercept)/100. The IAP values were divided into three categories; low (0–6), medium (7–12), and high (>12), based on natural breaks in data. IAP is calculated by multiplying Q (resistance factor), C (relative cover), and F (frequency). Cover is a vital component for calculating IAP. Recent studies are using advanced technologies to calculate correct lichen cover, for example in a study in Columbia the cover is estimated with acetate films which are digitized as TIF format without compression (Correa-Ochoaa et al., 2020). Site selection with respect to substrate in case of IAP is very important (Showman, 1988). Selection of phorophytes is also one of the important components of IAP to maintain homogeneity of substrate for lichens to grow. Selection of same tree or phorophyte ensures homogenous observation of lichens, else it needs to be seen that certain criteria are fulfilled such as bark character, DBH, and trunk angle from ground; although this is considered to be one of the challenges of IAP method (Conti and Cecchetti, 2001). Along with this, general bark condition, age of the tree, and bark pH should also be taken into consideration. Calculation of IAP with similar bark characteristics is emphasized, but if different phorophytes are used to calculate IAP, method suggested by Correa-Ochoaa et al., (2020) may be applicable, where ten random combination of eight phorophytes were generated for each zone resulting into ten IAP values in each location.

## 1.8 Conclusion

The IAP method was never out of criticism (Lopes, 2010). The main reason for this was the inclusion, in the same unique index, of species with contradictory responses to the stressor of interest (Kricke and Loppi, 2002). We need to be careful about the limitations of using IAP. For example, it is time consuming (Kularatne, 2012), requires lichen taxonomic understanding (Kularatne, 2012; Conti, 2008), and refers to specific group of environmental pollutant (Conti, 2008). Some researchers also suggest that IAP may not distinguish between impact on lichens of pollution and of climate (Wirth, 1988). Considering the fact that some authors believe, even if it has remained popular it is not an effective indicator of air quality (e.g., MacDonald and Coxson, 2013), IAP method has a high predictive power for ambient levels of mixtures of phytotoxic pollutants (Liebendörfer et al., 1988). Recent studies indicate that IAP is a diversity distribution technique which is economical and allow complex sampling grid (Abas, 2021). IAP helps us to understand changes in air quality to group of different environmental contaminants (Conti, 2008) and the method is the most commonly adopted in some European countries (Conti, 2008). But, it also requires regular validation and its usage has to be adapted to local conditions and lichen biota (Tanona and Czarnota, 2020). Finally, it is important that there should be streamlining of formulae for IAP calculation with more studies. Two primary and predominant ways of IAP use may be encouraged further; frequency based IAP ($IAP_F$) and companion factor based IAP ($IAP_Q$). Preliminary studies indicate that the former is more reliable for pollutants while the latter are useful for ecosystem functioning. More studies are required in different climatic conditions and different parts of the world to further validate this.

## References

Abas, A., 2021. A systematic review on biomonitoring using lichen as the biological indicator: a decade of practices, progress and challenges. Ecol. Indic. 121, 107197.

Agnan, Y., Probst, A., Nathalie, S.D., 2017. Evaluation of lichen species resistance to atmospheric metal pollution by coupling diversity and bioaccumulation approaches: a new bioindication scale for French forested areas. Ecol. Indic. 72, 99–110.

Alecsić, G., Stamenković, S., Ristić, S., Marković, M., 2019. Epiphytic lichens in the town of Zvečan and their bioindicators value. Publ. Nat. Sci. 9 (2), 1–5.

Ammann, K., Herzig, R., Liebendörfer, L., Urech, M., 1987. Multivariate correlation of deposition data of 8 different air pollutants to lichen data in a small town in Switzerland. Adv. Aerobiol. 51, 401–406.

ANPA (Agenzia Nazionale per la Protezione dell'Ambiente), 2001. I.B.L. Indice di Biodiversita Lichenica. Manuale ANPA. Dipartimento Stato dell'Ambiente, Controlli e Sistemi Informativi.

Asta, J., Erhardt, W., Ferretti, M., Fornasier, F., Kirschbaum, U., Nimis, P.L., Purvis, O.W., Pirintsos, S., Scheidegger, C., Van Haluwyn, C., Wirth, V., 2002. European Guideline for Mapping Lichen Diversity as an Indicator of Environmental Stress. British Lichen Society, London.

Attanayaka, A.N.P.M., Wijeyaratne, S.C., 2013. Corticolous lichen diversity, a potential indicator for monitoring air pollution in tropics. J. Natn. Sci. Foundation Sri Lanka 41 (2), 131–140.

Badin, G., Nimis, P.L., 1996. Biodiversity of epiphytic lichens and air quality in the province of Gorizia (NE Italy). Studio Geobotanica 15, 73–89.

Beckett, P.J., 1995. Lichens: sensitive indicators of improving air quality. In: Gunn, J.M. (Ed.), Restoration and Recovery of an Industrial Region. Springer Series on Environmental Management. Springer, New York, NY. https://doi.org/10.1007/978-1-4612-2520-1_6.

Braun-Blanquet, J., 1964. Pflanzensoziologie, Grundzüge der Vegetationskunde, 3rd ed. Springer, Wien-New York, pp. 865.

Case, J.W., Krouse, H.R., 1980. Variations in sulphur content and stable sulphur isotope composition of vegetation near a $SO_2$ source at Fox Ceek, Alberta, Canada. Oecologia 44 (2), 248–257.

Cislaghi, C., Nimis, P., 1997. Lichens, air pollution and lung cancer. Nature 387, 463–464. https://doi.org/10.1038/387463a0.

Conti, M.E., 2008. Lichens as bioindicators of air pollution. In: Conti, M.E. (Ed.). Biological monitoring: Theory & Applications. WIT Transactions on State of the Art in Science and Engineering, 30. WIT Press, pp. 111–162. doi:10.2495/978-1-84564-002-6/05.

Conti, M.E., Cecchetti, G., 2001. Biological monitoring: lichens as bioindicators of air pollution assessment—a review. Environ. Pollut. 114, 471–492.

Correa-Ochoa, M.A., Vélez-Monsalve, L.C., Saldarriaga-Molina, J.C., Jaramillo-Ciro, M.M., 2020. Evaluation of the index of atmospheric purity in an American tropical valley through the sampling of corticulous lichens in different phorophyte species. Ecol. Indic. 115, 106355.

Das, P., Joshi, S., Rout, J., Upreti, D.K., 2013. Lichen diversity for environmental stress study: application of index of atmospheric purity (IAP) and mapping around a paper mill in Barak Valley, Assam, northeast India. Trop. Ecol. 54 (3), 355–364.

Djekić, T., Ristić, S., Stamenković, S., Šajn, R., Engelman, M., 2020. Lichens as indicators of air quality in Balneological center Prolom Banja (Southern Serbia). J. Geogr. Inst. Cvijic 70 (2), 101–113.

Dymytrova, L., 2009. Epiphytic lichens and bryophytes as indicators or air pollution in Kyiv city (Ukraine), Folia Crypto. Estonica Fasc. 46, 33–44.

Estrabou, C., Filippini, E., Soria, J.P., Schelotto, G., Rodriguez, J.M., 2011. Air quality monitoring system using lichens as bioindicators in central Argentina. Environ. Monit. Assess. 182, 375–383.

Fatima, A, Messaoud, R, Takia, L., 2019. Relationship between lichen diversity and air quality in urban region in Bourdj Bou Arriridj. Algeria. Biodiversitas 20, 2329–2339.

Firdous, S.S., Naz, S., Shaheen, H., Dar, M.E.U.I., 2017. Lichens as bioindicators of air pollution from vehicular emissions in district poonch, Azad Jammu and Kashmir, Pakistan. Pak. J. Bot. 49 (5), 1801–1810.

Ghennam, K., Attou, F., Abdoun, F., 2021. Impact of atmospheric pollution on asthma and bronchitis based on lichen biomonitoring using IAP, IHI and GIS in Algiers Bay (Algeria). Environ. Monit. Assess. 193, 198.

Gibson, M.D., Heal, M.R., Li, Z., Kuchta, J., King, G.H., Hayes, A., Lambert, S., 2013. The spatial and seasonal variation of nitrogen dioxide and sulfur dioxide in Cape Breton Highlands National Park, Canada, and the association with lichen abundance. Atmos. Environ. 64, 303–311.

Gombert, S., Asta, J., Seaward, M.R.D., 2003. Correlation between the nitrogen concentration of two epiphytic lichens and the traffic density in an urban area. Environ. Pollut. 123 (2), 281–290.
Gombert, S., Asta, J., Seaward, M.R.D., 2004. Assessment of lichen diversity by index of atmospheric purity (IAP), index of human impact (IHI) and other environmental factors in an urban area (Grenoble, sourtheast France). Sci. Total Environ. 324 (1–3), 183–199.
Gombert, S., Asta, J., Seaward., M.R.D., 2005. The use of autecological and environmental parameters for establishing the status of lichen vegetation in a baseline study for a long-term monitoring survey. Environ. Pollut. 135, 501–514.
Gottardini, E., Cristofolini, F., Marchetti, F., 1999. Biomonitoraggio della qualità dell'aria della città di Trento tramite licheni epifiti. Acqua & Aria 30 (4), 67–71.
Granger, D., 1970. A study of the link between the index of atmospheric purity and epiphytic lichens. Water Air Soil Pollut. 30, 23–27.
Herben, T., Liška, J., 1986. A simulation study on the effect of flora composition, study design and index choice on the predictive power of lichen bioindication. Lichenologist 18 (4), 349–362. doi:10.1017/S0024282986000543.
Herzig, R., Liebendörfer, L., Urech, M., 1987. Flechten als Bioindikatoren der Luftverschmutzung in der Schweiz: Methoden-evaluation und Eichung mit wichtigen Luftschadstoffen. VDI-Berichte 60, 619–639.
Herzig, R., Liebendörfer, L., Urech, M., Ammann, K., Cuecheva, M., Landolt, W., 1989. Passive biomonitoring with lichens as a part of an integrated biological measuring system for monitoring air pollution in Switzerland. Intern. J. Environ. Anal. Chem. 35, 43–57.
Herzig, R., Urech, M., 1991. Flechten als Bioindikatoren, integriertes biologisches Messsystem der Luftverschmutzung für das Schweizer Mittelland. Bibliotheca Lichenol. 43, 1–283.
Jayalal, U., Oh, S.O., Park, J.S., Sung, J.H., Kim, S.H., Hur, J-S., 2016. Evaluation of air quality using lichens in three different types of forest in Korea. Forest Sci. Technol. 12 (1), 1–8. doi:10.1080/21580103.2014.1003983.
Jayalal, U., Ileperuma, O., Wolseley, P.A., Wijesundara, D.S.A., Karunaratne, V., 2017. Correlation of atmospheric purity index to the diversity of lichens in the Horton Plains National Park, Sri Lanka. Ceylon J. Sci. 46 (2), 13–29.
Jeran, Z., Jaćimović, R., Batić, F., Mavsar, R., 2002. Lichens as integrating air pollution monitors. Environ. Pollut. 120, 107–113.
Jeran, Z., Jaćimović, R., Smodiš, B., Batić, F., 1996. The use of lichens in atmospheric trace element deposition studies in Slovenia, Phyton. Ann. Rei Botanicae 36 (3), 91–94.
Käffer, M.I., Martins, S.M.D.A., 2014. Evaluation of the environmental quality of a protected riparian forest in Southern Brazil. Bosque (Valdivia) 35 (3), 325–336.
Käffer, M.I., Martins, S.M.D.A., Alves, C., Pereira, V.C., Fachel, J., Vargas, V.M.F., 2011. Corticolous lichens as environmental indicators in urban areas in southern. Brazil Ecol. Indic. 11 (5), 1319–1332.
Khairuddin, N.A., Muhammad, N., Hashim, N.H., Yusof, H., Jusoh, S., Abas, A., Talip, B.A., Abdullah, N., Din, L.B., 2017. Identification of most tolerant lichen species to vehicular traffic's pollutants at Batu Pahat area, The 2nd International Conference on Applied Science and Technology 2017 (ICAST'17) AIP Conf. Proc., 1891, 020078-1–020078-5. https://doi.org/10.1063/1.5005411.
Kommission Reinhaltung der Luft im VDI und DIN, 1995. Messen von Immissionswirkungen. Measurement of Immission Effects. Dusseldorf.
Kondratyuk, S.Ya., 1994. Lichen indication mapping of air pollution in Ukraine. Ukrainskyi Botanichnyi Zhurnal 51 (2–3), 148–153.
Kricke, R., Loppi, S., 2002. Bioindication: The IAP approach. In: Nimis, P.L., Scheidegger, C., Wolsley, P.A. (Eds.), Monitoring With Lichens—Monitoring Lichens. Kluwer Academic Publishers, Dordrecht, the Netherlands, pp. 21–37.

Kularatne, I., 2012. Performance of Lichen Species as Air Pollution Biomonitors in the Greater Auckland Urban Area, New Zealand PhD Thesis. The University of Auckland, New Zealand.

Kunze, M., 1971. Emittentenbezogene Flechtenkartierung auf Grund von Frequenzuntersuchungen. Oecologia 9, 123–133.

LeBlanc, F., De Sloover, J., 1970. Relation between industrialization and the distribution and growth of epiphytic lichens and mosses in Montreal. Can. J. Bot. 48, 1485–1496.

Liebendörfer, L., Herzig, R., Urech, M., Ammann, K., 1988. Evaluation und kalibrierung der Schweizer flechten-indikationsmethode mit wichtigen luftschadstoffen. Staub Reinhalt. Luft. 48, 233–238.

Lo Porto, A., Macchiato, M., Ragosta, M., 1992. Bioindicazione della qualità dell'aria tramite licheni epifiti nella provincia di Potenza. Acqua e Aria 1 (92), 11–18.

Lopes, P.A.P., 2010. Modeling Lichen Communities: Ecological Key Factors in a Changing Environment. Doutoramento em Biologia, Especialidade de Ecologia, Departamento de Biologia Vegetal Universidade de Lisboa, pp. 129.

Lopes, R.d.S., Marins, M.P., Oliveira, R.L., Amaral, F.P.G.d., 2019. Monitoring air pollution with living organisms. Case study use of lichens as bioindicators in the Miguel Pereira city, Rio de Janeiro, Brazil. Chem. Eng. Transact. 74, 253–258.

Loppi, S., 1996. Lichens as bioindicators of geothermal air pollution in Central Italy. Bryologist 99 (1), 41–48.

Loppi, S., 2000. Lichen biomonitoring as a tool for assessing air quality in geothermal areas, Proceedings World Geothermal congress, Kyushu - Tohoku, Japan May 28 – June 10.

Loppi, S., Corsini, A., 1995. Lichens as bioindicators of air quality in Montecatini Terme (central northern Italy). Ecol. Mediterr. 21 (3/4), 87–92.

MacDonald, A., Coxson, D.S., 2013. Lichen biomonitoring in the Prince George Area - Literature Review and Approaches for Future Studies. Report for the Fraser Basin Council 1st Floor, 470 Granville Street Vancouver, BC V6C 1V5.

Monaci, F., Bargagli, R., Gasparo, D., 1997. Air pollution monitoring by lichens in a small medieval town of central Italy. Acta Bot. Neerl. 46 (4), 403–412.

Moore, C.C., 1974. A modification of the "Index of Atmospheric Purity" method for substrate differences. Lichenologist 6, 156–157.

Muhammad, N., Hashim, N.H., Khairuddin, N.A., Yusof, H., Jusoh, S., Abas, A., Talip, B.A., Abdullah, N., Din, L., 2018. Identification of most tolerant lichen species to vehicular traffic's pollutants: a case study at Batu Pahat. J. Adv. Res. Fluid Mech. Therm. Sci. 42 (1), 57–64.

Mulgrew, A., Williams, P., 2000. Biomonitoring of Air Quality Using Plants, WHO Collaborating Centre for Air Quality Management and Air Pollution Control. WHO CC, Berlin. ISSN 0938-9822.

Neidoni, D.-G., Andres, L., Nicorescu, V., Lehr, C.B., Bucur, E., Sinitean, A., 2018. The use of corticolous lichen species to assess the level of air pollution with heavy metals, International symposium "The environment and the industry", SIMI 2018, Proceedings Book, Section Pollution Control and Monitoring, 442–450.

Rhzaoui, G.E., Divakar, P.K., Crespo, A., Tahiri, H., 2015. Biomonitoring of air pollutants by using lichens (*Evernia prunastri*) in areas between Kenitra and Mohammedia cities in Morocco. Lazaroa 36, 21–30.

Richardson, D.H.S., 1995. Metal uptake in lichens. Symbiosis 18, 119–127.

Sernander, R., 1926. Stockholm's Nature. Almqurst and Wiksells. Uppsala.

Showman, R., 1988Mapping air quality with lichens, the North American experience. In: Nash, T.H., Wirth, V., Cramer, J. (Eds.), Lichens, Bryophytes and Air Quality. Berlin.

Stamenković, S., Ristić, S.S., Dekić, T.L., Mitrović, T.L.J., Baošić, R.M., 2013. Air quality indication in Blace (Southern Serbia) using lichens as bioindicators. Arch. Biol. Sci. Belgrade 65 (3), 893–897.

Stamenković, S., Cvijan, M., Arandjelović, M., 2010. Lichens as bioindicators of air quality in Dimitrovgrad (South-eastern Serbia). Arch. Biol. Sci. Belgrade 62 (3), 643–648.
Stamenković, S., Djekić, T., Ristić, S., Novković, V., Mitrović, T., Marković, M., 2016. Air quality lichen monitoring at three selected urban areas in the Southern Serbia. Biol. Nyssana 7 (1), 19–29.
Tanona, M., Czarnota, P., 2020. Index of atmospheric purity reflects the ecological conditions better than the environmental pollution in the Carpathian forests. J. Mountain Sci. 17 (11). https://doi.org/10.1007/s11629-0206266-1.
Trass, H., 1973. Lichen sensitivity to the air pollution and index of poleotolerance (I. P.). United States: N. p., Web.
Tulumello, P.M., 2010. The Use of Lichens as Indicators of Ambient Air Quality in Southern Ontario MSc Thesis. Department of Earth Sciences, Bock University, St. Catharines, Ontario.
V. D. I., 1995. Richtlinie 3799. Blatt 1: Ermittlung und Beurteilung phytotoxischer Wirkungen von Immissionen mit Flechten: Flechtenkartierung. Düsseldorf.
Vitarana, M.C., 2013. Lichens as a Biomonitoring Tool for Detecting Heavy Metal Air Pollution Associated With Industrial Activities in Collie, South-Western Australia. Edith Cowan University. PhD Thesis. https://ro.ecu.edu.au/theses/679.
Will-Wolf, S., 1988. Quantitative approaches to air quality studies. In: Nash, T.H., Wirth, V., Cramer, J. (Eds.), Lichens, Bryophytes and Air Quality. Berlin.
Wirth, V., 1988. Phytosociological approaches to air pollution monitoring with lichens. In: Nash, T.H., Wirth, V., Cramer, J. (Eds.), Lichens, Bryophytes and Air Quality. Berlin, Stuttgart, pp. 91.
Yatawara, M., Dayananda, N., 2019. Use of corticolous lichens for the assessment of ambient air quality along rural–urban ecosystems of tropics: a study in Sri Lanka. Environ. Monit. Assess. 191, 179. https://doi.org/10.1007/s10661-019-7334-2.
Zocchi, A., Casarini, P.G., Guidetti, L., Roella, V., Borlandelli, C., Stefanetti, M.V., 1997. Air quality monitoring, using epiphytic lichens, in some northern-Italian areas (Lombardy and Piedmont). Lagascalia 19 (1-2), 505–512.

# CHAPTER 2

# Monitoring of airborne heavy metal using plants: Perspective and challenges

**Anoop Singh[a], Shiv Prasad[b], Dheeraj Rathore[c]**
[a]Department of Scientific and Industrial Research (DSIR), Ministry of Science and Technology, Government of India, Technology Bhawan, New Delhi, India
[b]Centre for Environment Science and Climate Resilient Agriculture, Indian Institute of Agricultural Research, New Delhi, India
[c]School of Environment and Sustainable Development, Central University of Gujarat, Gandhinagar, Gujarat, India

## 2.1 Introduction

Increasing atmospheric pollution is a serious global predicament affecting not only the living being but also the material. It influences economic efficiency, decreases agricultural productivity, damages property and modifies the ecology of the area which in turn increases the hazard of environmental disasters. Over expanding population and to fulfil the need of expanding population has worsened the situation. While air pollution signals are distinctly evident in large and megacities (Karagulian et al., 2015; Gargiulo et al., 2016), smaller cities, towns, and even small human settlements in remote are also experiencing these signals as well (Chaparro et al., 2020). Gases and particles released by all types of combustion into the air include sulfur and nitrogen oxides, carbon monoxide and soot particles, toxic metals, organic molecules, and radioactive isotope (Agbaire and Esiefarienrhe, 2009). Particles of varying size and shape, added to the air by natural and anthropogenic means and remained in the air for varying lengths of time depending on their size and weight. Particulate matter (PM) in the atmosphere with aerodynamic diameter < 10 μm diameter ($PM_{10}$) or < 2.5 μm diameter ($PM_{2.5}$) are of great apprehension for public health due to the presence of various pollutants, including heavy metals. While crustal material, suspended road dust, and long-range transport events are primarily identified as sources of the particles (Park and Kim, 2005; Vallius et al., 2005), vehicular emissions, agricultural activities generate local dust concentrations close to the source, which surpass environmental guideline values (Leys et al., 1998; Manins et al., 2001).

*New Paradigms in Environmental Biomonitoring Using Plants*
DOI: https://doi.org/10.1016/B978-0-12-824351-0.00006-7

PM is a complex mixture of the suspended solid and liquid particle in semi equilibrium with surrounding gases. Asian dust aerosol is the major mineral aerosol transported from central Asia to the Pacific, even to the western coast of America (Duce et al., 1991; Arimoto et al., 1997) and carries not only mineral-rich particles but also the anthropogenic pollutants from the eastern industrialized cities in Asia (Fu et al., 2010). Major pollutants suspended in the air include smoke, metal, and inert dust. Nakajima and Aryal (2018) reviewed the characteristics of urban dust and reported increased background concentration of heavy metal in and around the urban area due to dust and has been causing a severe threat to human and ecosystem health. Uptake of toxic metals present in soil by vegetables poses the potential health risks via foods' dietary consumption. The toxic elements in minerals, such as Cd and Pb in pyrite, pyrrhotite, and chalcopyrite (Zhuang et al. 2009) and Cr in the city and industrial areas, cause the highest health quotient (Liu et al., 2013; Zhao et al., 2018; Luo et al., 2018), and have potential health risks.

The use of plants as monitors of air pollution has long been recognized as plants are the initial acceptors of air pollution. They act as the scavengers for many airborne particulates in the atmosphere (Joshi and Swami, 2009). Garty (2001) recommended biomonitoring as a complementary or an alternative method for extensive detailed surveys whenever on-site use of instrumentation is costly or impractical. He has also explained biomonitoring as advantageous, which can substantiate the physical and chemical data by physiological data referring to this material's viability. Recently anthropogenic biodiversity of urban roadside plants is now increasingly being realized as an eco-sustainable tool for mitigation and monitoring air pollution. Various investigators have used these common roadside plant species to study the impact of PM pollutants and heavy metals on leaf morphology (stomata), biochemical (heavy metals, protein, and sugars) parameters, and enzyme activity (peroxidase and catalase), which are very useful in monitoring of these pollutants (Rai, 2015; Chaudhary and Rathore, 2018a, b, 2019).

## 2.2 Heavy metal in the air

According to Toy et al. (2002) and Field et al. (2010), dust is fine particulate material, removed from the land surface by wind erosion, and is small enough to be suspended in the atmosphere. Since the development of sophisticated instrumental techniques, dust composition has been explored and identified as combining minerals with metal pollutants based on the source. Dust is a mixture of solid particles and liquid droplets, varying microscopic in size, composition, and origin. According to USEPA (2013), air-suspended PM can be generated from either natural or manmade sources and can be composed of acid, organic, metal, soil, or mineral dust. The size of airborne particles ranges over a broad scale, from 0.005 μm to 100 μm in aerodynamic diameter, although the suspended portion is generally less than 40 μm (Nagar et a., 2014).

Additionally, fly ash and soot particles from coal combustion and other carbon-based combustion processes such as diesel fuel and wood burning add to the dust flux (Merefield and Stone, 1998). Heavy metals are elements having high densities, specific gravities, or atomic weights (Duffus, 2002) and were reported in air samples of different countries such as Brazil (Silvia et al., 2004), Denmark (Karl et al., 2002), Portugal (Vasconcelos and Tavares 1997), India (Khillare et al., 2004), Spain (Moreno-Grau et al., 1997, 2000; Mateu et al., 1999), Egypt (Abdel-shafy et al., 1992), Jordan (Gharaibeh et al., 2010), Greater Stockholm (Johansson et al., 2008), and Russia (Drobyshev and Emelina, 2001). Mn, Cu, Zn, Cd, Cr, Fe, Ni, K, Ca, Va, Ba, As, Se, and Sr are the most commonly found metals in the pollution sources and have been studied widely (Papoola et al., 2018). Two independent studies conducted by Godoi et al. (2008) and Quiterio et al. (2004) on PM in Brazil identified elements, including Zn, Cu, Cr, Co, Ca, Mg, K, Na, Al, Fe, Mn, Si, S, Cl, Ti, Pb, Cd in air-suspended dust.

### 2.2.1 Source of heavy metal

The sources of atmospheric trace elements in the atmosphere can be divided into two groups: (1) natural sources, e.g., sandstorms and volcanic eruptions, and (2) anthropogenic sources, e.g., industrial exhaust, fossil fuel combustion, and traffic exhaust emissions (Liu et al., 2013; Jiao et al., 2021). Heavy metals are in consideration of many studies since they are persistent and belong to the most widely dispersed industrial pollution. The significant contribution of many elements in the atmosphere is from the particle emission processes. Natural sources such as continental dust, salt spray, and biogenic particles further play a considerable role for trace metals (Pd, Cd, Zn, etc.) (Silvia et al., 2004; Katja et al., 1998; Figen et al., 2000; Munir and Shaheen 2008). A calculated Gaussian air quality model by Johansson et al. (2008) identified air traffic as the major contributor of atmospheric Cu, Cr, Zn, Pb, W, Hg (particulate), and Ni at Greater Stockholm. Csavina et al. (2011) suggested mining operations as the potential sources of airborne metal and metalloid contaminants through direct smelter emissions and wind erosion of mine tailings (Fig. 2.1).

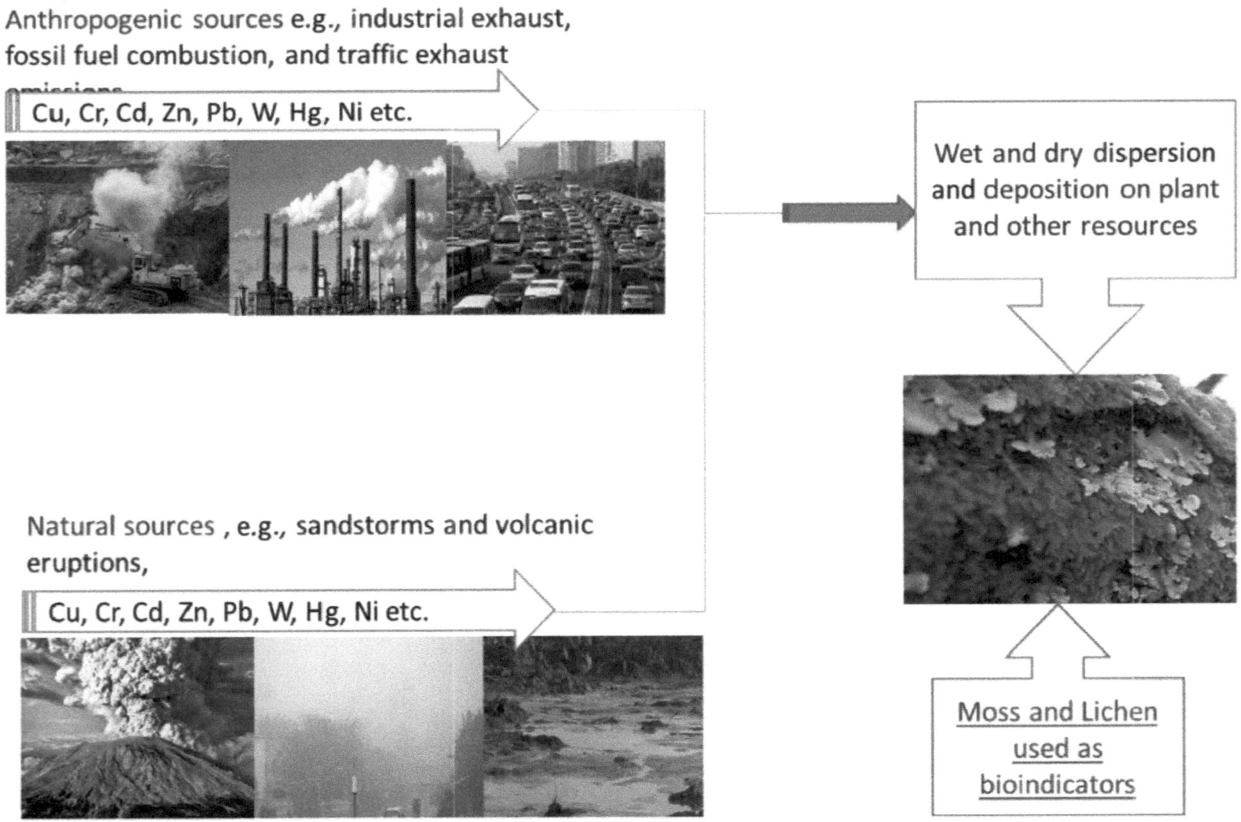


**Fig. 2.1** ***Source of atmospheric heavy metal and their dispersion and deposition and monitoring.***

### 2.2.2 Dispersion and deposition of heavy metals

The air-suspended dust is carried from one place to another through the blowing wind and the rate of transfer of dust from the air to vegetation surfaces which varies greatly with dust properties and the nature of the receiving environment (Grantz et al., 2003). The particle deposition depends on various conditions: viz, wind speed, particle size, temperature, humidity, and surface. Particle diameter affects the fate and transport of the aerosols. Small dust particles can be transported thousands of miles before being deposited on the surface (Csavina et al., 2011). Near to surface transport of particle size, less than 0.3 μM is mainly by Brownian motion and can be explained mathematically through molecular diffusive approach (Chamberlain, 1975) while, transport of particles larger than 0.3 μM is mostly by sedimentation, impaction, and interception (Martin and Coughtrey, 1982). Deposition velocities for fine particles to forest surfaces have been reported in the range of 1–15 cm $s^{-1}$ (Smith, 1990). Transfer resistance is affected remarkably by the surface roughness of the depositing surface. Martin and Coughtrey (1982) explain that the impaction velocity increased with turbulence velocity on the smooth surface, while for rough surfaces, roughness becomes a vital factor.

According to Walker and Everett (1987), the rates of decrease in surface dust load with increasing distance from highways were exponential. Keller and Lamprecht (1995) reported that the dust levels near the Dalton Highway in Alaska were relatively invariable over much of the summer growing season. Over 85% of the dust falling on vegetation surfaces may be removed. Johansson et al. (2008) suggested that metal particle measuring micrometer or sub-micrometer in size suspended in the air for several hours before deposition, which is much longer than the mixing and dilution. Unpaved roads produce higher dust levels than paved roads. Roberts et al. (1975) found that an unpaved dry gravel road with average daily traffic of 250 cars produced mean air concentrations of 584 g $m^{-3}$, while a paved road with an average daily traffic of 18000 produced mean concentrations of 463 g $m^{-3}$. Everett (1980) undertook a detailed study of an unpaved road in Alaska and found that in the summer, about 10 g $m^{-2}$ $day^{1}$ was deposited at the roadside and that there was a logarithmic decline in deposition away from the road, with deposition still occurring 1 km away.

## 2.3 Phytomonitoring of air pollutants

The development of instrumentation technology opens a new paradigm to monitor the various pollutants, although these methods are expensive and sometimes generate operational difficulty. Moreover, cost-effectiveness limits these technologies use in case of higher number of sites and at certain locations. To expand monitoring beyond a limited number of sites, it is necessary to develop simple, low-cost, and reliable techniques. Biomonitoring methods may satisfy these requirements (Chiarantini et al., 2016). Phytomonitors are defined as the indicator or accumulator plant species that can provide a quantitative description of the pollutant and illustrate temporal changes in pollutants (Garty, 2001). Biomonitoring of air pollution using plants can be traced back to as early as the mid-19th century (Hawksworth and Rose, 1976). Since then, the tree leaves have been used widely and effectively to monitor atmospheric pollution as an alternative to expensive instrumentation techniques (Baldantoni et al., 2014). The occurrence, character, or behavior of a plant species indicates the habitat's characteristics and can monitor the environmental condition prevailing in that area. Morphological alterations of a plant and the floristic composition of a plant community are commonly used to indicate environmental changes. Lichens and mosses are the primary and most studied pollution indicator; however, these plants are often missing in urban and industrial settlings (Anicic et al., 2011). In such cases, higher plants are gaining attention as primary biomonitoring tools (Chaudhary and Rathore, 2018a, 2019).

## 2.4 Mechanisms and factors that affect heavy metals foliar uptake

The foliar uptake and translocation mechanism of metal in plants occur after it absorbs by plant leaves. Several authors proposed various mechanisms of foliar transfer in plant tissues. According to Chamel et al. (1991), heavy metal enters through leaf cuticle essentially comprises four steps viz. (1) adherence to the cuticle, (2) penetration through the cuticle (possibly via endocytosis), (3) desorption in the apoplast, and (4) absorption by the subjacent cells. Schreck et al. (2012) explained two steps involved in this process: (1) adsorption through plant cuticle; and (2) insertion of metals through pores of stomata. Birbaum et al. (2010) reported that fine particles enter inside the leaves, while big particles stuck on the surface wax then diffused into leaves.

Heavy metal diffusion in the plant depends on various factors such as cuticle maturity, the morphology and surface area of the plant leaves, the physicochemical forms of the adsorbed metal, leaves surface texture, exposure of metal duration, environmental conditions like gas exchange (Beckett et al., 2000; Nair et al., 2010). Plants' metabolic processes also play a vital role in active metals transport in plant tissue. The deposition and adsorption of heavy metals on plant leave greatly vary with heavy metals in atmospheric PM fall-out. Rao and Dubey (1992) described that plant morphology like stomatal index, trichome density, and length affect the efficiency of dust collection by plants.

## 2.5 Plant as a monitoring tool for atmospheric heavy metal

Metals can be present in different matrices, such as airborne particulates, vegetation, river sediments, road dust, soils, etc. (Fujiwara et al., 2011), and deposited through the wet or dry method on plant or soil surface. Metals can be deposited directly on plant or soil surface or deposit through the transport of rainwater. Metals can also be absorbed through the roots after depositing on the soil surface (Gunawardena et al., 2013; Shahid et al., 2016). Some heavy metals such as Fe, Mn, Zn, and Cu are worked as micro-nutrient for plants and take part in biological activity however, become very toxic in high concentrations. Hg, and Pb have no known role in biological functioning and showed toxicity even at very low concentrations (Doganlar and Yurekli, 2009). Higher levels of heavy metals are linked with the high reactive oxygen species production, e.g., superoxide free radicals ($O^*$-), hydroxyl free radicals ($OH^*$-), or singlet oxygen ($O_2^*$-), hydrogen peroxide ($H_2O_2$) and cytotoxic methylglyoxal, which can cause oxidative stress by disordering the balance in prooxidant and antioxidant homeostasis inside the plant cells (Sytar et al., 2013). However, many studies have successfully demonstrated various plant species as monitoring tools for air-born heavy metal pollution with a comparable efficiency of expensive instrumentation techniques.

Monitoring using plants could be passive when native plants are analyzed to monitor the trace element concentration or active when specific accumulator or indicator plants are shifted from ecosystem with lower pollution level to specific location for monitoring.

### 2.5.1 Lower plants as indicator of aerial fallout of heavy metals

Lichens and mosses are mainly used for biomonitoring of heavy metal. Mosses (bryophytes), being lower plants without roots, vascular bundles, and cuticles, take up and store a considerable amount of water over the entire body surface. Heavy metals are accumulated on the moss surface by wet and dry deposition and pass through the plant's cell walls with high cation exchange capacity. Various forms of mosses such as naturally occurring epiphytic, moss bags, peat profile of sphagnum, and transplanted mosses have been reported for successful heavy metal monitoring (Onianwa, 2001; Kularatne and De Freitas, 2013; Maxhuni et al., 2016). Some bryophytes are metal tolerant and can withstand levels of heavy metals that are toxic to other species. Bryophytes are reliable indicators and monitors of air pollution as they are easy to handle and show a vast range of specific sensitivity and visible symptoms to pollutants greatly exceeding higher plants (Govindapyari et al., 2010). Bajpai et al., (2009) conducted a source identification study of metallic pollutant Al, Cd, Cr, Cu, Fe, Ni, and Zn using six lichen species viz. *Phaeophyscia hispidula*, *Parmotrema praesorediosum*, *Caloplaca subsoluta*, *Diploschistes candidissmus*, *Peltula euploca*, and *Phylliscum indicum*, concluded that the metals are originated from atmospheric pollution and the experimental species can be used for heavy metal monitoring. Airborne metals are incorporated into lichen thalli by the interception of soluble or PM. Soluble metals tend to settle in extra- or intracellular locations (Garty, 2001). Mosses were considered comparably efficient monitors to lichen and less sensitive to pollutants (Govindapyari et al., 2010). They were further considered a good monitor for mapping the extent and deposition of heavy metal (Onianwa, 2001; Maxhuni et al., 2016). Studies have also demonstrated the potential of mosses for monitoring radioactive materials too. Roos et al. (1994) has demonstrated monitoring of Pb-210, Cs-137, Pu-239+240, Pu-238, and Am-241 at the Antarctic Peninsula Area and Nifontova (1995) has monitored Sr-90 and Cs-137 levels at the Yamal Peninsula of Russia using mosses. Table 2.1 presented various lower plants used for heavy metal monitoring.

### 2.5.2 Higher plants as indicator of atmospheric heavy metals

Usually, higher plants are not considered suitable for monitoring (Jiang et al., 2018) however, wide spearheading of higher plants provide an edge over lower plants. Moreover, higher plants are preferable choice of heavy metal monitoring due to the non-suitability of lower plants at the sites with heavy pollution load. Another advantage of using higher plant is that they are long lived; therefore, experiment and sampling of the long interval are possible (Sawidis et al., 2011). Further, the ease of identification using leaves of higher plants over lower plants (Berlizov et al., 2007) can be used for international network monitoring deposition (Sawidis et al., 2011). Several researchers have successfully demonstrated the efficiency

**Table 2.1** Lower plants species used for atmospheric heavy metal monitoring.

| Plant name | Heavy metal | Study area | References |
|---|---|---|---|
| *Pseudocleropodium purum* and *Hypnum cupressiforme* (moss species) | Cd, Cr, Cu, Fe, Hg, Ni, Mn, Pb, and Zn | Kosovo, Balkan Peninsula | Maxhuni et al., 2016 |
| Active moss species | Cu, Pb, Cd, Zn, Cr, Mn, Ni, Co, and Fe | Elbasan city, Albania | Lazo et al. 2013 |
| Pleurozium schreberi, Papillaria crocea (moss species) | Cd, Cr, Pb, and Zn | Hailuogou Glacier foreland, Eastern Tibetan Plateau | Bing et al., 2016 |
| *Haplocladium angustifolium* (moss species) | Ag, As, Cd, Co, Cr, Cu, Mn, Mo, Ni, V, Pb, and Zn | Wuhan City in China | Jiang et al., 2018 |
| *Marchantia paleacea Bertol* (Liverwort) | Pb, Cu, Zn, and Ni | Nainital, India | Nath et al., 2009 |
| *Dirinaria pictaand* (lichen species) | Cr, Fe, Co, Ni, Zn, and Pb | Malaysia | Sulaiman et al., 2018 |
| *Parmotrema reticulatum* (lichen species) | Cr, Cu, Pb, and Zn. | greater Auckland region of New Zealand | Kularatne and De Freitas, 2013 |

of higher plants for heavy metal monitoring. Plant foliages are directly exposed to air pollution and provide a surface for air particle deposition (Chaudhary and Rathore, 2018a, 2019). The characteristics of leaf such as leaf geometry, phyllotaxy, leaf pubescence, micro-morphological leaf surfaces (e.g., the presence of stomata, trichomes, waxy cuticular layers, and cell wall appendages of epidermal cells), on both the adaxial and abaxial sides, and the height and canopy size of the plant determine the rate of deposition or accumulation of PM-bound toxic metals (Singh et al., 2005; Simon et al., 2016; Chaudhary and Rathore, 2018b). A study by Simon et al. (2014) proved Acer negundo and Celtis occidentalis to be suitable biomonitors for atmospheric metal pollution due to their foliage which is covered by a large number of trichomes. Gajbhiye et al. (2019) reported 12 airborne metals, including Pb, Cd, Cu, Ni, and Zn, on the surface of the leaf of *Clostridium procera,* which supported the use of this species as a phytomonitoring tool for a tropical condition to assess size-fractionated quantification of airborne PM-bound toxic metals in urban areas. Gawel et al. (1996) asses the phytochelatins, and intercellular metal complexing agent, obtained from foliage needle of red spruce to indicate the heavy metal toxicity in the region. Kord et al. (2010) conducted a source appropriation study of atmospheric heavy metals using needles of *Pinus Eldarica*. Study identified traffic emission as the major source of Pb, Ni, and Cu, whereas industrial areas are the major source of Zn and Ni in the atmosphere.

Tree bark has also been reported for its monitoring efficiency. Using bark overleaf is beneficial as it can retain the metal for a longer duration without affecting the health of the plant (Sawidis et al., 2011). A study by Chiarantini et al. (2016) indicates that bark Hg concentrations are comparable with values reported for lichens in the same areas, suggesting that tree barks may be an additional helpful tool for biomonitoring airborne Hg. Similarly, standardized grass culture (*Lolium multiflorum*) and many tropical tree species, including *Psidium guajava* cv. Paluma and *Tibouchina pulchra* have been studied for their potential as bioaccumulation (Bulbovas et al., 2015; Domingos et al., 2003; Klumpp et al., 2009; Moraes et al., 2002; Nakazato et al., 2015). The study of Zhai et al. (2016) indicated that the bark followed by the leaf of *Cinnamomum camphora* (L) Presl has a good potential for heavy metal accumulation. The plant can be grown along the highway for monitoring heavy metal originated from vehicles. Study of Barnes et al. (1976) indicated that rough barks are potential metal accumulator than smooth barks as the heavy metals is not merely accumulated after settling but the surface of the bark influences it due to cation exchange capacity of bark (Kreiner and Härtel, 1986; Farmer et al., 1991). Plant bark has also been reported for source appropriation of heavy metal pollution. El-Hasan et al. (2002) successfully investigated the biomonitoring potential of bark of the *Cupressus semervirens* L. at Amman city of Jordan. They identified traffic emission as the main source of Pb, Zn, Mn, Cr, Ni, Cd, and Cu pollution in the atmosphere.

Several epiphytic bromeliads have also been used as biomonitors of trace metals due to their high ability to accumulate toxic elements, mostly without showing visible damage, with the advantages of absorbing these elements only from the air showing a higher capacity of accumulation. Biological monitoring using bromeliads allows the assessment of hundreds of sites at the same time. It integrates exposure over longer periods, very low cost, and giving an idea of bioavailability (Filho et al., 2002). Table 2.2 presented various plants and their parts used for the atmospheric heavy metal monitoring or as a suitable biosindicator of aerial fallout of heavy metals.

**Table 2.2** Various higher plants species used for atmospheric heavy metal monitoring.

| Tree species | Heavy metal | Study area | Used plant part | References |
|---|---|---|---|---|
| Alpine tree species *Salix rehderiana, Populus purdomii, Betula albosinensis* | Cd, Pb, and Zn | Eastern Tibetan Plateau | Leaves and bark | Bing et al., 2016 |
| *M. indica, P. sylvestris, E. globulus, D. sissoo, A. marmelos,* and *F. benghalensis* | As, Cr, Cu, Mn, Zn, and Pb | Rajnandgaon district (Chhattisgarh, India) | Bark | Patel et al., 2020 |
| Pinus eldarica Medw | Pb, Zn, Cu, Ni, and Cr | Tehran City | Bark | Kord and Kord, 2011 |
| *Schinus molle L.* | Pb, Cu, and Zn | Adana, Mediterranean region of Turkey | Leaves | Doganlar et al., 2012 |
| *Nerium oleander L.* | Fe, Al, and Ba. | Adana, Turkey | Leaves | Doganlar et al., 2012 |
| *Pinus eldarica* | Zn, Cu, Pb, and Cd | Yazd city, Iran | Leaves | Miri et al. 2017 |
| *Melia azedarach, Morus alba, and Ficus benghalensis* | Cu, Cd, Pb, and Zn | Mansoura city | Leaves | El-Amier and Alghanem, 2018. |
| *Araucaria heterophylla* | Ba, Zn, Cu, Cr, Pb, and Co | Andean city Quito, Ecuador | Needle | Alexandrino et al., 2020 |
| *Platanus orientalis L.* and *Pinus nigra Arn* | Cr, Cu, Fe, and Pb | European cities (Salzburg, Belgrade and Thessaloniki) | Leaves and bark | Sawidis et al., 2011 |

## 2.6 Present scenario, challenges, and future perspective

The lower and higher plant has significant advantages for biomonitoring of heavy metals. They are widely distributed and can be used over large areas for monitoring and evaluating heavy metal at a lower cost than direct pollution measurement methods since no collection or measurement equipment has to be installed in the observation area. The use of an individual part of an ecosystem for determining the latter's heavy metal status has the advantage that it permits conclusions going beyond the biomonitor itself.

By occupying a niche in the ecosystem, biomonitors make it possible to integrate the analysis results into an overall biological system. Through biotic interactions, they allow ecological sustainability within the whole community of organisms.

Plant biomonitoring challenges come up against its limits when the environment's quality to be quantified. There is a linear relation in the concentration of metals in the atmosphere, allowing organisms to supply unequivocal statements about their environment. But many plants are accumulators, i.e., they store substances at high accumulation factors, especially at low environmental concentrations; at high environmental concentrations, the accumulation factor decreases considerably. On the other hand, many organisms react as excluders. This means that they can keep the concentration low despite a high level in the environment. The concentration in organisms does not rise until a level in the air is reached, which causes the regulatory mechanisms to break down. These limitations need further investigation to fine-tune these hurdles. Limitation of the range of the species and the favoritism to other environmental factors restrict the study area of biomonitoring (Stafanut et al., 2019). Mosses, lichens, many trees, and plant species have been identified as the successful biomonitor of heavy metals. They have great future potential and sustainable and eco-friendly solution in monitoring air-suspended heavy metal in the environment.

## 2.7 Conclusion

Metal particles emitted from road transport and various industries have been seriously concerned with society's wellbeing. The increasing metal pollution and settling of wet and dry deposition in the atmosphere have forced us to find an adequate monitoring system that allows reliable assessments of the environment's condition. Lower and higher plant species accumulate toxic metals and are also influenced morphologically, anatomically, and physiologically changes and can be used as biomonitor. However, to control heavy metal atmospheric pollution, more studies, surveys, and analyses on heavy metals should be conducted to effective environmental risk management.

## References

Abdel-Shafy, H.I., Hindy, K.T., Abdel-Sabour, M.F., 1992. Heavy metal levels in the environment of Ismailia city, Egypt. Int. J. Environ. St. 39 (4), 279–289. doi:10.1080/00207239208710703.

Agbaire, P.O., Esiefarienrhe, E., 2009. Air Pollution Tolerance Indices (APTI) of some plants around Otorogun Gas Plant in Delta State, Nigeria. J. Appl. Sc. Environ. Manag. 13 (1), 11–14.

Alexandrino, K., Viteri, F., Rybarczyk, Y., Andino, J.E.G., Zalakeviciute, R., 2020. Biomonitoring of metal levels in urban areas with different vehicular traffic intensity by using Araucaria heterophylla needles. Ecol. Ind. 117, 106701.

Arimoto, R., Ray, B.J., Lewis, N.F., Tomza, U., 1997. Mass-particle size distributions of atmospheric dust and the dry deposition of dust to the remote ocean. J. Geophys. Res. 102 (D13), 15867–15874.

Anicic, M., Spasic, T., Tomasevic, M., Rajsic, S., Tasic, M., 2011. Trace elements accumulation and temporal trends in leaves of urban deciduous trees (*Aesculus hippocastanum* and *Tilia* spp.). Ecol. Indic. 11, 824–830.

Bajpai, R., Upreti, D.K., Dwivedi, S.K., et al., 2009. Lichen as quantitative biomonitors of atmospheric heavy metals deposition in central India. J Atmos. Chem. 63, 235–246. https://doi.org/10.1007/s10874-010-9166-x.

Baldatoni, D., Nicola, F.D., Alfani, A., 2014. Air biomonitoring of heavy metals and polycyclic aromatic hydrocarbons near a cement plants. Atmos. Pollut. Res. 5, 262–269.

Barnes, D., Hammadah, M.A., Ottaway, J.M., 1976. The lead, copper, and zinc contents of tree rings and barks, a measurement of local pollution. Sci. Total Environ. 5, 63–67.

Beckett, K.P., Freer-Smith, P.H., Taylor, G., 2000. The capture of particulate pollution by trees at five contrasting urban sites. Arboricult. J. 24, 209–230.

Berlizov, A.N., Blum, O.B., Filby, R.H., Malyuk, I.A., Tryshyn, V.V., 2007. Testing applicability of black poplar (*Populus nigra* L.) bark to heavy metal air pollution monitoring in urban and industrial regions. Sci. Total Environ. 372, 693–706.

Bing, H., Wu, Y., Zhou, J., Sun, H., 2016. Biomonitoring trace metal contamination by seven sympatric alpine species in Eastern Tibetan Plateau. Chemosph 165, 388–398.

Birbaum, K., Brogioli, R., Schellenberg, M., Martinoia, E., Stark, W.J., Günther, D., Limbach, L.K., 2010. No evidence for cerium dioxide nanoparticle translocation in maize plants. Environ. Sci. Technol. 44, 8718–8723.

Bulbovas, P., Camargo, C.Z., Domingos, M., 2015. Ryegrass cv. Lema and guava cv. Paluma biomonitoring suitability for estimating nutritional contamination risks under seasonal climate in southeastern Brazil. Ecotox. Environ. Saf. 118, 149–157.

Chamberlain, A.C., 1975. Pollution in plant canopies. In: DeVries, D.A., Afgan, N.H. (Eds.), Heat and Mass Transfer in the Biosphere. Part 1. Transfer Processes in the Plant Environment. Scripta Book Company, New York, pp. 561–582.

Chamel, A., Pineri, M., Escoubes, M., 1991. Quantitative determination of water sorption by plant cuticles. Plant Cell Environ. 14, 87–95.

Chaparro, M.A.E., Chaparro, M.A.E., Castañeda-Miranda, A.G., Marié, D.C., Gargiulo, J.D., Lavornia, J.M., Natal, M., Böhnel, H.N, 2020. Fine air pollution particles trapped by street tree barks: In situ magnetic biomonitoring. Environ. Pollut. 226, 115229. https://doi.org/10.1016/j.envpol.2020.115229.

Chaudhary, I.J., Rathore, D., 2019. Dust pollution: Its removal and effect on foliage physiology of urban trees. Sust. Cities Soc. 51, 101696. https://doi.org/10.1016/j.scs.2019.101696.

Chaudhary, I.J., Rathore, D., 2018a. Phytomonitoring of dust load and its effect on foliar micro morphological characteristics of urban trees. Plant. J. Plant Sc. 2 (3), 170–179.

Chaudhary, I.J., Rathore, D., 2018b. Suspended particulate matter deposition and its impact on urban trees. Atmos. Pollut. Res. 9 (6), 1072–1082. https://doi.org/10.1016/j.apr.2018.04.006.

Chiarantini, L., Rimondi, V., Benvenuti, M., Beutel, M.W., Costagliola, P., Gonnelli, C., Lattanzi, P., Paolieri, M., 2016. Black pine (*Pinus nigra*) barks as biomonitors of airborne mercury pollution. Sci. Tot. Environ. 569-570, 105–113. https://doi.org/10.1016/j.scitotenv.2016.06.029.

Csavina, J., Landázuri, A., Wonaschütz, A., Rine, K., Rheinheimer, P., Barbaris, B., Conant, W., Sáez, A.E., Betterton, E.A., 2011. Metal and metalloid contaminants in atmospheric aerosols from mining operations. Water Air Soil Pollut. 221 (1-4), 145–157. doi:10.1007/s11270-011-0777-x.

Doganlar, B.Z., Doganlar, O., Erdogan, S., Onal, Y., 2012. Heavy metal pollution and physiological changes in the leaves of some shrub, palm and tree species in urban areas of Adana, Turkey. Chem. Specia. Bioavail. 24 (2), 65–78.
Doğanlar, Z.B.P., Yürekli, F., 2009. Interactions between cadmium and phytochelatin accumulation in two different sunflower cultivars. Fresen. Environ. Bull. 18, 304–310.
Domingos, M., Klumpp, A., Rinaldi, M.C.S., Klumpp, G., Modesto, I.F., Delitti, W.B.C., 2003. Combined effects of air and soil pollution by fluoride emissions on *Tibouchina pulchra* Cogn., at Cubatão, SE Brazil, and their relations with aluminium. Plant Soil 249, 297–308.
Duce, R.A., et al., 1991. The atmospheric input of trace species to the world ocean. Global Biogeochem. Cycles 5, 193–259.
Drobyshev, A.I., Emelina, O.I., 2001. Analysis of atmospheric aerosols by atomic emission spectrometry with electrical discharge sampling. J. Analyt. Chem. 56, 577–580. https://doi.org/10.1023/A:1016697130936.
Duffus, J.H., 2002. Heavy metals: a meaningless term? Pure Appl. Chem. 74 (5), 793–807.
El-Amier, Y.A., Alghanem, S.M., 2018. Tree leaves as bioindicator of heavy metal pollution from soil and ambient air in urban environmental. Plant Arch 18 (2), 2559–2566.
El-Hasan, T., Al-Omari, H., Jiries, A., Al-Nasir, F., 2002. Cypress tree (*Cupressus semervirens* L.) bark as an indicator for heavy metal pollution in the atmosphere of Amman City. Jordan. Environ. Int. 28, 513–519.
Everett, K.R., 1980. Distribution and properties of road dust along the northern portion of the Haul Road. In: Brown, J., Berg, R. (Eds.), Environmental Engineering and Ecological Baseline Investigations Along the Yukon River–Prudhoe Bay Haul Road. U.S. Army Cold Regions Research and Engineering Laboratory, Fairbanks, Alaska, pp. 101–128.
Farmer, A.M., Bates, J.W., Bell, J.N.B., 1991. Seasonal variations in acidic pollutant inputs and their effects on the chemistry of stemflow, bark and epiphyte tissues in three oak (*Quercus petraea*) woodlands in NW Britain. New Phytol. 119, 567–570.
Field, J.P., Belnap, J., Breshears, D.D., Neff, J.C., Okin, G.S., Whicker, J.J., Reynolds, R.L., 2010. The ecology of dust. Front. Ecol. Environ. 8 (8), 423–430. https://doi.org/10.1890/090050.
Figen, V., Yasushi, N., Shigeru, T., 2000. The concentration trend and seasonal variation of metals in the atmosphere in 16 Japanese cities shown by the results of National Air Surveillance Network (NASN) from 1974 to 1996. Atmos. Environ. 34, 2755–2770.
Filho, A.G.M., Andrade, L.R., Farina, M., Malm, O., 2002. Hg localization in *Tillandsia usneoides* L. (Bromeliaceae), an atmospheric biomonitor. Atmos. Environ. 36, 881–887.
Fu, Q., Zhuang, G., Li, J., Huang, K., Wang, Q., Zhang, R., Chen, Y., 2010. Source, long-range transport, and characteristics of a heavy dust pollution event in Shanghai. J. Geophys. Res. Atmosph. 115 (D7). https://doi.org/10.1029/2009JD013208.
Fujiwara, F., Jimenez Rebagliati, R., Dawidowski, L., Gomez, D., Polla, G., Pereyra, V., Smichowski, P., 2011. Spatial and chemical patterns of size fractionated road dust collected in a megacity. Atmos. Environ. 45, 1497–1505.
Gajbhiye, T., Pandey, S.K., Lee, S.S., Kim, K.H., 2019. Size fractionated phytomonitoring of airborne particulate matter (PM) and speciation of PM bound toxic metals pollution through *Calotropis procera* in an urban environment. Ecol. Indi 104, 32–40.
Gargiulo, J.D., Kumar, S.R., Chaparro, M.A.E., Chaparro, M.A.E., Natal, M., Rajkumar, P., 2016. Magnetic properties of air suspended particles in thirty eight cities from south India. Atmos. Pollut. Res. 7, 626–637. http://dx.doi.org/10.1016/j.apr.2016.02.008.
Garty, J., 2001. Biomonitoring atmospheric heavy metals with lichens: theory and application. Crit. Rev. Plant Sc. 20 (4), 309–371.
Gawel, J.E., Ahner, B.A., Friedland, A.J., Morel, F.M.M., 1996. Role for heavy metals in forest decline indicated by phytochelatin measurements. Nature 381 (6577), 64–65. doi:10.1038/381064a0.

Gharaibeh, A.A., El-Rajoob, A.O., Harb, M.K., 2010. Determination of selected heavy metals in air samples from the northern part of Jordan. Environ. Monit. Assess. 160, 425–429. doi:10.1007/s10661-008-0706-7.
Godoi, R.H.M., Potgieter-Vermaak, S., Godoi, A.F.L., Stranger, M., Van Grieken, R., 2008. Assessment of aerosol particles within the Rubens House Museum in Antwerp, Belgium. X-Ray Spectr 37 (4), 298–303.
Govindapyari, H., Leleeka, M., Nivedita, M., Uniyal, P.L., 2010. Bryophytes: indicators and monitoring agents of pollution. NeBIO 1 (1), 35–41.
Grantz, D.A., Garner, J.H.B., Johnson, D.W., 2003. Ecological effects of particulate matter. Environ. Inter. 29 (2-3), 213–219. doi:10.1016/S0160-4120(02)00181-2.
Gunawardena, J., Egodawatta, P., Ayoko, G.A., Goonetilleke, A., 2013. Atmospheric deposition as a source of heavy metals in urban storm water. Atmos. Environ. 68, 235–242.
Hawksworth, D.L., Rose, F., 1976. Lichens as Pollution Monitors. Edward Arnold Ltd, London, UK.
Jiang, Y., Fan, M., Hu, R., Zhao, J., Wu, Y., 2018. Mosses are better than leaves of vascular plants in monitoring atmospheric heavy metal pollution in urban areas. Inter. J. Environ. Res. Public Health 15 (6), 1105. https://doi.org/10.3390/ijerph15061105.
Jiao, X., Dong, Z., Kang, S., Li, Y., Jiang, C., Rostami, M., 2021. New insights into heavy metal elements deposition in the snowpacks of mountain glaciers in the eastern Tibetan Plateau. Ecotoxicol. Environ. Saf. 207, 111228.
Johansson, C., Norman, M., Burman, L., 2008. Road traffic emission factors for heavy metals. Atm. Environ. 43, 4681–4688. doi:10.1016/j.atmosenv.2008.10.024.
Joshi, P.C., Swami, A., 2009. Air pollution induced changes in the photosynthetic pigment of selected plant species. J. Environ. Biol. 30, 295–298.
Karagulian, F., Dora, C., Belis, C.A., Dora, C.A.C., Prüss-Ustün, A., Bonjour, S., Adair-Rohani, H., Amann, M., 2015. Contributions to cities' ambient particulate matter (PM): a systematic review of local source contributions at global level. Atm. Environ. 120, 475–483. https://doi.org/10.1016/j.atmosenv.2015.08.087.
Karl, X.Y., Swami, K., Liaquat, H., 2002. Determination of trace 264 metals in atmospheric aerosols with a heavy matrix of cellulose by 265 microwave digestion-inductively coupled plasma mass spectroscopy. Spectrochim. Acta 57, 73–84. doi:10.1016/S0584-8547(01)00354-8.
Katja, V.V., Claude, B., Christophe, F., Tania, B., Carlo, B., Sergey, R., et al., 1998. Seasonal variations of heavy metals in the 1960s Alpine ice: sources versus meteorological factors. Earth Planet. Sc. Lett. 164, 521–533. doi:10.1016/S0012-821X(98)00243-X.
Keller, J., Lamprecht, R., 1995. Road dust as an indicator for air pollution transport and deposition: an application of SPOT imagery. Remote Sens. Environ. 54 (1), 1–12.
Khillare, P.S., Balachandran, S., Bharat, R.M., 2004. Special and temporal variation of heavy metal in atmospheric aerosol of Delhi. Environ. Monit. Assess. 90, 1–21. doi:10.1023/B:EMAS.0000003555.36394.17.
Klumpp, A., Ansel, W., Klumpp, G., Breuer, J., Vergne, P., Sanz, M.J., Calatayud, V., 2009. Airborne trace element pollution in 11 European cities assessed by exposure of standardized ryegrass cultures. Atmos. Environ. 43, 329–339.
Kord, B., Mataji, A., Babaie, S., 2010. Pine (*Pinus Eldarica* Medw.) needles as indicator for heavy metals pollution. Int. J. Environ. Sci. Tech. 7 (1), 79–84.
Kord, B., Kord, B., 2011. Heavy metal levels in pine (*Pinus eldarica*Medw.) tree barks as indicators of atmospheric pollution. BioRes 6 (2), 927–935.
Kreiner, W., Härtel, O., 1986. Spruce bark influenced by $SO_2$. Eur. J. For. Pathol. 16, 171–181.
Kularatne, K.I.A., De Freitas, C.R., 2013. Epiphytic lichens as biomonitors of airborne heavy metal pollution. Environ. Exp. Bot. 88, 24–32.
Lazo, P., Bekteshi, L., Shehu, A., 2013. Active moss biomonitoring technique for atmospheric deposition of heavy metals in Elbasan city, Albania. Fresenius Environ. Bull. 22 (1a), 213–218.

Leys, J.F., Larney, F.J., Muller, J.F., et al., 1998. Anthropogenic dust and endosulfan emissions on a cotton farm in northern New South Wales, Australia. Sci. Total Environ. 220, 55–70.

Liu, X., Song, Q., Tang, Y., Li, W., Xu, J., Wu, J., Brookes, P.C., 2013. Human health risk assessment of heavy metals in soil-vegetable system: a multi-medium analysis. Sci. Environ. 463-464, 530–540.

Luo, L., Shen, Y., Wang, X., Chu, B., Xu, T., Liu, Y., Liu, J., 2018. Phytoavailability, bioaccumulation, and human health risks of metal(loid) elements in an agroecosystem near a lead-zinc mine. Environ. Sci. Pollu. Res. 25 (24), 24111–24124. https://doi.org/10.1007/s11356-018-2482-4.

Manins, P., Allan, R., Beer, T., et al., 2001. Atmosphere. Australia State of the Environment Report 2001 (Theme Report). CSIRO Publishing, Melbourne, Australia.

Martin, M.H., Coughtrey, P.J., 1982. The use of vegetation for monitoring airborne heavy metal deposition. Biol. Monit. Heavy Metal Pollu. 60–150. doi:10.1007/978-94-009-7352-7_3.

Mateu, J., Mirabo, F.B., Forteza, R., Cerda, V., Colom, M., Oms, M., 1999. Heavy metals in the aerosols collected at two stations in Mallorca (Spain). Water Air Soil Pollu. 112, 349–363. doi:10.1023/A:1005091027762.

Maxhuni, A., Lazo, P., Kane, S., Qarri, F., Marku, E., Harmens, H., 2016. First survey of atmospheric heavy metal deposition in Kosovo using moss biomonitoring. Environ. Sci. Pollu. Res. 23 (1), 744–755.

Merefield, J., Stone, I., Roberts, J., Jones, J., Barron, J., 1998. Airborne particulate characterisation for environmental regulation. In: Bennett, M.R., Doyle, P. (Eds.), Environmental Geology: A British Perspective. Geological Society, London, pp. 277–289.

Miri, M., Ehrampoush, M.H., Ghaffari, H.R., Aval, H.E., Rezai, M., Najafpour, F., Fathabadi, Z.A., Aval, M.Y., Ebrahimi, A., 2017. Atmospheric heavy metals biomonitoring using a local *Pinus eldarica* tree. Health Scope 6 (1), 39241.

Moraes, R.M., Klumpp, A., Furlan, C.M., Klumpp, G., Domingos, M., Rinaldi, M.C.S., Modesto, I.F., 2002. Tropical fruit trees as bioindicators of industrial air pollution in southeast Brazil. Environ. Int. 28, 367–374.

Moreno-Grau, S., Perez-Tornell, A., Moreno-Grau, J., Moreno-Clavel, J., 1997. Determination of lead in aerosol sample collected on glass fiber filters by an improved atomic absorption spectrometry method. Water Air Soil Pollu. 96, 145–153.

Moreno-Grau, S., Tornell, A., Bayo, J., Moreno, J., Angosto, J.M., Moreno-Clavel, J., 2000. Particulate matter and heavy metals in the atmospheric aerosol from Cartagena, Spain. Atmos. Environ. 34, 5161–5167. doi:10.1016/S1352-2310(00)00349-6.

Munir, H.S., Shaheen, N., 2008. Annual and seasonal variations of trace metals in atmospheric suspended particulate matter in Islamabad. Pakistan Water Air Soil Pollut. 190, 13–25. doi:10.1007/s11270-007-9575-x.

Nagar, J.K., Akolkar, A.B., Kumar, R., 2014. A review on airborne particulate matter and its sources, chemical composition and impact on human respiratory system. Int. J. Environ. Sci. 5 (2), 447–463.

Nair, R., Varghese, S.H., Nair, B.G., Maekawa, T., Yoshida, Y., Kumar, D.S., 2010. Nanoparticulate material delivery to plants. Plant Sci. 179, 154–163.

Nakajima, F., Aryal, R., 2018. Heavy metals in urban dust. In: Hosam, El-Din M.S., Refaat, F.A. (Eds.), Heavy Metals. IntechOpen. doi:10.5772/intechopen.74205. Available from. https://www.intechopen.com/books/heavy-metals/heavy-metals-in-urban-dust.

Nakazato, R.K., Rinaldi, M.C.S., Domingos, M., 2015. Will technological modernization for power generation at an oil refinery diminish the risks from air pollution to Atlantic Rainforest in Cubatão, SP, Brazil? Environ. Pollut. 196, 489–496.

Nath, V., Sahu, V., Asthana, A.K., Yunus, M., 2009. Heavy metal indicator *Marchantia paleacea* Bertol. (Liverwort): a case study of Nainital (Uttarakhand). Proc. Nat. Acad. Sci. Ind.: Biol. Sci. 79, 147–152.

Nifontova, M., 1995. Radionuclides in the moss-lichen cover of tundra communities in the Yamal Peninsula. Sci. Total Environ. 160-161, 749–752.
Onianwa, P.C., 2001. Monitoring atmospheric metal pollution: a review of the use of mosses as indicators. Environ. Monit. Assess. 71, 13–50.
Park, S.S., Kim, Y.J., 2005. Source contributions to fine particulate matter in an urban atmosphere. Chemosphere 59 (2), 217–226.
Patel, K.S., Yadav, A., Sahu, Y.K., Lata, L., Milosh, H., Corns, W.T., Martín-Ramos, P., 2020. Tree bark as a bioindicator for arsenic and heavy metal air pollution in Rajnandgaon district, Chhattisgarh, India. J. Hazard. Toxic Radio. Waste 24 (1), 05019006.
Popoola, L.T., Adebanjo, S.A., Adeoye, B.K., 2018. Assessment of atmospheric particulate matter and heavy metals: a critical review. Int. J. Environ. Sci. Technol. 15, 935–948. https://doi.org/10.1007/s13762-017-1454-4.
Quiterio, S.L., Escaleira, V., Sousa, C.R.S., Maia, L.F.P.G., Arbilla, G., 2004. Metals in airborne particulate matter in downtown Rio de Janeiro, Brazil. Bull. Environ. Contam. Toxicol. 72 (5), 916–922.
Rai, P.K., 2015. Multifaceted health impacts of particulate matter (PM) and it's management: an overview. Environ. Skeptics Critics 4, 1–26.
Rao, M.V., Dubey, P.S., 1992. Occurrence of heavy metals in air and their accumulation by tropical plants growing around an industrial area. Sci. Total Environ. 126, 1–16.
Roberts, J.W., Watters, H.A., Mangold, C.A., Rossano, A.T., 1975. Cost and benefits of road dust control in Seattle's industrial valley. J. Air Poll. Control Assoc. 25 (9), 948–952.
Roos, P., Holm, Z., Persson, R.B.R., Aarkrog, A., Nielsen, S.P., 1994. Deposit of 210Pb, 137Cs, 239+240Pu and 241Am in the Antartic Peninsula Area. J. Environ. Radioact. 24 (3), 235–251.
Sawidis, T., Breuste, J., Mitrovic, M., Pavlovic, P., Tsigaridas, K., 2011. Trees as bioindicator of heavy metal pollution in three European cities. Environ. Poll. 159 (12), 3560–3570. doi:10.1016/j.envpol.2011.08.008.
Schreck, E., Foucault, Y., Sarret, G., Sobanska, S., Cécillon, L., Castrec-Rouelle, M., Uzu, G., Dumat, C., 2012. Metal and metalloid foliar uptake by various plant species exposed to atmospheric industrial fallout: mechanisms involved for lead. Sci. Total Environ. 427–428, 253–262.
Shahid, M., Dumat, C., Khalid, S., Schreck, E., Xiong, T., Niazi, N.K., 2016. Foliar heavy metal uptake, toxicity and detoxification in plants: a comparison of foliar and root metal uptake. J. Hazard. Mater. 325, 36–58.
Silvia, M.S., Annibal, D.P.N.A.D., Pereira, N., Emmanoel, V.S.F., Martha, T.A., 2004. Shortterm and spatial variation of selected metals in the atmosphere of Niteroi City, Brazil. Microchem. J. 78, 85–90. doi:10.1016/j.microc.2004.03.015.
Simon, E., Baranyai, E., Braun, M., Cserháti, C., Fábián, I., Tóthmérész, B., 2014. Elemental concentrations in deposited dust on leaves along an urbanization gradient. Sci. Total Environ. 490, 514–520.
Simon, E., Harangi, S., Baranyai, E., Fábián, I., Tóthmérész, B., 2016. Influence of past industry and urbanization on elemental concentrations in deposited dust and tree leaf tissue. Urban For. Urban Green 20, 12–19.
Singh, A., Agrawal, S.B., Rathore, D., 2005. Amelioration of Indian urban air pollution phytotoxicity in Beta vulgaris L. by modifying NPK nutrients. Environ. Pollu. 134, 385–395.
Smith, W.H., 1990. Series on Environmental Management. Springer-Verlag, New York, USA, pp. 147–180.
Ştefănuţ, S., Öllerer, K., Manole, A., Ion, M.C., Constantin, M., Banciu, C., Maria, G.B., Florescu, L.I., 2019. National environmental quality assessment and monitoring of atmospheric heavy metal pollution—a moss bag approach. J. Environ. Manag. 248, 109224. https://doi.org/10.1016/j.jenvman.2019.06.125.

Sulaiman, N.O., Fuzy, S.F., Muis, S.I., Sulaiman, N., Ismail, B.S., 2018. Use of lichens as bioindicators for determining atmospheric heavy metal concentration in Malaysia. Pak. J. Bot. 50 (1), 421–428.

Sytar, O., Kumar, A., Latowski, D., Kuczynska, P., Strzałka, K., Prasad, M.N.V., 2013. Heavy metal-induced oxidative damage, defense reactions, and detoxification mechanisms in plants. Acta Physiol. Plant. 35 (4), 985–999.

Toy, T.J., Foster, G.R., Renard, K.G., 2002. Soil Erosion: Processes, Prediction, Measurement, and Control. Wiley Publishing.

U.S. EPA, 2013. Report on the 2013 U.S. Environmental Protection Agency (EPA) International Decontamination Research and Development Conference. Research Triangle Park, NC, November 05 - 07, 2013. U.S. Environmental Protection Agency, Washington, DC EPA/600/R-14/210, 2014.

Vallius, M., Janssen, N.A., Heinrich, J., et al., 2005. Sources and elemental composition of ambient PM (2.5) in three European cities. Sci. Total Environ. 33 (1-3), 147–162.

Vasconcelos, M.T.S.D., Tavares, H.M.F., 1997. Atmospheric metal pollution (Cr, Cu, Fe, Mn, Ni, Pb and Zn) in Oporto City derived from results for low-volume aerosol samplers. Sci. Total Environ. 212, 11–20. doi:10.1016/S0048-9697(97)00322-7.

Walker, D.A., Everett, K.R., 1987. Road dust and its environmental impact on Alaskan Taiga and Tundra. Arctic Alpine Res 19 (4), 479–489. https://doi.org/10.2307/1551414.

Zhai, Y., Dai, Q., Jiang, K., et al., 2016. Traffic-related heavy metals uptake by wild plants grow along two main highways in Hunan Province, China: effects of soil factors, accumulation ability, and biological indication potential. Environ. Sci. Pollut. Res. 23, 13368–13377. https://doi.org/10.1007/s11356-016-6507-6.

Zhao, Q., An, M., Chen, H., 2018. Research on geochemical characteristics of soil in a chemical industrial factory site in Jinan City. J. Rock Miner. Anal. 37, 201–208. https://doi.org/10.15898/j.cnki.11-2131/td.201708240135.

Zhuang, P., Zou, B., Li, N.Y., et al., 2009. Heavy metal contamination in soils and food crops around Dabaoshan mine in Guangdong, China: implication for human health. Environ. Geochem. Health 31, 707–715. https://doi.org/10.1007/s10653-009-9248-3.

# CHAPTER 3

# The importance and effectiveness of aquatic biomonitoring

**Didem Gökçe**
Department of Biology, İnönü University, Malatya, Turkey

## 3.1 Introduction

Natural ecosystems quickly change as a result of many human-induced activities. Anthropogenic impacts, together with naturally occurring fluctuations, cause environmental problems. These changes are seen as the degradation of habitat structure and changes in species composition. In other words, there is a rapid change from the current situation in the form of deterioration of the ecosystem balance. It causes more severe fluctuations than naturally occurring effects on the ecosystem.

Aquatic ecosystems have been the most exposed to anthropogenic impacts from past to present. Many factors such as temperature increase, irregularities in the precipitation regime, climatic change, groundwater withdrawal, lake water level fluctuations, and changes in stream flow influence biological diversity. While the effect of invasive species on the community structure is gradually increasing, the number of species in that community decreases and losses of species are observed. Water is of critical importance due to some main reasons; first, water has biochemical and physiological importance for living creatures and secondly, it is an ecosystem as freshwater and marine systems. The function and balance of the ecosystem are ecological integrity that affects all living creatures. Another reason for importance of water is that it is used for waste removal and transport and also as energy resources in the industry (Barinova and Mamanazarova, 2021). Although water is considered as a renewable resource, it is a non-renewable resource and/or asset when considered as ecosystem integrity.

Consequently, energy flow between trophic levels deteriorates through the ecosystem due to spatial and temporal differences. At the level of the ecosystem, functional components play crucial roles ranging from decomposers to consumers, thereby bottom-up and top-down control processes balance the ecosystem functions. Primary producers have the function

*New Paradigms in Environmental Biomonitoring Using Plants*
DOI: https://doi.org/10.1016/B978-0-12-824351-0.00007-9

of energy storage and provide the energy needed for the heterotrophic communities. The permanency of the aquatic ecosystem is limited by the availability of nutrients. In this respect, phytoplankton and algae are vital organisms for the food web.

Algae, which are polyphyletic group as eukaryotic and cyanobacteria, can live in many habitats as benthic and pelagic waters on earth and they are also found as epiphyte, epilith and zoophyte. According to Huisman (2000) and Wu et al. (2017), algae were recorded during the ancient Chinese Empire in 3000 BC for the first time. At the present time, in addition to classical taxonomy, algae, whose molecular identification has been made, provide reliable information in the water quality assessment and biomonitoring of the ecosystems.

Interannual variability of algae/phytoplankton diversity is based on climatic change, water quality, water level fluctuations, and pollution. These changes have occurred due to main principles (Borics et al., 2021):

- species in algae community show different behaviors as sensitive and tolerant species.
- alteration in community structure and community parameters is caused by changed ecological situations.
- changed ecological situations as spatial and temporal qualifications change the annual pattern of species.

Phytoplankton and algae are sensitive organisms that immediately respond to environmental quality degradation by changing their population structure and community composition. Since phytoplankton/algae is the primary producer and they directly reflect any point and nonpoint changes in the aquatic ecosystem due to the effect of anthropogenic and non-anthropogenic stressors, a change in community composition makes them an important component in the food web. It reveals more accurate ecosystem quality than chlorophyll-*a* and nutrient concentrations. It is possible to explain the periodicity in abundance and composition of phytoplankton and algae assemblages at the integrity of biotic responses, from the molecular level to the growth and reproduction, and to the community level of functional groups.

## 3.2 Population dynamics and life strategies

In phytoplankton ecology, two functional approaches are taken: (1) determination of population dynamics is a major factor to identify morphological, physiological and behavioral characteristics; (2) the identification of distinctive functional units within phytoplankton communities leads to the improvement of the functional group (ecological groups) concepts. Ecological similarities among taxa reflect more resemblance between community composition and the ecological gradients.

Temporal differences of major factors such as light, temperature, and nutrients in natural ecosystems upon the biomass and productivity determine and alter the dominance of species. Critical factors are responsible for altering dominant vernal diatom populations and their replacement by species that are representative of other algal groups. The properties of aquatic habitats (e.g., circulation pattern, morphometry, and water level fluctuations) allow to identify the quality and quantity of primary producer composition. Therewithal, the loss in the different gross planktonic primary production rates of loss influences the algae to compete for dominance. Nutrients (nitrogen, phosphorus and dissolved silica), thermal stratification, and grazing by zooplankton can be considered as impacts.

MacArthur and Wilson (1967) stated the concept of r and K selected species which are placed on a continuum based on their life-history traits. Margalef's Mandala model combines the interactive effects of habitat mixing and nutrient conditions on the selection of phylogenetic morphotypes and their seasonal succession, and realizes this along a pattern of r versus K strategies (Margalef, 1978; Margalef et al., 1979). The Margalef's model was applied to red tide blooms to predict selection of harmful algal bloom (HAB) species in nutrient-rich waters with low turbulence (situations which except diatoms). While r-selected species are small and have high surface area to volume ratios, K-selected species have large dimensions, consists of large cells or large colonies, are resistant to grazing, and often motile. Reynolds (2006) classified freshwater phytoplankton within three principal adaptive strategies involving morphologies, growth rates, and behaviors. First, r-selected species is an invasive strategy such as *Chlorella* in which organisms encountering favorable resource and energy fluxes can embark upon the rapid resource processing, biosynthesis and reproduction that constitute growth; briefly, the invasive, r-selected, small, fast-growing, high surface-to-volume colonist (C) species. The second set of strategic adaptations has variously combined specific mechanisms for obtaining scarce or limiting resources in short supply. K-selected species presents a high surface-to-volume ratio because a relatively large size is essential for obtaining the diminishing resources. *Microcystis aeruginosa* is a good model of this strategy type. r and K-selected species are widely applied in marine and freshwater phytoplankton and algal ecology and biomonitoring studies.

Algal populations develop three life strategies: (1) rapid consumption of the resources available (competitors), C-strategists; (2) tolerance of nutrient and ecological factors stress, by efficient obtaining of the limited resource, S-strategists; and (3) tolerance of disturbance and improving situations of transient habitats and resources into biomass, R-strategists. Many phytoplankton and algal species present intermediate properties. Morphological and physiological properties are compatible with their ecologies. Reynolds (2006) explained that *Dinobryon*, *Dictyosphaerium*, *Coenochloris*, *Pseudosphaerocystis*, *Eudorina*, *Volvox*, *Aphanocapsa*, and *Aphanothece* are reflected in C–S strategy. Algae in the C–R strategy comprise predominantly centric diatoms of varying tolerance of turbidity and the *Scenedesmus* and *Pediastrum*. The R–S strategy is described by the slow-growing, long-surviving species such as *Planktothrix rubescens* and *Lyngbya limnetica*. Salmaso et al. (2015) emphasized that large diatoms are found in the prevernal-vernal period as R strategists, small flagellates are seen as C strategists at the end of the vernal period, and dinoflagellate and cyanobacteria were dominant as S strategists in the serotinal period in European lakes.

Because chemical and nutrient levels of water, temperature, and light are control factors on photosynthesis, they affect the algal physiological and morphological structure. Sevindik et al. (2017) explained that phytoplankton/algae functional groups (FGs) and their morphology are indicators of water quality. C strategists, dominant *Stephanodiscus neoastraea*, *Lindavia ocellata*, *Cryptomonas ovata*, and *Cryptomonas caudata* have been noted in different groups of FGs in vertical heterogeneity. S-strategist *Oscillatoria limosa* was found in deep waters. The results are convenient to the Padisák et al. (2010) where the C→R→S strategy structure in phytoplankton communities in Mediterranean basin lakes was determined.

## 3.3 Bioindicator species

The distribution of the species with a wide range of tolerance to cosmopolitan and ecological factors covers wide limno-geographic areas. In other words, the species can survive to the extent permitted by ecological factors of the habitat in which it exhibits its metabolic activities and successfully lives. Depending on time (temporal and spatial variability) and environmental degradation (natural and/or anthropogenic effects), limiting factors determine population size, species diversity, and community composition. Accordingly, the concept of Baas Becking comes in; "everything is everywhere: but the environment selects" (Becking and Canfield, 2015).

However, it is seen clearly that the species reflects environmental qualifications when the distribution of a certain population is restricted to a narrow ecological range. Cyanobacteria, diatoms, and chrysophytes are excellent bioindicators for physical and chemical composition (especially temperature, N and P concentrations) (Gökçe, 2016). Changes in ecological conditions depending on the time result in alterations in the bioindicator group. Characteristic species or communities that provide information about environmental factors such as water quality parameters in an ecosystem are included in the definition of bioindicators. In the aquatic ecosystem, the natural trophic states of structural arrangements and functional properties of primary producers are regulated significantly by two factors: nutrient concentration and temperature. Moreover, bioindicators are used to show the effects of direct and indirect anthropogenic environmental changes. The N: P ratio identifies which taxa are dominant or exist in these nutrient-influenced freshwater habitats. Total phosphorus (TP) and nitrogen (TN) concentrations are used to describe the trophic state in the freshwater ecosystem. Redfield (1958) explained in detail that the 16N:1P stoichiometric ratio reflects productivity in the aquatic ecosystem and is the most important limiting factor on phytoplankton and algae growth. The algal composition and abundance are dependent on P when $N:P > 17$, and on N when $N:P < 10$. In addition to these ratios, the amount of silica is particularly effective on the abundance and diversity of Bacillariophyta and other community representatives with a frustule structure.

This relationship has been discussed by Gökçe and Gülbenk (2019) concerning Tohma Stream ecosystem of the transboundary Euphrates River tributary. Because Bacillariophyta contains a significant amount of silica in the frustule, it can alter the silica concentration in the water. $SiO_2$ consumption by the dominant algae population causes temporal and spatial decreases in biodiversity. In this study, Bacillariophyta was the dominant group of total dominance (98.57%) and redundancy analysis (RDA) indicated that they were affected by silica ($SiO_2$) concentrations. The algal composition and abundance were dependent on phosphorus and the amount of nitrogen was the control factor on the abundance of cyanobacteria as well as epilithic diatom. Especially algal community components are dependent on P when $N:P > 17$ and on N when $N:P < 10$.

In the temperate latitudinal zone, Bacillariophyta has two maximum periods of abundance and diversity: the highest peak in the cold season and the lower second peak in the early summer (Gökçe and Gülbenk, 2019). The most abundant taxa *Gomphonema subclavatum*, *Synedra ulna*, and

*Cocconeis placentula* were recorded in December and January. In the second temporal peak, *Achnanthes minutissima*, *Navicula tripunctata*, and *Cymbella minuta* were recorded. The highest chlorophyll-*a* and chlorophyll-*b* values were observed due to the abundance of Bacillariophyta in the cold period (Gökçe and Gülbenk, 2019). Furthermore, the dynamics of dominant species and abundance in cyanobacteria community were strongly associated with the $SiO_2$ concentration.

In particular, the population increase seen in certain algae-like cyanobacteria reflects nutrient excess and eutrophication. Fish deaths can be observed when toxic algae increase (Díaz-Torres et al., 2021; Mardones et al., 2021). In cause–effect relationship, water temperature, light penetration, pH, dissolved oxygen, chemical oxygen demand, TN, TP amount, as well as zooplankton population abundance are seen as determinants of excessive phytoplankton/algae growth. *Microcystis*, *Dolichospermum*, *Aphanizomenon*, *Planktothrix*, *Nodularia*, *Cylindrospermopsis*, *Phormidium*, *Anabaenopsis*, *Gloeotrichia*, and *Oscillatoria* are the most common cyanobacterial bloom genus in freshwaters (Benayache et al., 2019).

Environmental conditions are effectively monitored by the use of bioindicator species due to their rapid response to ecological variability (Fig. 3.1).

Han et al. (2015) and Parmar et al. (2016) indicated standards on bioindicator selection as following:

- to be well-known taxonomic category and ecology,
- to be distributed in geographically widespread area,
- to reflect prominent habitat properties,
- to be directly affected by physical and chemical factors,
- to give early warning for a change,
- to be easy and economically beneficial for the research,
- to have many independent populations,
- to indicate the reaction of other species,
- to represent the ecology change caused by the anthropogenic pressure,
- to reflect on climate change,
- to be easy to monitor, to appear for a long time,
- to be important socially, economically and culturally.

Dokulil (2003) explained that the term of bioindicator is evaluated as biomass, density, and/or population growth due to presence/absence of algae as a result of environmental influences or specific stress responses. The bioindicator species reflects environmental degradation as a biosensor. Bioindicator species accurately represent the aquatic environment quality

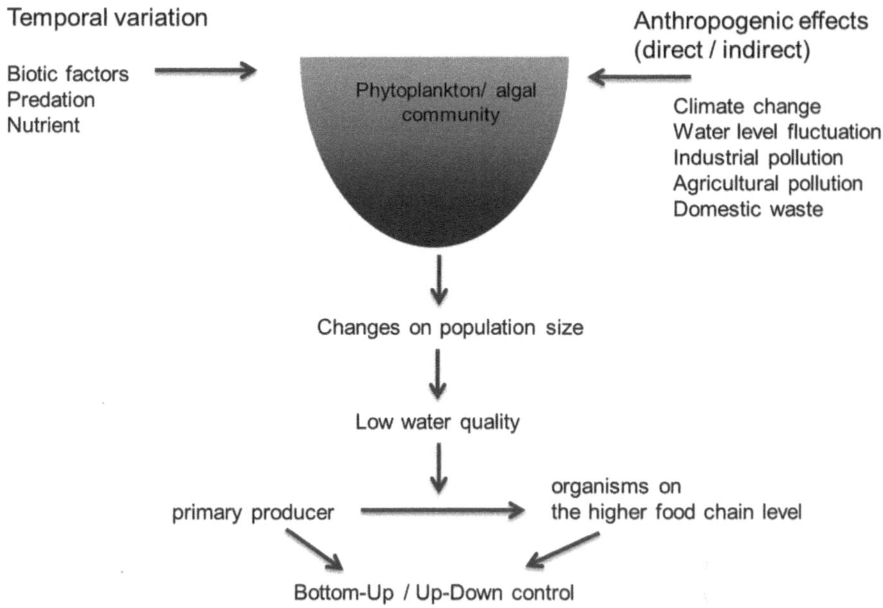


**Fig. 3.1** ***The energy is transferred to the upper levels due to bottom-up cascade feeding steps in aquatic ecosystems.*** However, environmental and anthropogenic limiting factors affecting water quality can disrupt energy transfer pathways or contaminants are transmitted to grazing invertebrates by accumulating. Therefore, primary producers play a major role.

and allow quality of ecosystem to be followed in long periods. Primary producer algae species are widely preferred as bioindicators. The compositions of chemical and physical conditions that are decisive in a freshwater determine the quality and quantity of the algal community. On the other hand, they play the role of bioaccumulators that accumulate pollutants and transfer them to the upper levels.

In monitoring study made using algae between 2009 and 2015, Barinova and Mamanazarova (2021) studied the water quality and diatoms of the Zarafshan River (Uzbekistan). *Cymbella*, *Navicula* and *Nitzschia* were reported as the richest species. Saprobic index, Water Ecosystem State Index, River Pollution Index, and Watanabe pollution indices were applied on diatom species. Diatoms were found to be tolerant of pollution of the river ecosystem. *Pleurosira laevis*, a mesohalobic species, has been described for the first time in Uzbekistan and Central Asia. As a result of the multimetric analysis, it was determined by diatom biomonitoring that the deterioration of the river's ecosystem structure resulting from climate change and drought was more effective than pollution. According to the study, diatoms can be good indicators of aquatic ecosystem quality in a semi-arid region and they demonstrate the climatic change and anthropogenic degradation levels.

Bioindicator species or upper taxonomic categories demonstrate impacts of abiotic/ biotic conditions in the aquatic ecosystem due to environmental changes. Therefore, bioindicator systems are advantageous than chemical analysis (Stevenson and Pan, 1999; Solak et al., 2007; Gökçe, 2016):

- Bioindicator organisms provide information about aquatic habitat change over time. Environmental physical and chemical analyses also measure and detect this change at the moment (snap-shots), but cannot detect environmental stress factors and changes on populations.
- The quality of the aquatic ecosystem is evaluated by considering the relationship between temporal and regional population size changes in organisms identified as indicators.
- Low concentration contaminants can be detected by chemical analysis. However, bioindicators show harmful effects by accumulating toxic substances in their tissues.
- It is difficult to determine the indirect effect of bioaccumulation by chemical analysis. Most pollutants are transported in the food web, increasingly from the lower level to the upper level through energy transfer (bottom-up mechanism). The changes in population parameters of bioindicator give more precise information than chemical analysis. Life tables, reproduction dynamics and cellular response levels (reactive oxygen system enzymes) provide this information.

Bioindicator species are moderately tolerant compared to rare and cosmopolitan species. This characteristic provides bioindicators with sensitivity in order to reflect the variability of environmental factors (Barlas, 1995; Reynolds, 2006; Holt and Miller, 2011; Sanal and Demir, 2018). Narrow tolerant species are very sensitive to ecological change. It is also seldom preferred to reflect the environmental response. Broad-tolerant species are less sensitive to disturbing ecological changes in the habitat (Holt and Miller, 2011).

While the quality of the ecosystem is directly measured with chemical and physical analysis, the use of bioindicators is safely preferred to evaluate the multicanonical effects of chemical pollutants depending on time and habitat quality (Barlas, 1995; Yorulmaz and Ertaş, 2021). Pollutants and environmental degradation may occur at very low concentrations or at irregular intervals. In this case, the evaluation of bioindicators reflects the past, present and future quality of the ecosystem. As shown in Fig. 3.1, existing pollutants are transported to the upper levels in the food web by means of bioaccumulation (bottom-up effects). Fluctuations in population dynamics display this situation and are monitored. Bioaccumulators are widely preferred for freshwater monitoring because of their quantitative response to environmental degradation. In addition, they provide reliable responses in the detection of point sources and intermittent pollution. Correctly selected bioaccumulator phytoplankton/algae is important for detecting environmental pollutants that cannot be determined by other indicators. Bellino et al. (2020) stated that the macrophytic algae *Chara gymnophylla* accumulated elements (Al, As, Ca, Cd, Co, Cr, Cu, Fe, K, Mg, Mn, Na, Ni, P, Pb, S, Si, V, Zn) for the experimental duration (7 months) and the algae can be used as a biomonitor.

Environmental indicators for detecting changes in environmental quality; ecological indicators for determining the stress factors affecting biota and its level; and biodiversity indicators for the long-term monitoring of taxon changes in the habitat (Gerhardt, 2011) are used. Through these methods, problems are determined with (1) diagnostic and (2) early warning indicators (Füreder and Reynolds, 2003; Hamza-Chaffai, 2014). Subgroup indicators such as environmental, ecological, and biodiversity indicators are observed in long-term monitoring in environmental deteriorations such as climatic causes and pollutants (Han et al., 2015; Parmar et al., 2016).

Another important role of algae in environmental quality indicators is that it is monitor-species in paleolimnology. Lentic habitat sediments reflect records of past physical, chemical and biological structure and habitat properties of lakes and their surrounding environments. Comparison of past fossil species with present-day diatom species gives sufficient information about environmental changes such as climate changes, eutrophication and pollution that the ecosystem has been exposed to from past to present (Dönmez et al., 2021; Ruwer et al., 2021; Shibabaw et al., 2021). Many paleolimnological records use diatoms due to their sensitivity to limnological characteristics and well-preserved silica cell walls (Avendaño et al., 2021; Levi et al., 2016; Maslennikova, 2020). The frustule structure of diatoms, one of the representatives of benthic algae, provides information about the consequences of anthropogenic stress factors and climate change from past to present. Diatom fossils in sediment samples directly reflect ecosystem changes. Therefore, they are used as long-term biomonitor fossil organisms in paleolimnological studies.

### 3.3.1 Assessment of indices

The evaluation and monitoring of water quality using algae populations and community structures were first studied by Cohn in 1853 (Krstić et al., 2007). Kolkwitz and Marsson (1902) defined the relationship of organisms with aquatic ecosystem quality and the concept of saprobic organisms. The concept of saprobic, which refers to the degree of organic matter pollution that reduces the oxygen content of a stream (mostly from domestic wastewater), was developed by researchers (Kolkwitz, 1950; Liebmann, 1962; Sládeček, 1973; Friedrich, 1990). Water quality was evaluated in terms of the presence/absence and abundance of the species or community quality. Algae communities vary depending on physical variables such as temperature, light, chemical composition and nutrients (ecosystem integrity). Due to this canonical relationship, many factors determine water quality and algae composition (Dokulil, 2003). For this reason, the necessity of developing different indices has emerged.

Plankton species are considered as indicators of nutrients, eutrophication, pollution, climatic warming, and aquatic ecosystem degradation (Özdemir et al., 2021). Enumeration and taxonomic classification of algae and phytoplankton community (quantitative and qualitative assessment) provide many indices. Determining the effects of factors that cause deviations and differences from the natural balance between algae communities and the ecosystem is important in determining the degree of threat. Therefore, qualitative and quantitative data are evaluated in biotic index and biomonitoring (Dalu et al., 2020). Table 3.1 illustrates the indicators used in algae-based biological monitoring programs. In water quality assessments, phytoplankton and algae are considered as indicator groups. Phytobenthos such as diatoms and filamentous algae are evaluated by multimetric indices in monitoring processes worldwide (Poikane et al., 2016).

In addition to general algal metrics, diatoms are widely preferred in water quality assessment and monitoring studies due to their seasonal sensitivity to the amount of nitrogen, phosphorus and silica salts (Gökçe and Gülbenk, 2019). Indices are obtained by following the diatom communities in lakes and rivers. Qualitative and quantitative observations and sampling are performed by standardized procedures from the points that identify the habitat for the accuracy of the indices used. On the other side, while cyanobacteria play a dominant role in the formation of eutrophication conditions, some species are characteristic of clean waters (Mantzouki et al., 2018). Some species are used as indicators in routine monitoring programs reflecting water quality according to their ecophysiological tolerance values and N and P concentrations.

**Table 3.1** Indices used in water quality assessment of phytoplankton/algae-based metric associations according to saprobic, trophic, metabolism, and community variables (adapted from Dokulil (2003); Maznah and Omar (2010); Poikane et al. (2011); Salmaso et al. (2015); Dalu et al. (2020); Solak et al. (2020)).

| | Indices | Abbreviation | References |
|---|---|---|---|
| **Pollution indices** | | | |
| | Algal genus/species Pollution Index | | Palmer and Adams (1977) |
| | Artois-Picardie Diatom Index | IDAP | Prygiel and Coste (1996) |
| | Biological Diatom Index | BDI | Lenoir and Coste (1996) |
| | Commission for Economical Community Metric—European Index | CEE | Descy and Coste (1991) |
| | Descy's Index | DES | Descy (1979) |
| | Generic Diatom Index | IDG | Coste and Ayphassorho (1991) |
| | Specific Pollution Sensitivity Index | IPS | CEMAGREF (1982) |
| | Pampean Diatom Index | IDP | Gómez and Licursi (2001) |
| | Biological Index of Water Quality | BIWQ | Lobo et al. (2004) |
| | Index of Biotic Integrity | IBI | Wang et al. (2005) |
| | Pollution Tolerance Index | | Lange-Bertalot (1979); Kelly and Whitton (1995) |
| **Saprobity** | | | |
| | Diatomic Index | IDES | Leclerq and Maquet (1987) |
| | Rott's Saprobic Metric | ROTTs | Rott et al. (1997) |
| | Saprobic Index | SLA | Sládeček (1986) |
| | Watanabe Index (Diatom Assemblage Index) | WAT-DAI | Watanabe et al. (1986) |
| **Trophic** | | | |
| | Brettum Index | | Brettum (1989) |
| | Carlson's Trophic State Index | TSI | Carlson (1977) |
| | Eutrophication/Pollution Index | EPI-D | Dell'Uomo (1996) |
| | The Mediterranean Phytoplankton Trophic Index | MedPTI | Marchetto et al. (2009) |
| | Proportion of taxa tolerant to organic pollution | %PT | Kelly and Whitton (1995) |

*(Continued)*

**Table 3.1** Cont'd.

| | Indices | Abbreviation | References |
|---|---|---|---|
| | Phytoplankton Trophic Lake Index | PTSI | Riedmüller et al. (2006) |
| | Steinberg and Schiefele's Index | SHE | Steinberg and Schiefele (1988) |
| | Rott's Trophic Metric | ROTTt | Rott et al. (1999) |
| | Swiss Diatom Index | DI-CH | Hürlimann and Niederhauser (2002) |
| | Trophic Diatom Index | TDI | Kelly and Whitton (1995) |
| | Trophic-Saprobic index | LOBO | Lobo et al. (2004) |
| | Photosynthetic capacity | | Rier and King (1996) |
| **Growth potential** | | | |
| | Catalán Index | | Catalan et al. (2003) |
| | Growth | | Fujimoto et al. (1997) |
| | Net production | | Ho (1976); Keithan and Lowe (1985) |
| | Ash-free-dry-weight | AFDW | Ho (1976); Hill et al. (2000) |
| | Chlorophyll-*a* | | Ho (1976); Hill et al. (2000) |
| | Cell biovolume | | Stevenson and Lowe (1986) |
| | Autotrophic Index | AFDW: chlorophyll-*a* | Putz (1997); Bourassa and Cattaneo (1998) |
| **Functional and morphometric attributes** | | | |
| | r and K selection | r/K | Margalef (1978; Reynolds); Reynolds (2006) |
| | Competitive, Stress-tolerant and Ruderal strategists | CSR | Reynolds (2006) |
| | Biomass size spectrum; | BSS, NBS | Platt and Denman (1978) |
| | Traditional Taxonomic Size Spectrum | TTSS | Kamenir et al. (2006) |
| **Community** | | | |
| | species richness | | Margalef (1958) |
| | diversity index | | Wiener (1948); Shannon and Weaver (1949); Menhinick (1964) |
| | evenness index | | Mulder et al. (2004) |

In Austria, Germany, Czech Republic, Poland, and Turkey, cyanobacteria are included in algal monitoring programs as well as Diatoms (Kelly, 2013; Mateo et al., 2015; Beklioğlu et al., 2020; Çelekli and Lekesiz, 2021). Furthermore, composition of both cyanobacteria and diatoms is recorded in monitoring programs in America (USA), Israel, Russia, and Iran. Additionally to these groups, chrysophytes and green algae are evaluated in terms of water quality as indicator organisms (Gökçe, 2016).

The Shannon diversity index (H′) is used to express the species diversity in the habitat. The H′ value reflects the number of species, as well as how the abundance of species is distributed among all species in the community. The Shannon index also reflects the pollution level of the aquatic habitat (H′ = 3.0–4.5 very light, 2.0–3.0 light, 1.0–2.0 medium, and 0.1–1 pollution level; Datta et al., 2010). Different pollution levels have different effects on algae diversity.

Using few indices remains insufficient to explain the stress factors; therefore, it is necessary to apply multiple indices for describing ecosystem integrity (Wu et al., 2017). Mao et al. (2018), Pandey et al. (2018) and Gocer et al. (2019) demonstrated that algae, especially diatoms, are considered as good indicators of water quality and respond rapidly to habitat degradation. This species diversity has been noted to decline in case of anthropogenic stressors.

Species composition and seasonal changes of phytoplankton and algae were classified by Reynolds (1997). The change of nutrient concentration (nitrogen and phosphorus) and temperature (its effect on metabolism and water circulation-stratification structure) that are effective on water quality control population dynamics and composition has been stated as an environmental factor (Fakıoğlu and Demir, 2011). Reynolds et al. (1997) explained in detail functional groups defined according to lake and stream qualities and coded each group based on their characteristics. These identified functional algae groups were revised by Dokulil (2003) and Padisák et al. (2003), and were present in Table 3.2.

Because of knowing bio-ecology of phytoplankton and algae, and predicting their responses to environmental variables using multimetric indices, different phytoplankton and algae associations have been evaluated in water quality studies worldwide. Dalu et al. (2020) followed the warm temperate Bloukrans River ecosystem quality (South Africa) in 2020. One hundred eighteen diatom taxa belonging to 36 genera have been identified. The stable isotope δ15N in phytobenthos provides a reliable short-term index of δ15N for monitoring in the primary producer signal. Δδ15N data

**Table 3.2** The variations of phytoplankton and algae compositions depending on the seasonal and trophic structure were discussed by Reynolds (1997) and revised by Dokulil (2003); Padisák et al. (2003).

| Codes | Habitat and trophic types | | Indicator species |
|---|---|---|---|
| A | Oligotrophic | Clear water, well-mixed lakes | *Urosolenia, Cyclotella comensis* |
| B | Oligotrophic | Mixed, mesotrophic small to medium sized lakes | *Aulacoseira subarctica, Aulacoseira islandica* |
| C | Eutrophic | Mixed, eutrophic, small to medium sized lakes | *Asterionella formosa, Aulacoseira ambigua, Stephanodiscus rotula* |
| D | Hertropic | Shallow, nutrient-rich, turbid waters (including rivers) | *Synadra acuis, Nitzschia* spp. *Stephanodiscus hantzschii* |
| E | Mesotrophic | Oligotrophic nutrient-rich, small lakes or heterotrophic pools | *Dinobryon, Mallomonas (Synura)* |
| F | Oligotrophic | Clear epilimnion | *Colonial Chlorophytes, Botryococcus, Pseudosphaerocystis, Coenochloris, Oocystis lacustris* |
| G | Eutrophic | Shallow, nutrient-rich waters | *Eudorina, Volvox* |
| $H_1$ | Eutrophic | Nitrogen-fixing Nostocalean species | *Anabaena flos-aquae, Aphanizomenon* |
| $H_2$ | | Nitrogen-fixing Nostocalean species, large mesotrophic lakes | *Anabaena lemmermanni, Gloeotrichia echinulata* |
| J | Eutrophic | Shallow, nutrient-rich lakes, ponds, streams | *Pediastrum, Coelastrum, Scenedesmus, Golenkinia* |
| K | Eutrophic | Nutrient-rich lakes | *Aphanothece, Aphanocapsa* |
| $L_0$ | Oligo-mesotrophic | Summer epilimnion, mesotrophic lakes | *Peridinium, Woronichinia, Merismopedia* |
| $L_M$ | Meso-eutropic | Summer epilimnion, eutrophic lakes | *Ceratium, Microcystis* |
| M | | Low latitude small eutrophic lakes with diurnal mixing | *Microcystis, Sphaerocavum* |
| $M_P$ | | Continuous mixed shallow lakes | *Surirella, Campylodiscus, Fragilaria construens* |
| N | Oligo-mesotrophic | Mesotrophic epilimnion | *Tabellaria, Cosmarium, Staurodesmus* |
| $N_A$ | Eutrophic | Oligo-mesotrophic lakes | *Cosmarium, Staurodesmus, Staurastrum* |

| Codes | Habitat and trophic types | | Indicator species |
|---|---|---|---|
| P | | Eutrophic epilimnion | *Fragilaria crotonensis, Aulacoseira granulata, Closterium aciculare, Staurastrum pingue* |
| Q | | Small humic lakes | *Gonyostomum* |
| R | | Mesotrophic mesolimnion | *Planktothrix rubescens, P. mougeotii* |
| $S_1$ | Eutrophic | Turbid mixed layers | *Planktothrix agardhii, Limnothrix redekei, Pseudoanabaena* |
| $S_2$ | | Turbid shallow mixed layers | *Spirulina, Arthrospira, Raphidiopsis* |
| $S_N$ | | Temperate mixed layers | *Cylindrospermopsis, Anabaena minutissima* |
| T | | Deep, well-mixed epilimnion | *Geminella, Mougeotia, Tribonema* |
| $T_B$ | | Stream | *Didymosphaenia geminata, Gomphonema, Fragilaria, Achnantes, Melosira varians* |
| $T_C$ | | Eutrophic lakes, slow flowing streams | *Oscillatoria, Phormidium, Lyngbya, Rivularia* |
| $T_D$ | | Mesotrophic lakes, slow-flowing streams and riparian macrophytes | Epiphyte diatoms and filamentous chloropyhtes |
| U | | Summer epilimnion | *Uroglena* |
| V | Meso-eutropic | Eutrophic metalimnion | *Chromatium, Chlorobium* |
| $W_1$ | | Ponds | *Euglenoids, Synura, Gonium* |
| $W_2$ | | Shallow mesotrophic lakes | Phytobenthos, *Trachelomonas* |
| $X_1$ | Eutrophic | Shallow, nutrient-rich mixing layers | *Chlorella, Ankyra, Monoraphidium* |
| $X_2$ | Eutrophic | Clear water, mixing shallow meso-eutrophic lakes | *Plagioselmis, Chrysochromulina* |
| $X_3$ | Oligotrophic | Clear water, shallow mixing layers | *Koliella, Chrysococcus,* Eukaryotic picoplankton |
| Z | Oligotrophic | Clear water, well-mixed layers | *Synechococcus*, Prokaryotic picoplankton |
| $Z_{MX}$ | | Deep oligotrophic lakes | *Synechococcus* spp., *Ceratium hirundinella* |

show anthropogenic nitrogen (N) inputs, identification, monitoring, and spatial effect of nonpoint organic pollution. The amount of phytobenthos δ15N was found to be significantly associated with the Trophic Diatom Index, Specific Pollution Sensitivity Index and Artois-Picardie Diatom Index (Dalu et al., 2020).

Çelekli and Lekesiz (2021) stated that *Lagerheimia genevensis* was considered as an indicator for slightly to moderately trophic-rich in European ecosystems. Also, species was only recorded in the brackish ecosystems and it was correlated with electrical conductivity values in the West Mediterranean basin in Anatolia. Epipelic diatom communities are widely used as an indicator of water quality and pollution (Akbulut, 2003; Solak et al., 2012; Solak et al., 2020).

## 3.4 Biomonitoring

In general, evaluation of the biological response of certain species or certain communities to monitor and evaluate changes in the ecosystem due to anthropogenic effects is included in the concept of biomonitoring. Biomonitoring can be considered qualitatively, semi-quantitatively or quantitatively. Aquatic macrophytes, algae, phytoplankton, benthic macroinvertebrates, and fish species and population dynamics are used for monitoring organisms. As described above, the presence or absence of an indicator species or community exhibits ecological conditions. The absence of many species from different species with similar tolerance levels previously found in the same habitat is more important than the absence of a single species, indicating the degree of impact of the pollution. A whole algal community survey can contain the different parameters such as taxonomy, abundance, dominance and diversity, thus providing a more defined ecosystem status (Sagova-Mareckova et al., 2020). The different indices used for phytoplankton and algae community composition describe how to survive under ecosystem conditions (Çelekli and Lekesiz, 2021; Padisák et al., 2009; Sanal and Demir, 2018; Solak et al., 2007). The utilization of phytoplankton/algae indices in the assessment of water quality has undertaken a critical role in freshwater management.

Environmental monitoring includes biological test methods, bioanalytics, and biomonitoring (Szczerbińska and Gałczyńska, 2015). Biological monitoring is an implementation of biological response for the assessment of ecosystem change of using organisms in the quality control programs (Maznah and Omar, 2010). Biomonitoring aims at evaluating the state of the natural ecosystems and levels of pollution.

Basically, the concept of monitoring covers reliable observations from nature in order to measure, evaluate, and draw conclusions about how species and ecosystems change in time and space by natural or human influence. Monitoring has been defined by Hellawell (1991) as intermittent (regular or irregular) surveillance to determine the degree of compliance to a predetermined standard or the degree of an expected deviation from normal. Elzinga et al. (2001) gave the monitoring concept referring to collection and analysis of repeated observations or measurements to evaluate changes in conditions and progress towards the management goal in the ecosystem. In line with these definitions, population, community, and ecosystem changes can be monitored with repetitive standard data in temporal and spatial terms.

According to (Tucker et al., 2005), a monitoring survey could include:

- determining the standards that include monitoring,
- detecting changes in population/community and defining responses,
- identifying the reasons for these changes,
- establishing measures to maintain standards or ensure that negative changes are improved,
- evaluation of management plans/strategies with the results of monitoring programs.

Biological monitors can be divided into three groups depending on their physiological characteristics, life time, and ecosystem sensitivities: bioindicators, biomonitors, and test organisms. Some biomonitor organisms can be considered "passive" according to their life cycle; some of organisms and populations are "active" exposed to the monitored pollutants and their responses are recorded (Szczerbiñska and Gałczyñska, 2015). For these two groups, "sensitive" or "cumulative" indicators can be used. The sensitive approach is more important in ecosystem quality monitoring and continuity. Active and passive biomonitors metabolically accumulate pollutants in their tissues in certain periods. In other words, they reflect water quality by providing both sensitive and accumulative response.

It can be asserted that bioindicators, which basically have different characteristics, are descriptive at distinct levels in monitoring. Sensitive organisms that accumulate environmental pollution in their metabolism respond to stress and act as an early warning system. These indicators can respond at the organism or cellular level. For this reason, they are considered as biomarkers. The quality and quantity of all kinds of pollutants or factors that damage environmental quality lead to multiple complex effects on ecosystem dynamics. Therefore, monitoring of multiple parameters in cellular response (multi-biomarker) is critical in long-term biomonitoring. As previously mentioned, organisms, which are ecological, biodiversity, and environmental indicators, are used in monitoring studies from species level to community-ecosystem level. Monitoring from cellular response to population structure determines that ecosystem has characteristic with the same organism. Ecosystem or landscape-level indicators are identified according to sensitive stress response patterns (Fig. 3.2).

While the changes in the ecosystem are reflected by the qualitative change of bioindicators, biomonitors are used in environmental monitoring studies with quantitative structure (Jongman et al., 1995; Markert et al., 2003; Parmar et al., 2016; Sanal and Demir, 2018; Tokatlı et al., 2020). These data evaluate precisely the level of current and past change of ecosystem health by assessing temporal and spatial changes. Biomonitoring is the main element of studies on water quality and water pollution. It plays an important role in anthropogenic effects and taking precautions.

According to phytoplankton/algae community structure in ecosystem monitoring shows spatial and temporal variations. Remote sensing is one of the important monitoring tools used in recent years in monitoring according to pigment quality and concentration. Remote sensing of bloom is the measurement of changes in algae or cyanobacteria biomass, changes in the color of the water that can be detected by optical sensors by satellites (Kutser et al., 2006; Binding et al., 2020; Zilius et al., 2021). Depending on the different pigment contents of phytoplankton and algae groups, absorption levels of chlorophyll-*a*, and phycocyanin pigment concentrations showed that cyanobacteria caused a more intense algal bloom (Mantzouki

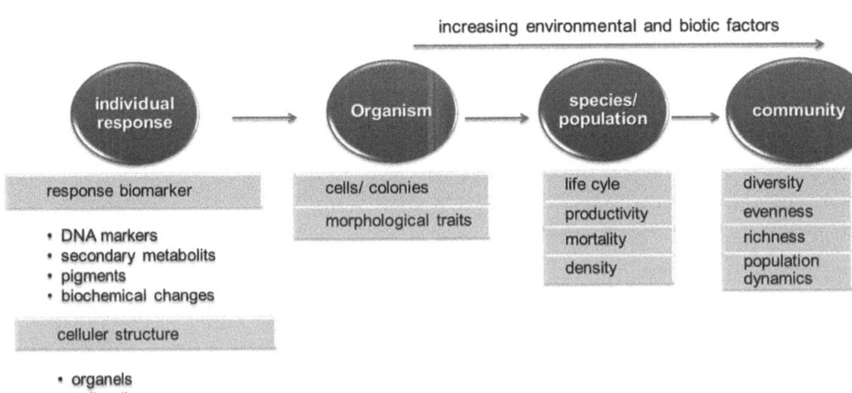


**Fig. 3.2** ***Stress responses are observed in organisms from cellular level to population structure and community composition.***

et al., 2018). Phycocyanin is a commonly used for cyanobacteria marker pigment and forms the basis of many proposed remote sensing algorithms for detecting cyanobacteria (Kutser et al., 2006; Pirasteh et al., 2020). Excessive algae growth depends on other water quality parameters, especially N and P, in that ecosystem. Cyanobacteria harmful algal blooms often have severe harmful effects on lacustrine habitats, drinking water reservoirs and recreational areas. For this reason, remote detection of blooms in the Great Lakes and Lake Erie in Canada and the USA is routinely applied (Matthews et al., 2020; Binding et al., 2020). Therefore, monitoring of pigment concentration by remote sensing provides information about the amount of algae biomass so that water quality is monitored. This provides important information about the management plans including eutrophication and water quality improvement studies.

## 3.5 Water management

Biomonitoring is an important component of sustainable and effective management plans for water quality improvement and protection. An acceptable management approach would be more favorable for species that allow relatively easy data collection and are abundant. Endangered and vulnerable species are often found at low densities and data about them are more limited. Therefore, sampling studies for both unethical and rare species in feasible management plans are likely to be misleading. Monitoring is performed based on data related to population abundance and density.

Fluctuations in population dynamics vary depending on the characteristics of that habitat (microhabitat structure), nutrients, and seasons. These results also differ over the years due to the changes in environmental factors. Population estimates for a species may vary depending on the timing of data collection (Fig. 3.3). Therefore, area, duration, and sampling time are important for obtaining reliable results in regular biomonitoring studies. Thus, it may be possible to determine environmental quality. In order for only one bioindicator/biomonitor organism to be sufficient in evaluating the system in the ecosystem, it should be evaluated by different indices in the long term.

In Europe, 60% of surface water ecosystems do not have a good ecological status and are exposed to environmental pollutant factors. For this reason, it has become necessary to prepare and develop sustainable integrated water management policies by evaluating bioindicator and biomonitor organisms (Baattrup-Pedersen et al., 2019). Because the protection and

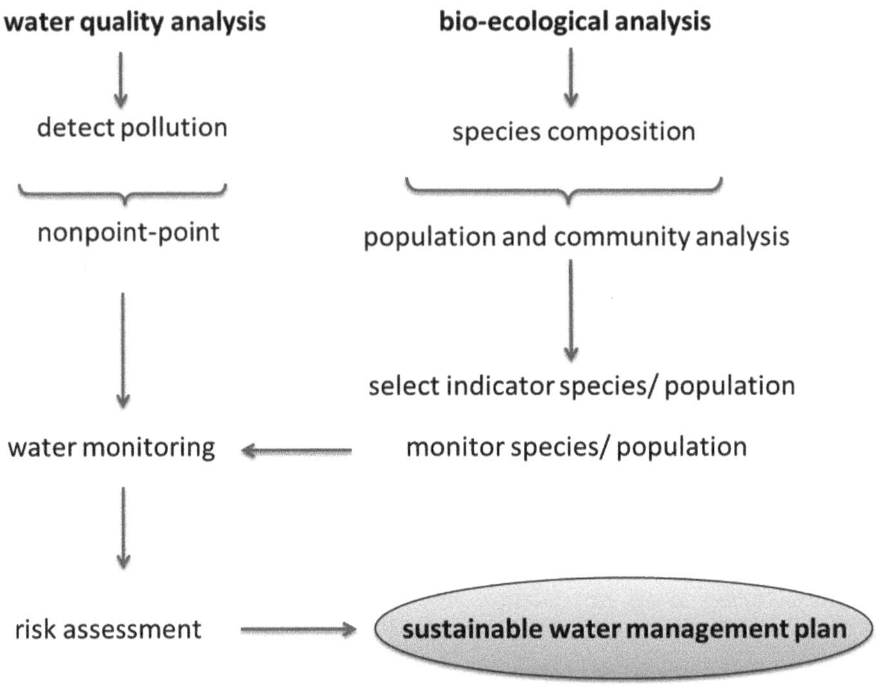


**Fig. 3.3** ***Allochthonous pollutant has an effect on the aquatic ecosystem affects the physical and chemical quality of that system.*** Organisms living in that habitat (bioindicator, biomonitor) are directly and indirectly related to ecosystem components. For example, contaminants (organic, metal, plastic, other industrial wastes, etc.) affect primary producer algae and thus other consumer organisms. In this case, water quality variables as well as organisms and species associations reflecting pollution provide valuable information about water monitoring. The deviation from the homeorhesis equilibrium state is evaluated through risk assessment and management plans are prepared against environmental degradation.

sustainability of water resources are the priority area of the EU water policy. The Water Framework Directive (WFD; 2000/60/EC) was adopted in 2000 to implement assessment systems for all major groups of aquatic organisms (Poikane et al., 2011). Biological monitoring is the most significant monitoring method for water quality monitoring according to WFD and includes living groups (such as phytoplankton, phytobenthos, macrophyte, macroalgae, angiosperm, macroinvertebrates, fish) are considered as monitor organisms (Krstić et al., 2007; Poikane et al., 2011). Biological Quality Elements within the scope of WFD monitoring provisions are evaluated with standard index methods. When phytoplankton composition, abundance and biomass data are analyzed; the frequency of monitoring in this group is determined as 6 months in EUWFD. The Directive states that it would be beneficial for countries having insufficient data for monitoring studies to carry out biological monitoring every year and three seasons a year until sufficient data are obtained.

The US Environmental Protection Agency (EPA) implements the water quality protocol on aquatic ecosystem quality monitoring. Clean Water Act regulations were made in 1972 in order to ensure chemical, physical, and biological integrity (Hill et al., 2020). Accordingly, chemical (e.g., N, P, dissolved oxygen), physical and biological indicators are used as quality indicators (WQI) in ecosystem monitoring. WQI indicators reflect the ecosystem integrity of water resources. For this purpose, multimetric indices and observed-to-expected (O/E) taxa compositions are evaluated in EPA National Aquatic Resources Surveys (NARS) (Hill et al., 2020). The richness, evenness, diversity, pollution tolerance taxa, and taxonomic composition are used in the multimetric monitoring of aquatic habitats (Stoddard et al., 2008). Limiting factors in the environment affect in-situ water and habitat quality so that in-situ O/E scores are evaluated. The watershed ecosystem is evaluated by determining reference conditions and habitat. The same approach is used in EUWFD and good condition status is targeted.

## 3.6 Conclusion

Freshwater ecosystems have been unprotected against anthropogenic influence all over the world. Qualifications of pollutants from agricultural, industrial, domestic or urban inputs are significant for aquatic organisms and ecosystem quality. In addition, climate change also affects water ecosystems. Organisms adapt to environmental factors and changes. Within this framework, the definition of "natural" spatial heterogeneity should be determined by reliable analysis. The objective of water management is to assess the abovementioned responsiveness using biomonitoring results, with many ecosystem components.

Therefore, phytoplankton and algae are widely used in water quality research as an indicator and monitor organism. Algal indexes are sensitive to pollution parameters. In general, the use of indices in taxonomic classification at the species level reflects ecological integrity and ecosystem health status. Biological monitoring is performed to appreciate these impacts. Cyanobacteria and especially diatoms are the most widely used bioindicators. Biomonitoring and bioindicator not only reflect the polluting concentration (e.g., nanoparticles, microplastics, heavy metals, pesticide, wastewaters) in the habitats and particularly in the organisms living in the environment but also determine the effect of the previous pollution load even if there are no pollutants entering into the habitat during that time. Other ecological stressors such as drought or eutrophication and management practices such as irrigation also have a strong effect on aquatic habitats.

Ecosystem health is the desired goal in environmental management. However, this concept has not been easy to use because of the contradiction between water ecology and environmental economics. Due to climate change and anthropogenic pressures, freshwater resources are needed more than ever. It is clearly seen that aquatic monitoring and ecosystem health are very important values in the concept of environmental sustainability in the future.

## References

Akbulut, A., 2003. Planktonic diatom (Bacillariophyceae) flora of Sultan Sazlığı Marshes (Kayseri). Turk. J. Bot. 27, 285–301.

Avendaño, D., Caballero, M., Vázquez, G., 2021. Ecological distribution of *Stephanodiscus niagarae* Ehrenberg in central Mexico and niche modeling for its last glacial maximum habitat suitability in the Nearctic realm. J. Paleolimnol. 66, 1–14. https://doi.org/10.1007/s10933-021-00178-w.

Baattrup-Pedersen, A., Larsen, S.E., Rasmussen, J.J., Riis, T., 2019. The future of European water management: demonstration of a new WFD compliant framework to support sustainable management under multiple stress. Sci. Total Environ. 654, 53–59. https://doi.org/10.1016/j.scitotenv.2018.11.008.

Barinova, S., Mamanazarova, K., 2021. Diatom algae-indicators of water quality in the Lower Zarafshan River, Uzbekistan. Water 13 (3), 358. https://doi.org/10.3390/w13030358.

Barlas, M., 1995. Biological and chemical evaluation and criteria of river pollution (in Turkish). Eastern Anatolia Region I. and II, Fisheries Symposium, 465–479.

Becking, L.G.M.B., Canfield, D.E., 2015. Baas Becking's Geobiology. John Wiley & Sons, Ltd, Chichester, UK.

Beklioğlu, M., Bucak, T., Levi, E.E., Erdoğan, Ş., Özen, A., Filiz, N., et al., 2020. Influences of climate and nutrient enrichment on the multiple trophic levels of Turkish shallow lakes. Inland Waters 10 (2), 173–185. https://doi.org/10.1080/20442041.2020.1746599.

Bellino, A., Alfani, A., Riso, L. de, Gregorio, R., Pellegrino, T., Baldantoni, D., 2020. Long-established and new active biomonitors jointly reveal potentially toxic element gradients across spatial scales in freshwater ecosystems. Ecol. Indic. 118, 106742. https://doi.org/10.1016/j.ecolind.2020.106742.

Benayache, N-Y., Nguyen-Quang, T., Hushchyna, K., McLellan, K., Afri-Mehennaoui, F-Z., Bouaïcha, N., 2019. An overview of cyanobacteria harmful algal bloom (CyanoHAB) issues in freshwater ecosystems Ed. In: Gökçe, D. (Ed.), Limnology - Some New Aspects of Inland Water Ecology. IntechOpen.

Binding, C.E., Stumpf, R.P., Shuchman, R.A., Sayers, M.J., 2020. Advances in remote sensing of great lakes algal blooms. In: Crossman, J., Weisener, C. (Eds.). Contaminants of the Great Lakes. The Handbook of Environmental Chemistry, 101. Springer, Cham. https://doi.org/10.1007/698_2020_589.

Borics, G., Abonyi, A., Salmaso, N., Ptacnik, R., 2021. Freshwater phytoplankton diversity: models, drivers and implications for ecosystem properties. Hydrobiologia 848, 53–75. https://doi.org/10.1007/s10750-020-04332-9.

Bourassa, N., Cattaneo, A., 1998. Control of periphyton biomass in Laurentian streams (Québec). J. North Am. Benthol. Soc. 17 (4), 420–429.

Brettum, P., 1989. Alger som indikator på vannkvalitet i norske innsjøerPlanteplankton. NIVA, Trondheim, Norway.

Carlson, R.E., 1977. A trophic state index for lakes. Limnol. Oceanogr. 22 (2), 361–369.

Catalan, J., Ventura, M., Munné, A., Godé, L., 2003. Desenvolupament d'un index integral de qualitat ecològica i regionalització ambiental dels sistemes lacustres de Catalunya. Agencia Catalana del Aigua. Generalitat de Catalunya. Agencia Catalana del Aigua Generalitat de Catalunya. Agencia Catalana del Aigua.

CEMAGREF, 1982. Etude des Méthods Biologiques Quantatatives d'Appréciation de la Qualité des Eaux. Rapport Division Qualité des Eaux Lyon. Lyon, French. Agence Financiére de Bassin Rhône, Mediterranée-Corse.

Coste, M., Ayphassorho, H., 1991. Etude de la qualité des eaux du Bassin Artois-Piccardie á l' aide des communautés des diatomées benthiques (application des indices diatomiques). Douai, France Rapport CEMAGREF, Bordeaux; Agence de l'Eau Artois-Picardie.

Çelekli, A., Lekesiz, Ö., 2021. Limno-ecological assessment of lentic ecosystems in the western Mediterranean basin (Turkey) using phytoplankton indices. Environ. Sci. Pollut. Res. Int. 28 (3), 3719–3736. https://doi.org/10.1007/s11356-020-10697-0.

Dalu, T., Cuthbert, R.N., Taylor, J.C., Magoro, M.L., Weyl, O.L.F., Froneman, P.W., et al., 2020. Benthic diatom-based indices and isotopic biomonitoring of nitrogen pollution in a warm temperate Austral river system. Sci. Total Environ. 748, 142452. https://doi.org/10.1016/j.scitotenv.2020.142452.

Datta, S.N., Chakraborty, S.K., Jaiswar, A.K., Ziauddin, G., 2010. A comparative study on intertidal faunal biodiversity of selected beaches of Mumbai coast. J. Environ. Biol. 6, 981–986.

Dell'Uomo, A., 1996. Assessment of Water Quality of an Apennine River as a Pilot Study for Diatom-Based Monitoring of Italian Watercourses. The Use of Algae for Monitoring Rivers II. Institut für Botanik Universität Innsbruck, Innsbruck, Austria.

Descy, J.P., 1979. A new approach to water quality estimation using diatoms. Nova Hedwig 64, 305–323.

Descy, J.P, Coste, M., 1991. A test of methods for assessing water quality based on diatoms. Verh. Intern. Limnol. 24, 2112–2216.

Díaz-Torres, O., Anda, J. de, Lugo-Melchor, O.Y., Pacheco, A., Orozco-Nunnelly, D.A., Shear, H., et al., 2021. Rapid changes in the phytoplankton community of a subtropical, shallow, hypereutrophic lake during the rainy season. Front. Microbiol. 12, 617151. https://doi.org/10.3389/fmicb.2021.617151.

Dokulil, M.T., 2003. Algae as ecological bio-indicatorsBioindicators & Biomonitors - Principles, Concepts and Applications. Elsevier, pp. 285–327.

Dönmez, E.O., Ocakoğlu, F., Akbulut, A., Tunoğlu, C., Gümüş, B.A., Tuncer, A., et al., 2021. Vegetation record of the last three millennia in central Anatolia: Archaeological and palaeoclimatic insights from Mogan Lake (Ankara, Turkey. Quat. Sci. Rev. 262, 106973. https://doi.org/10.1016/j.quascirev.2021.106973.

Elzinga, C.L., Salzer, D.W., Willoughby, J.W., Gibbs, D.P., 2001. Monitoring Plant and Animal Populations. Blackwell Scientific Publications, Abingdon.

Fakıoğlu, Ö., Demir, N., 2011. The use of phytoplankton communities in the assessment of the ecological state of lakes. Ankara Üniv Çevrebilimleri Derg 3 (1), 99–105.

Friedrich, G., 1990. Eine Revision des Saprobiensystems. Zeit. Wasser- und Abw.-Forrsch. 23, 141–152.

Fujimoto, N., Sudo, R., Sugiura, N., Inamori, Y., 1997. Nutrient-limited growth of *Microcystis aeruginosa* and *Phormidium tenue* and competition under various N:P supply

ratios and temperatures. Limnol. Oceanogr. 42 (2), 250–256. https://doi.org/10.4319/lo.1997.42.2.0250.
Füreder, L., Reynolds, J.D., 2003. Is *Austropotamobius pallipes* a good bioindicators? Bull. Fr. Pêche Piscic. 370-371, 157–163. https://doi.org/10.1051/kmae:2003011.
Gerhardt, A., 2011. Environmental Monitoring – Vol. I – Bioindicator Species and Their Use in Biomonitoring.
Gómez, N., Licursi, M., 2001. The Pampean Diatom Index (IDP) for assessment of rivers and streams in Argentina. Aquat Ecol 35 (2), 173–181. https://doi.org/10.1023/A:1011415209445.
Gocer, M., Koc, D., Sıvacı, R.E., 2019. Examining the effects of herbicide echo-5 EC on epilithic diatoms. Fresenius Environ. Bull. 28, 10096–10100.
Gökçe, D., 2016. Algae as an indicator of water quality Eds. In: Thajuddin, N., Dhanasekaran, D. (Eds.), Algae - Organisms for Imminent Biotechnology. InTech, pp. 1203–1217.
Gökçe, D., Gülbenk, H., 2019. Assessment of water quality and epilithic algae community structure of Tohma River, Turkey. Fresenius Environ. Bull. 28, 1203–1217.
Hamza-Chaffai, A., 2014. Usefulness of bioindicators and biomarkers in pollution biomonitoring. Int. J. Biotech. Well Indus. 3 (1), 19–26. https://doi.org/10.6000/1927-3037.2014.03.01.4.
Han, Y-G., Kwon, O., Cho, Y., 2015. A study of bioindicator selection for long-term ecological monitoring. J. Ecol. Environ. 38 (1), 119–122. https://doi.org/10.5141/ecoenv.2015.013.
Hellawell, J., 1991. Development of a rationale for monitoring editor. In: Goldsmith, F.B. (Ed.), Monitoring for Conservation and Ecology: Development of a Rationale for Monitoring. Chapman and Hall, London, pp. 1–14.
Hill, B.H., Herlihy, A.T., Kaufmann, P., Stevenson, R.J., McCormick, F.H., Burch Johnson, C., 2000. Use of periphyton assemblage data in an index of biotic integrity. J. North Am. Benthol. Soc. 19 (1), 50–67.
Hill, R., Moore, C., Doyle, J., Leibowitz, S.G., Ringold, P., Rashleigh, B., 2020. Valuing Aquatic Ecosystem Health at a National Scale: Modeling Biological Indicators Across Space and Time. U.S. EPA, National Center for Environmental Economics Working Paper 20-04.
Ho, S.C., 1976. Periphyton production in a tropical lowland stream polluted by inorganic sediments and organic wastes. Arch. Hydrobiol. 77 (4), 485–494.
Holt, E.A., Miller, S.W., 2011. Bioindicators: using organisms to measure environmental impacts. Nat. Educ. Knowl. 2 (2), 8.
Huisman, J.M., 2000. Marine Plants of Australia. University of Western Australia Press, Australia.
Hürlimann, J., Niederhauser, P., 2002. Méthode d'Analyse et d'Appréciation des Cours d'eau en Suisse, Diatomées, Niveau R Région. Office Fédéral de l'Environnement, des Forêts et du Paysage OFEFP, Berne, Switzerland.
Jongman, R.H., Braak, C.F. ter, van Tongeren, O.F., 1995. Data Analysis in Community and Landscape Ecology. Cambridge University Press, Cambridge.
Kamenir, Y., Dubinsky, Z., Zohary, T., 2006. The long-term patterns of phytoplankton taxonomic size-structure and their sensitivity to perturbation: a Lake Kinneret case study. Aquat. Sci. 68 (4), 490–501. https://doi.org/10.1007/s00027-006-0864-z.
Keithan, E.D., Lowe, R.L., 1985. Primary productivity and spatial structure of phytolithic growth in streams in the Great Smoky Mountains National Park, Tennessee. Hydrobiologia 123 (1), 59–67. https://doi.org/10.1007/BF00006615.
Kelly, M., 2013. Data rich, information poor? Phytobenthos assessment and the Water Framework Directive. Eur. J. Phycol. 48 (4), 437–450. https://doi.org/10.1080/09670262.2013.852694.

Kelly, M.G., Whitton, B.A., 1995. The Trophic Diatom Index: a new index for monitoring eutrophication in rivers. J. Appl. Phycol. 7 (4), 433–444. https://doi.org/10.1007/BF00003802.

Kolkwitz, R., 1950. Ökologie der Saprobien. Schri Vere Wass Boden und Luft 4, 1–64.

Kolkwitz, R., Marsson, M., 1902. Grundsätze für die biologische Beurteilung des Wassers nach seiner Flora und Fauna. Mitt Prüfun Wasser und Abwasser 1, 33–72.

Krstić, S, Svirčev, Z, Levkov, Z, Nakov, T., 2007. Selecting appropriate bioindicator regarding the WFD guidelines for freshwaters—a Macedonian experience. Int. J. Algae 9 (1), 41–63.

Kutser, T, Metsamaa, L, Strömbeck, N, Vahtmäe, E., 2006. Monitoring cyanobacterial blooms by satellite remote sensing. Estuar Coast Shelf Sci 67 (1-2), 303–312. https://doi.org/10.1016/j.ecss.2005.11.024.

Lange-Bertalot, H., 1979. Pollution tolerance of diatoms as a criterion for water quality estimation. Nova Hedwig Beih 64, 285–303.

Leclerq, L, Maquet, B., 1987. Deux nouveaux índices chimique et diatomique de qualité d'eau courante: comparaison avec diérents indices existants. Cah Biol Mar 28, 303–310.

Lenoir, A, Coste, M., 1996. Development of a Practical Diatom Index of Overall Water Quality Applicable to the French National Water Board Network. Use of Algae for Monitoring Rivers II. Universität Innsbruck, Schmitten, Austria.

Levi, E.E., Bezirci, G., Çakıroğlu, A.İ., Turner, S., Bennion, H., Kernan, M., et al., 2016. Multi-proxy palaeoecological responses to water-level fluctuations in three shallow Turkish lakes. Palaeogeogr. Palaeoclimatol. Palaeoecol. 449, 553–566. https://doi.org/10.1016/j.palaeo.2016.02.052.

Liebmann, H., 1962. Handbuch der Frischwasser- und Abwasserbiologie. Fischer Verlag Bd. 2. Aufl. G.

Lobo, EA, Callegaro, V, Hermany, G, Bes, D, Wetzel, CA, Oliveira, MA., 2004. Use of epilithic diatoms as bioindicator from lotic systems in southern Brazil, with special emphasis on eutrophication. Acta Limnol. Bras. 16, 25–40.

MacArthur, R, Wilson, EO., 1967. The Theory of Island Biogeography, 2001 reprint ed. Princeton University Press, Princeton.

Mantzouki, E, Campbell, J, van Loon, E, Visser, P, Konstantinou, I, Antoniou, M, et al., 2018. A European Multi Lake Survey dataset of environmental variables, phytoplankton pigments and cyanotoxins. Sci. Data 5, 180226. https://doi.org/10.1038/sdata.2018.226.

Mao, S, Guo, S, Deng, H, Xie, Z, Tang, T., 2018. Recognition of patterns of benthic diatom assemblages within a river system to aid bioassessment. Water 10 (11), 1559. https://doi.org/10.3390/w10111559.

Marchetto, A, Padedda, BM, Mariani, MA, Luglie, A, Sechi, N., 2009. A numerical index for evaluating phytoplankton response to changes in nutrient levels in deep Mediterranean reservoirs. J. Limnol. 68 (1), 106–121.

Mardones, JI, Paredes, J, Godoy, M, Suarez, R, Norambuena, L, Vargas, V, et al., 2021. Disentangling the environmental processes responsible for the world's largest farmed fish-killing harmful algal bloom: Chile, 2016. Sci. Total Environ. 766, 144383. https://doi.org/10.1016/j.scitotenv.2020.144383.

Margalef, R., 1958. Information theory in ecology. Gen. Syst. 3, 36–71.

Margalef, R., 1978. Life-forms of phytoplankton as survival alternatives in an unstable environment. Oceanol. Acta 1 (4), 493–509.

Margalef, R., Estrada, M., Blasco, D, 1979. Functional morphology of organisms involved in red tides, as adapted to decaying turbulence. In: Taylor, D., Seliger, H. (Eds.), Toxic Dinoflagellate Blooms. Elsevier, New York, pp. 89–94.

Markert, BA., Breure, AM., Zechmeister, HG., 2003. Definitions, strategies and principles for bioindication/biomonitoring of the environmentBioindicators & Biomonitors - Principles, Concepts and Applications. Elsevier, pp. 3–39.

Maslennikova, AVE., 2020. Development and application of an electrical conductivity transfer function, using diatoms from lakes in the Urals, Russia. J. Paleolimnol. 63 (2), 129–146. https://doi.org/10.1007/s10933-019-00106-z.

Mateo, P., Leganés, F., Perona, E., Loza, V., Fernández-Piñas, F., 2015. Cyanobacteria as bioindicators and bioreporters of environmental analysis in aquatic ecosystems. Biodivers. Conserv. 24 (4), 909–948. https://doi.org/10.1007/s10531-015-0903-y.

Matthews, MW., Bernard, S., Evers-King, H., Lain, LR., 2020. Distinguishing cyanobacteria from algae in optically complex inland waters using a hyperspectral radiative transfer inversion algorithm. Remote Sens. Environ. 248, 111981. https://doi.org/10.1016/j.rse.2020.111981.

Maznah, W., Omar, W., 2010. Perspectives on the use of algae as biological indicators for monitoring and protecting aquatic environments, with special reference to Malaysian freshwater ecosystems. Trop. Life Sci. Res. 21 (2), 51–67.

Menhinick, E.F., 1964. A comparison of some species-individuals diversity indices applied to samples of field insects. Ecology 45 (4), 859–861.

Mulder, C., Bazeley-White, E., Dimitrakopoulos, P.G., Hector, A., Scherer-Lorenzen, M, Schmid, B., 2004. Species evenness and productivity in experimental plant communities. Oikos 107, 50–63.

Özdemir, C.D., Saygı, Y., Gündüz, E., Demirkalp, F.Y., Karacaoğlu, Ç., 2021. Assessment of the zooplankton community structure of the coastal Uzungöl Lagoon (Kızılırmak Delta, Turkey) based on community indices and physicochemical parameters. Turk. J. Zool. 45 (1), 33–45. https://doi.org/10.3906/zoo-2006-9.

Padisák, J., Borics, G., Fehér, G., Grigorszky, I., Oldal, I., Schmidt, A., et al., 2003. Dominant species, functional assemblages and frequency of equilibrium phases in late summer phytoplankton assemblages in Hungarian small shallow lakes. Hydrobiologia 502 (1-3), 157–168. https://doi.org/10.1023/B:HYDR.0000004278.10887.40.

Padisák, J., Crossetti, L.O., Naselli-Flores, L., 2009. Use and misuse in the application of the phytoplankton functional classification: a critical review with updates. Hydrobiologia 621 (1), 1–19. https://doi.org/10.1007/s10750-008-9645-0.

Padisák, J., Hajnal, É., Naselli-Flores, L., Dokulil, M.T., Nõges, P., Zohary, T., 2010. Convergence and divergence in organization of phytoplankton communities under various regimes of physical and biological control. Hydrobiologia 639 (1), 205–220. https://doi.org/10.1007/s10750-009-0021-5.

Palmer, C.M., Adams, S., 1977. Algae and Water Pollution. Office of Research and Development, Ohio.

Pandey, L.K., Sharma, Y.C., Park, J., Choi, S., Lee, H., Lyu, J., et al., 2018. Evaluating features of periphytic diatom communities as biomonitoring tools in fresh, brackish and marine waters. Aquat. Toxicol. 194, 67–77. https://doi.org/10.1016/j.aquatox.2017.11.003.

Parmar, T.K., Rawtani, D., Agrawal, Y.K., 2016. Bioindicators: the natural indicator of environmental pollution. Front. Life Sci. 9 (2), 110–118. https://doi.org/10.1080/21553769.2016.1162753.

Pirasteh, S., Mollaee, S., Fatholahi, S.N., Li, J., 2020. Estimation of phytoplankton chlorophyll-*a* concentrations in the Western Basin of Lake Erie using sentinel-2 and sentinel-3 Data. Can. J. Remote Sens. 46 (5), 585–602. https://doi.org/10.1080/07038992.2020.1823825.

Platt, T., Denman, K., 1978. The structure of pelagic ecosystem. Rapports et Proc'es-Verbaux des R´eunions, Conseil Perma- nent International pour l'Exploration de la. Mer 173, 60–65.

Poikane, S., van den Berg, M., Hellsten, S., Hoyos, C. de, Ortiz-Casas, J., Pall, K., et al., 2011. Lake ecological assessment systems and intercalibration for the European Water Framework Directive: aims, achievements and further challenges. Proc. Environ. Sci. 9, 153–168. https://doi.org/10.1016/j.proenv.2011.11.024.

Poikane, S., Kelly, M., Cantonati, M., 2016. Benthic algal assessment of ecological status in European lakes and rivers: challenges and opportunities. Sci. Total Environ. 568, 603–613. https://doi.org/10.1016/j.scitotenv.2016.02.027.

Prygiel, J., Coste, M., 1996. Recent Trends in Monitoring French Rivers Using Algae, Especially Diatoms. Institut für Botanik Universität Innsbruck, Schmitten, Austria.

Putz, R., 1997. Periphyton communities in Amazonian black- and whitewater habitats: community structure, biomass and productivity. Aquat. Sci. Res. Across Bound. 59 (1), 74–93.

Redfield, A., 1958. The biological control of chemical factors in the environment. Am. Sci. 46 (3), 230A–2221.

Reynolds, C.S., 1997. Vegetation Processes in the Pelagic. Ecology Institute, Oldendorf-Luhe, Germany.

Reynolds, CS., 2006. The Ecology of Phytoplankton. Cambridge University Press.

Riedmüller, U, Hoehn, E, Nixdorf, B, Mischke, U., 2006. Bewertung der natürlichen Seen Deutschlands anhand des Phytoplanktons. Deutsche Gesellschaft für Limnologie (DGL), Berlin, Germany.

Rier, ST, King, DK., 1996. Effects of inorganic sedimentation and riparian clearing on benthic community metabolism in an agriculturally-disturbed stream. Hydrobiologia 339 (1-3), 111–121. https://doi.org/10.1007/BF00008919.

Rott, E, Hofmann, G, Pall, K, Pfister, P, Pipp, E., 1997. Indikationslisten für Aufwuchsalgen in Österreichischen Fliessgewässern, Teil 1: Saprobielle Indikation Wasserwirtschaftskataster. Bundeministerium für Land- und Forstwirtschaft, Wienna, Astralia.

Rott, E, Pfister, P, van Dam, H, Pipp, E, Pall, K., 1999. Indikationslisten für Aufwuchsalgen. Bundeministerium für Land- und Forstwirtschaft, Wienna, Astralia.

Ruwer, DT, Osório, NC, Costa-Böddeker, S, Rodrigues, L., 2021. Influence of temporal changes on a diatom species in a floodplain lake: paleolimnological and biomonitoring approach. Diatom. Res. 36 (1), 1–9. https://doi.org/10.1080/0269249X.2020.1848927.

Sagova-Mareckova, M, Boenigk, J, Bouchez, A, Cermakova, K, Chonova, T, Cordier, T, et al., 2020. Expanding ecological assessment by integrating microorganisms into routine freshwater biomonitoring. Water Res. 191, 116767. https://doi.org/10.1016/j.watres.2020.116767.

Salmaso, N, Naselli-Flores, L, Padisák, J., 2015. Functional classifications and their application in phytoplankton ecology. Freshwater Biol. 60 (4), 603–619. https://doi.org/10.1111/fwb.12520.

Sanal, M, Demir, N., 2018. Use of the epiphytic diatoms to estimate the ecological status of Lake Mogan. Appl. Ecol. Environ. Res. 16 (3), 3529–3543. https://doi.org/10.15666/aeer/1603_35293543.

Sevindik, TO, Çelik, K, Naselli-Flores, L., 2017. Spatial heterogeneity and seasonal succession of phytoplankton functional groups along the vertical gradient in a mesotrophic reservoir. Ann. Limnol. Int. J. Lim. 53, 129–141. https://doi.org/10.1051/limn/2016040.

Shannon, CE, Weaver, W., 1949. The Mathematical Theory of Communication. University of Illinois Press, Urbana.

Shibabaw, T, Beyene, A, Awoke, A, Tirfie, M, Azage, M, Triest, L., 2021. Diatom community structure in relation to environmental factors in human influenced rivers and streams in tropical Africa. PLoS One 16 (2), e0246043. https://doi.org/10.1371/journal.pone.0246043.

Sládeček, V., 1973. System of water quality from the biological point of view. Arch. Hydrobiol. 7, 1–218.

Sládeček, V., 1986. Diatoms as indicators of organic pollution. Acta Hydrochim. Hydrobiol. 14 (5), 555–566. https://doi.org/10.1002/aheh.19860140519.

Solak, CN, Fehér, G, Barlas, M, Pabugcu, K., 2007. Use of epilithic diatoms to evaluate water quality of Akçay Stream (Büyuk-Menderes River) in Mugla/Turkey. Large Rivers 17 (3-4), 327–338. https://doi.org/10.1127/lr/17/2007/327.

Solak, CN, Ector, L., Wojtal, AZ, Ács, É, Morales, EA., 2012. A review of investigations on diatoms (Bacillariophyta) in Turkish inland waters. Nova Hedwigia 141, 431–462.

Solak, CN, Peszek, Ł, Yilmaz, E, Ergül, HA, Kayal, M, Ekmekçi, F, et al., 2020. Use of diatoms in monitoring the Sakarya River Basin, Turkey. Water 12 (3), 703. https://doi.org/10.3390/w12030703.

Steinberg, C, Schiefele, S., 1988. Biological indication of trophy and pollution of running waters. Z. Wasser Abwasser-Forsch 21, 227–234.

Stevenson, RJ, Lowe, RL., 1986. Sampling and Interpretation of Algal Patterns for Water Quality Assessment. Rationale for Sampling and Interpretation of Ecological Data in the Assessment of Freshwater Ecosystems. American Society for Testing and Materials, Philadelphia, USA.

Stevenson, R.J., Pan, Y., 1999. Assessing environmental conditions in rivers and streams with diatoms. In: Stoermer, E.F., Smol, J.P. (Eds.), The Diatoms. Cambridge University Press, Cambridge, pp. 11–40.

Stoddard, J.L., Herlihy, A.T., Peck, D.V., Hughes, R.M., Whittier, T.R., Tarquinio, E., 2008. A process for creating multimetric indices for large-scale aquatic surveys. J. North Am. Benthol. Soc. 27 (4), 878–891. https://doi.org/10.1899/08-053.1.

Szczerbiñska, N., Gałczyñska, M., 2015. Biological methods used to assess surface water quality. Arch. Polish. Fish 23 (4), 185–196. https://doi.org/10.1515/aopf-2015-0021.

Tokatlı, C., Solak, C.N., Yılmaz, E., 2020. Water quality assessment by means of bio-indication: a case study of Ergene River using Biological Diatom Index. ASE 35 (2), 43–51. https://doi.org/10.26650/ASE2020646725.

Tucker, G., Bubb, P., Heer, M. de, Miles, L., Lawrence, A., Bajracharya, S.B., et al., 2005. Guidelines for Biodiversity Assessment and Monitoring for Protected Areas, KMTNC, Kathmandu, Nepal. UNEP, Cambridge.

Wang, Y-K., Stevenson, R.J., Metzmeier, L., 2005. Development and evaluation of a diatom-based Index of Biotic Integrity for the Interior Plateau Ecoregion, USA. J. North Am. Benthol. Soc. 24 (4), 990–1008. https://doi.org/10.1899/03-028.1.

Watanabe, T., Asai, K., Houki, A., 1986. Numerical estimation to organic pollution of flowing water by using the epilithic diatom assemblage—diatom assemblage index (DAIpo). Sci. Total Environ. 55, 209–218. https://doi.org/10.1016/0048-9697(86)90180-4.

Wiener, N., 1948. Cybernetics. Wiley, New York.

Wu, N., Dong, X., Liu, Y., Wang, C., Becking -Pedersen, A., Riis, T., 2017. Using river microalgae as indicators for freshwater biomonitoring: review of published research and future directions. Ecol. Indic. 81, 124–131. https://doi.org/10.1016/j.ecolind.2017.05.066.

Yorulmaz, B., Ertaş, A., 2021. Water quality assessment of Selendi Stream and comparative performance of the indices based on benthic macroinvertebrates and physicochemical parameters. Biologia 76, 2599–2607. https://doi.org/10.1007/s11756-021-00756-3.

Zilius, M., Vybernaite-Lubiene, I., Vaiciute, D., Overlingė, D., Grinienė, E., Zaiko, A., et al., 2021. Spatiotemporal patterns of $N_2$ fixation in coastal waters derived from rate measurements and remote sensing. Biogeosciences 18 (5), 1857–1871. https://doi.org/10.5194/bg-18-1857-2021.

CHAPTER 4

# Urban roadside trees as eco-sustainable filters of atmospheric pollution: A review of recent evidence from atmospheric trace elements deposition

**Hamza Badamasi**
Department of Chemistry, Federal University Dutse, Dutse, Jigawa State, Nigeria

## 4.1 Introduction

Atmospheric pollution has been recognized among the most significant environmental challenges, especially in urban areas, owing to its direct negative consequences on ecosystems and public health (Zhang et al., 2018). It is accounting for roughly one out of every nine deaths in the world annually (WHO, 2016). Increased population explosion, rapid industrial development, urban sprawl, and traffic congestion have contributed to the release of large amounts of harmful atmospheric trace elements (ATEs) into the atmosphere in recent times (Kumar et al., 2013; Tibbetts, 2015; Kiliç, 2019). ATEs have received considerable attention because of their toxic effects, nondegradability, and persistence in the nature (Ali et al., 2019; Ouro-Sama et al., 2020). Some ATEs present in the atmosphere like copper, zinc and molybdenum are necessary for optimum functions of plants at lower doses when absorbed as soil solution (Rai, 2019). However, they are reported to be harmful at higher concentration levels (Ali et al., 2019). Other ATEs like As, Cd, Cr, and Hg are recognized to be harmful even at low levels and have no known biological function (Harguinteguy et al., 2016). ATEs are usually passed to living organisms from the abiotic environment and accumulated as the final link of food chain in the human body (Zhang et al., 2014; Olajumoke and Ojo, 2020), and can also easily reach the human respiratory organ, causing a broad range of pulmonary and cardiovascular illnesses (Rehman et al., 2017). Although the source of ATEs could be natural (geogenic), it has been reported to be predominantly derived from human (anthropogenic) activities (Sawidis et al., 2011; Popoola et al., 2018).

*New Paradigms in Environmental Biomonitoring Using Plants*
DOI: https://doi.org/10.1016/B978-0-12-824351-0.00004-3

The geogenic sources include mineral dust, volcano etc. (Venter et al., 2017), while the anthropogenic sources are automobile emissions, mining and smelting, biomass, and waste burning (Fang and Huang, 2011). Among the anthropogenic sources of ATEs emissions, automobile emission has been identified as one of the most serious challenges of urban air quality in recent years, responsible for between 90% and 95% of air pollution (Duong and Lee, 2011; Moreira et al., 2016). ATEs had been reported to be emitted through various components of automobiles like tailpipe, tire scrape, corrosion, lubricating oil, roadside soil resuspension, and brake abrade either through the dry or wet deposition (Khan et al., 2011; Deepalakshmi et al., 2014; Charron et al., 2018; Alexandrino et al., 2020; Sharma et al., 2020). Owing to the significant rise in human exposure to atmospheric contamination and the deterioration of air quality caused by increased ATEs emissions over the last decade, there is a need for better air quality monitoring and assessment strategies (Rai, 2019). Atmospheric pollution studies have been consistently constrained by the exorbitant cost of instruments and the complexity of conducting a comprehensive sampling in space and time (Tomašević and Aničić, 2010; Norouzi and Khademi, 2015). For these reasons, there is a rising demand for using vegetation to control atmospheric pollution (Fujiwara et al., 2011; Tomaševića et al., 2011).

Using plants for atmospheric biomonitoring is economically viable, efficient, and eco-friendly compared to conventional instrumental methods (Rai, 2013; Pandey et al., 2015; Badamasi, 2017; Roy et al., 2020). Numerous plant species such as mosses, lichen, trees, etc., have been applied for atmospheric pollution monitoring (Matin et al., 2016; Miri et al., 2016). Using trees for biomonitoring activities is becoming increasingly popular in recent time (Rai, 2016a; Mazari and Filippelli, 2020). The key advantages of trees compared to smaller vegetation are large quantities of biological matter, very large surface area, and the ease of species identification and sampling, and longevity (Sawidis et al., 2011; Rai, 2016b). Lower plants, such as mosses and lichens, are characterized by an inconsistent and uneven spread and should be sampled by experts capable of distinguishing among similar species (Norouzi et al., 2015). The aforementioned weaknesses get more noticeable in manufacturing and heavily populated areas where increased pressure from anthropogenic activities might cause shortages or perhaps even the absence of biological indicators at certain sampling locations (Rai, 2016a). Tree species commonly used for biomonitoring of traffic-related atmospheric contaminants include; *Aesculus hippocastanum* for Cu and Pb (Tomašević and Aničić, 2010), *Betula pendula* for Zn, Cd, and Cr

(Petrova et al., 2014), *Nerium oleander* L. for Al, Ba, and Fe (Doganlara et al., 2012), and *Poincianella pluviosa* for Cd, Fe, Ni, Mn, Pb, and Zn (de Souza et al., 2019). Trees growing on the side of the road had been demonstrated to display many physiological, anatomical, and biochemical changes due to the high accumulation of ATEs in their bodies (Rai, 2016b; Singh et al., 2020). Many trees had been reported to tolerate elevated levels of ATEs by modifying their morphological, physiological, and biochemical characteristics, and the more sensitive ones will act as bio-indicators of atmospheric pollution (Achakzai et al., 2017; Sen et al., 2017; Singh et al., 2020).

Several factors had been reported to influence ATEs distribution and accumulation in trees (Miri et al., 2016; Sharma and Mandal, 2017; Allahnouri et al., 2018; Khalid et al., 2018). For example, Deepalakshmi et al. (2014) reported traffic volume and distance to roadside as the main factors influencing the ATEs deposition and accumulation in plants. The accumulation of ATEs also depends on trees species and types of the element being investigated (Luo et al., 2011; Enche-Constantinescu et al., 2015). This was confirmed by Turkyilmaz et al. (2018) who reported significant differences between conifer trees species (*Abies bornmulleriana, Pinus nigra, Pinus sylvestris,* and *Picea pungens*) for the accumulation of ATEs. This chapter seeks to report the role of urban roadside trees as eco-sustainable sinks of atmospheric pollution with respect to ATEs emissions. The chapter would serve as reference guidelines for a proper understanding of the significance of urban trees in the sustainable control and management of atmospheric pollution.

## 4.2 ATEs deposition on urban roadside trees

Atmospheric deposition is the movement of contaminants from the atmosphere to aquatic and terrestrial habitats (Amodio, 2014). It is an important process that eliminates the pollutants from the atmosphere and, is a vital aspect controlling the biogeochemical cycles of ATEs. However, because of concerns about public health and global warming, atmospheric deposition is widely regarded as one of the world's most serious environmental issues (Lynam et al., 2015; Pan and Wang, 2015; Wu et al., 2018). Fig. 4.1 shows the pathways for dry and wet depositions of ATEs on trees. Dry deposition is the process of removing pollutants from the atmosphere through nonprecipitation under the influence of gravity (Mohan, 2016). Dry deposition is caused by a variety of mechanisms, including turbulent diffusion, sedimentation, Brownian diffusion, interception, inertial forces, electrical migration,

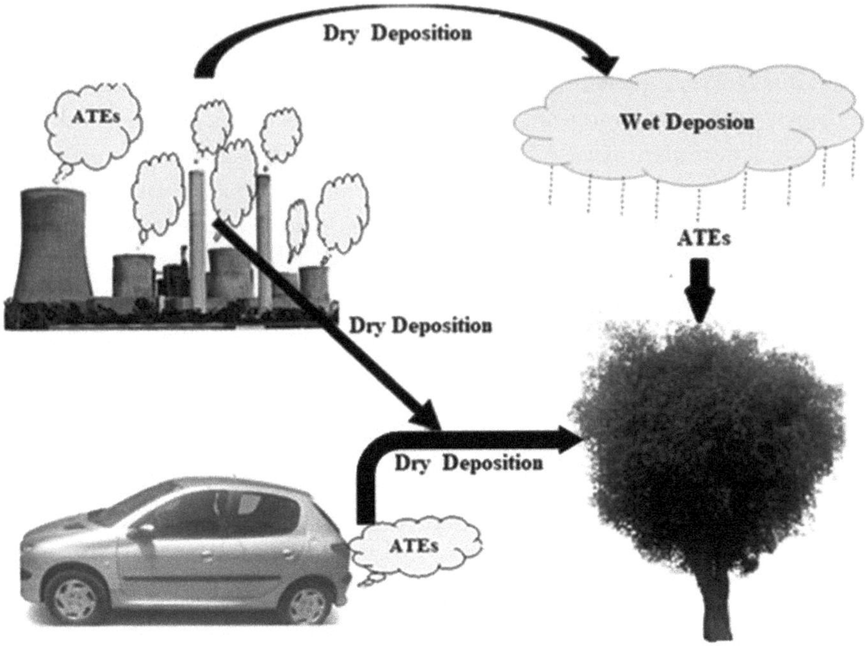


Fig. 4.1 *Dry and wet deposition pathways for ATEs.*

thermophoresis, and diffusiophoresis (Amodio et al., 2014). On a regional scale, the dry deposition process may be especially important close to urban and industrial areas where particle concentrations and sizes are large (Yang et al., 2012). Whereas wet deposition is a process in which atmospheric gases are combined with floated water in the atmosphere and then washed away by precipitation, snowfall, or smog (Selmi et al., 2016). ATEs removal from the atmosphere through wet deposition is commonly seen as an important natural regulating mechanism in atmospheric purification (Yang et al. 2012). Wet deposition is based on two different approaches: rain out and wash out (Chabbi et al., 2018). The rainout approach is the outcome of the absorption of soluble particulate matter and gasses by droplets of water in the same cloud. The wash-out approach refers to the leaking of the atmosphere under the clouds by rain that dissolves the particles and gasses during their falling (Chabbi et al., 2020).

Several factors are reported to influence the dry and wet depositions of ATEs (Schaubroeck et al., 2014; Selmi et al., 2016; Nie et al., 2017; Lu et al., 2018). These factors include seasonal variations, meteorological parameters such as wind speed and relative humidity, particle size and shape, and surface properties like friction speed, roughness and temperature and deposition velocity, etc. (Langner et al., 2011). For instance, Kim et al. (2012) investigated the wet deposition fluxes of ATEs in a rural area of Korea. Based on their findings, the wet deposition flux was significantly higher during the summer, due to large precipitation depths and snow was more effective scavenger of ATEs than rainfall. Sakata and Asakura (2011) also observed that dry deposition of ATEs are determined by their dry deposition rates and coarse particles derived from local sources. Zhu et al. (2016) observed that ATEs wet deposition was strongly associated with increasing industrialization, traffic density and burning of fuel. Chen et al. (2014) also noticed that the re-suspension of soils in the atmosphere may be favored by strong winds and the absence of vegetation, leading to more increase in dry deposition. In another study, Chabbi et al. (2018) reported that there was no association between lead, copper, cadmium, and Zinc and the amount of rainfall due to wet deposition in remote and urban areas of Monastir Region, Eastern Tunisia. A study of the total deposition (dry and wet) of ATEs in Daya Bay, South China Sea carried out by Wu et al. (2018) confirmed that, wet deposition flux of ATEs was significantly greater than a dry deposition. In an area with low precipitation, such as the Mediterranean regions, dry deposition is more important than wet deposition (Pan and Wang, 2015).

## 4.3 Biodiversity of urban roadside trees and their roles in ATEs deposition

As an important part of urban greenery, roadside trees are of great significance to biodiversity, leisure, and eco-sustainable atmospheric pollution filtration (Barwise and Kumar, 2020). Understanding tree diversity in urban roads can allow policymakers to make better decisions on the mitigation of atmospheric pollution and the preservation of urban biodiversity (Sawidis et al. 2011; Rai, 2016a; Mutlu et al., 2017). Several urban roadside trees such as *Artocarpus heterophyllus, Alstonia schlolaris, Azadirachta indica, Bombax ceiba, Bauhinia variegate, Bougainvillea spectabilis, Coreopsis auriculat, Cassia fistula, Cassia siamea, Delonix regia, Ficus religiosa, Ficus benghalensis, Hibiscus rosasinensis, Jacaranda mimosifolia, Lagerstroemia indica, Mangifera indica, Lagerstroemia speciose, Psidium. guajava,* and *Plumeria rubra* have been reported to be suitable for biomonitoring of ATEs (Alahabadi et al., 2017; Drava et al., 2017b; Kumar et al., 2013; Matin et al., 2016; Rai, 2016; Singh et al., 2020; Tak and Kakde, 2019).

Leaves and barks of urban roadside trees have been reported by numerous authors to be more efficient bioindicators than other organs of trees (Aničić et al., 2011). This was because, leaves and barks offer more surface area for proper deposition of ATEs than other parts of the tree (Matin et al. 2016; El-Khatib et al., 2020a). A study of ATEs accumulation conducted by Tomašević and Aničić (2010) between 2002 and 2006 indicated that *A. hippocastanum* accumulated more ATEs than *Tilia spp.*, and therefore more appropriate for monitoring specifically for lead and copper. Terekhina and Ufimtseva (2020) investigated trace elements in *Ulmus laevis*, *Tilia cordata*, *Populus sp.*, and *Rosa rugosa*. They recommended *Tilia cordata* as the most suitable bioindicator of ATEs pollution. Ukpebor et al. (2010) also evaluated the suitability *Delonix regia* and *Casuarina equisetifolia*, the two commonly grown ornamental trees in Nigeria. Their findings confirmed the effectiveness of the trees to be used for biomonitoring of ATEs, with *D. regia* moderately better in ATEs accumulation than *C. equisetifolia*. Bozdoğan (2016) determine the level of ATEs concentration in leaves of *Melia azedarach* that is commonly planted along the streets of Antakya, the central district of Hatay (Turkey). His result confirmed *M. azedarach* as a potential candidate ATEs biomonitoring. Olowoyo et al. (2010) assessed the suitability of *J. mimosifolia*, a tree found widespread in Tshwane Region, South Africa, for biomonitoring of ATEs (Pb, Cu, Zn, Cr. Cd, and V). Their findings confirmed that *J. mimosifolia* was suitable for biomonitoring ATEs. El-Khatib et al. (2020b) reported *Eucalyptus globulus* to be a good candidate for ATEs biomonitoring.

### 4.3.1 Evidence from foliar deposition

The foliar surface of the plant acts as an essential receiver, in which the leaves capture atmospheric particulate matter containing ATEs by wet and dry deposition (Simon et al., 2011; Wu et al., 2018). After wet or dry deposition, ATEs may be engrossed through leaves of trees, which are then distributed to other parts of the trees through the phloem in a very similar fashion as the photosynthetic process (Shahida et al., 2016; Alatou and Sahli, 2019).

The accumulation of ATEs through leaf systems of trees largely depends on multiple factors, such as cuticular nature, foliar morphology and surface area, speciation of the metals, pubescence, and roughness of leaves, life cycle of trees, and length of the exposure to atmospheric pollutants (Kardel et al., 2010; Simon et al., 2014; Shahid et al., 2016; Pace and Grote, 2020). It has been reported that the size of the tree leaves did not seem to affect the ATEs accumulation (Chen et al., 2017; Zhang et al., 2017; Muhammad et al., 2019). This was supported by Kumar et al. (2013) who discovered that *Tamarindus indicus* with shorter compound leaves collected more ATEs than trees with broader leaves. Shafiq et al. (2012) analyzed Pb and Cd in the leaves of some roadside trees of Karachi, Pakistan. Their findings revealed that *Pongamia pinnata* was better in capturing ATEs than *Peltophorum pterocarpum* due to its vast surface area.

Patel et al. (2015) investigated the concentrations some ATEs in the leaves of *Butea monosperma, Azadirachta indica, Mangifera indica, Eucalyptus, Ficus religiosa, Tectona grandis*, and *Mangifera indica*. Their results showed that *Azadirachta indica* leaves accumulated more trace metals than other trees. Zare et al. (2016) analyzed Cu, Ni, and Zn contents of *Cupressus arizonica* green leaves from traffic areas of Isfahan, Iran. The concentrations of studied ATEs were significantly higher in the leaves than the roots, indicating a considerable input of atmospheric deposition. Turkyilmaz et al. (2018) discovered that the leaves of *Abies bornmülleriana* accumulated a considerable amount of iron, lead, and zinc. This was also described by El-Khatib et al. (2020a) who established that *E. globulus* and *Ficus nitida* leaves trapped high concentrations of Cd and Pb from the atmosphere.

Scanning electron microscope (SEM) has been recognized as a promising tool for assessing the deposition and accumulation ATEs on trees leaves and barks. SEM helps in examining the role of morphology of leaves in the deposition of ATEs (Sæbø et al., 2015; Gajbhiye et al., 2019). For example, based on SEM imaging, it was evident that *C. siamea* accumulated the toxic ATEs from deposited particulate matter (Gajbhiye, et al., 2016). Roy et al. (2020) also confirmed from SEM study that *Ficus bengalensi, F. religiosa,* and *M. indica* exhibited the best ATEs capturing ability due to their coarse surface, a wide foliage surface area, and broad canopy system. Krupnova et al. (2020) observed considerable accumulation of Zn in birch tree leaves as confirmed by SEM image analysis. Based on SEM-EDX imaging report from Kocic′ et al. (2014), about 50%–60% of ATEs deposited on to the leaves of *T. spp and Aesculus hippocastrum*.

Numerous studies indicated that washing of leaves significantly reduces the amount of toxic ATEs present in urban roadside trees (Tomaševića et al., 2011; Serbula et al., 2012). This indicated that most of the deposited metals in tree leaves are attaching to the leaves surface instead of being absorbed into the leaf tissues (Serbula et al., 2012). Solgi et al. (2020) witnessed a significant reduction (20%–46%) of copper, lead, and zinc in the leaves of *Fraxinus excelsior* and *Pinus eldarica* after washing before analysis. Norouzi et al. (2015) confirmed that ATEs in the washed leaves of *Platanus orientalis* L. had been considerably reduced. The chronological list of some selected tree species whose leaves were used for ATEs biomonitoring is presented in Table 4.1.

### 4.3.2 Evidence of tree bark accumulation

Tree bark is usually very effective for the accumulation and retention of particulate matter due to its porous nature, making it an excellent organ for atmospheric contamination monitoring studies, especially for ATEs (Cocozza et al., 2016; Costagliola et al., 2017; Drava et al., 2017b). The benefit of the use of tree bark for monitoring of ATEs is that its structure

**Table 4.1** List of some selected tree species whose leaves were used for ATEs biomonitoring.

| Tree species | Location | Trace elements | References |
|---|---|---|---|
| *Aesculus hippocastanum* | Serbia | As, Cd, Cu, Ni, Pb, Ti, V | Tomašević and Aničić (2010) |
| *Azadirachta indica* | Katsina, Nigeria | Ni, Zn, Mn, Cu, Cr, Pb, | Lawal et al. (2011) |
| *Pongamia pinnata (L.) and Peltophorum pterocarpum* | Karachi, Pakistan | Pb, Cd | Shafiq et al. (2012) |
| *Quercus ilex* L. | Florence, Italy | As, Cd, Pb, Zn, Cu, Fe, Mn, Ba | Ugolini et al. (2013) |
| *Aesculus hippocastanum L., Acer platanoides, and Betula pendula* | Plovdiv, Bulgaria | Cr, Fe, Zn, Pb, Cu, Cd | Petrova et al. (2014) |
| *Tilia spp.* | Western Romania, Romania | Pb, Ni, Cr, Mn, Cu, Fe | Enche-Constantinescu et al. (2015) |
| *Pinus eldarica* | Yazd, Iran | Fe, Cu, Zn, Pb, Cd | Miri et al. (2016) |
| *Euonymus japonicus* | Beijing, China | Cr, Cu, Ni, Pb, Zn | Zhang et al. (2017) |
| *Athyrium esculentum, Chromolaena odorata, and Lantana camara* | Jengka, Malaysia | Cd, Cu, Fe, and Pb | Sulaiman and Hamzah (2018) |
| *Poincianella pluviosa* | Rio de Janeiro, Brazil | Cd, Cu, Fe, Mn, Ni, Pb, Zn | de Souza et al. (2019) |
| *Araucaria heterophylla* | Quito, Equador | Al, Ba, Cr, Co, Cu, Fe, Mn, Pb, Zn | Alexandrinoa et al. (2020) |

and the large surface area retains pollutants for a long period and is readily accessible without compromising the health of trees (Škrbić et al., 2012; Chiarantini et al., 2016; Chrabascz and Mróz, 2016). Ukpebor et al. (2010) and Chrabąszcz and Mróz (2016) reported that the bark of the trees has been found to accumulate more metals relative to the leaves and is commonly available all year round as opposed to leaves of the deciduous trees. Janta et al. (2016) confirmed that ATEs accumulated on the external layer of bark were certainly derived from air pollution as a result of direct exposure to the environment. de Souza et al. (2019) reported that the barks *P. pluviosa* tree accumulated high levels of cadmium, copper, manganese, and iron. Ejidike and Onianwa (2015) examined seven tree species namely, *A. indica, Gmelina arborea, Gliricidia sepium, M. indica, Musa spp, Prosopis juliflora,* and *Terminalia catappa* for their potential use in ATEs (Cd, Co, Cr, Cu, Pb, and Zn) biomonitoring. Their findings indicated that all the seven trees were potential candidates for ATEs biomonitoring. A significant concentration of copper, lead, cadmium, and zinc were reported to have been accumulated in the bark of *Cassia fistula* from road traffic flow in the city of Chiang Mai, Thailand (Janta and Chantara, 2017). Patel et al. (2020) recorded a significant accumulation of ATEs (As, Cu, Cr, Mn, Pb, and Zn) in the bark *of Aegie marmelos, Dalbergia sissoo, E. globulus, F. benghalensis, M. indica,* and *P. sylvestris* from the Indian district of Rajnandgaon. Table 4.2 lists some of the tree species whose barks were used for ATEs biomonitoring over a 10-year period.

## 4.4 Emissions sources of ATEs

Vehicular activity in roadside settings has been recognized as a significant source of ATEs emissions (Bhandarkar, 2013). Approximately 60%–70% of atmospheric pollution in the world is considered to be caused by the vehicle (Abhishek, 2018). ATEs bound Particulate matter linked to vehicular activities can be classified into three groups. Group one contains ATEs emission from vehicle tailpipe, i.e., exhaust emissions. Group two comprises non-tailpipe emissions, i.e., emissions from tire and brake wear, road abrasion, the motion of automobile parts, and lubricating oils. The final group contains resuspended soil particles caused by traffic movement (Winther and Slentø, 2010; Zhang et al., 2012; Crilley et al., 2017). Tailpipe (exhaust) and non-tailpipe emissions are the major sources of ATEs (Zhang et al., 2012). Burning of fuel, tire, and brake wear had been identified as the main factors that contribute to metals pollution in roadside soils (Pulles et al., 2012).

**Table 4.2** List of some selected tree species whose barks were used for ATEs biomonitoring.

| Tree species | Location | Trace elements | References |
|---|---|---|---|
| *Jacaranda, mimosifolia, Fagus sylvatica* | Shwane, South Africa, France & Germany | As, Cd, Co, Cr, Cu, Fe, Mn, Mo, Ni, Pb,V, Zn, Pb, Sb, Sn,V, Zn | Olowoyo et al. (2010), Guéguen et al. (2011) |
| *Pinus sylvestris* L. | Krakow, Poland | Cd, Pb, Zn | Chrzan and Marko-Worlowska (2012) |
| *Carica papaya* | Ebonyi, Nigeria | Cu, Pb, Zn | Chukwuka, and Uka, (2013) |
| *Populus nigra* L. | Romania | Ni, Cu, Zn | Barbes et al. (2014) |
| *Cupressus arizonica Green* | Tehran, Iran | Cu, Cr, Ni, Pb, Zn | Kord et al. (2015) |
| *Tipuana tipu* | Brazil | Al, Ba, Cu, Fe, Mn, Zn | Moreira T.C.L. et al. (2016) |
| *Quercus ilex* L.) | Liguria, Italy | As, Cd, Co, Cu, Fe, Mn, Ni, Pb, V, Zn | Drava et al. (2017b) |
| *Celtis australis* L.) | Istanbul, Turkey | Pb, Cd, Cu, Zn | Ozturk et al. (2018) |
| *Cupressus sempervirens var. fastigiata, Eucalyptus cladocalyx F. Muell* | Constantine, Algeria | Cd, Cu, Pb, Zn | Alatou and Sahli (2019) |
| *Acer rubrum* | Toronto, Canada | Cu, Hg, Ni, Mn, Pb, Zn | Yousaf et al. (2020) |

Tire corrosion and lubricants leakage may lead to release of Zn (Adamiec et al., 2016), and abrasion of tires may emits Cr, Cd, Cu, Zn, Pb, Hg, Ni, and Co (Apeagyei et al., 2011; Sahu et al., 2011; Hetem and Andrade, 2016). Brake scrape discharges Ba, Cr, Cu, Fe, Zn, Fe, and Pb to soil (Harrison et al., 2012). Vehicular exhaust emits Cd, Cu, Mn, Pb, Sr, and Zn (Song and Gao, 2011). ATEs released both by the exhaust pipe and non-exhaust channels endanger public and environmental health after they have been dispersed out from the atmosphere, primarily through soil and plant contamination (Khalid et al., 2018). Moreira et al. (2016) analyzed 171 tree bark samples for Al, Ba, Cu, Fe, and Mn, their findings revealed that the main source of ATEs was vehicular non-exhaust emissions primarily released from brakes and tires abrasion, and the re-suspension from roadside dust. Alexandrino et al. (2020) revealed using Pearson and principal component analyses that road traffic emissions were the main source of Cr, Cu, Co, Ba, Pb, and Zn. Janta and Chantara (2017) investigated the levels of some ATEs in the bark of *Cassia fistula* from the street of Chiang Mai, Thailand. The cluster and principal component analyses results showed that iron (Fe) and aluminum (Al) were released from re-suspended soil adhered to the surface of the road and automobile wheels whereas Cr, Cu, Pb, and Zn were emitted from vehicles directly.

## 4.5 Methods of selecting urban roadside trees for ATEs deposition

Atmospheric pollution resulting from rapid industrial development, urban sprawl, and, population growth has disrupted the primeval urban ecosystems (Rai, 2016). Identifying species of trees that can monitor air pollution biologically and can withstand toxic ATEs-induced pollution is very critical for successful and sustainable green space management around any contaminated sites (Sen et al., 2017; Singh et al., 2020). Choosing suitable tolerant tree species for environmentally sustainable growth is a reliable and cost-effective process (Chaudhary and Rathore, 2019). For selecting appropriate trees for biomonitoring purposes, trees should be rapidly propagated, easily obtainable, moderate-growing, economically and environmentally highly appropriate, pest and disease-resistant, and need minimal management (Kumar et al., 2013). These criteria should be given top priority. Other important criteria are that tree species should be identified in massive quantities in the monitoring sites, have a broad geographic spread, be able to distinguish between airborne and soil-borne trace metals, be easily sampled, and should have no identification problems (Zhang et al. 2012). Mori et al. (2015) and Chaudhary and Rathore (2018) proposed that a tree species with a plain and coarse surface of the leaves was best in ATEs capturing than a tree with a smooth leaf surface. Sett (2017) described that the tree species with horizontal leaf positioning are better in trapping particulate matter bound ATEs than a tree with vertically oriented smooth leaf. Chrabąszcz and Mróz (2016) cited some conditions for the selection of suitable tree barks for the monitoring of air pollution. According to them, a tree bark should be selected from the species commonly found in the region and should have familiar biology, so that sufficient sample size can be obtained. A species with rough and extremely porous bark, according to Moreira et al. (2018) is the most desirable for biomonitoring purposes. In assessing the resistance of plants to air pollution, plant morphology also plays a significant role. Some features, such as size stomata and cuticle, thickness of cell walls are, therefore, of particular significance for mitigating atmospheric pollution.

Another important approach for selecting tolerant and sensitive tree species for biomonitoring purposes is the use of tree's physiology and biochemistry to study the responses of trees to ATEs pollution (Rai, 2019). This can be accomplished by calculating the air pollution tolerance index (APTI) and anticipated pollution index (API) (Rai, 2016a). According to APTI and API, tolerant trees selected are used for urban green-belt development, while sensitive ones are used as biomonitors of environmental pollution (Roy, 2020; Sahu et al., 2020).

In APTI method, tree species are classified and rank base on their tolerance to atmospheric contamination (Rai, 2013). APTI is usually determined using the following relation (Sen et al., 2017; Hajizadeha et al., 2019):

$$APTI = \frac{A(T+P)+R}{100} \tag{4.1}$$

where; A = ascorbic acid, mg/g, P = pH, R = RWC,%, T = Chlorophyll (total), mg/g, APTI is classified as follows (Sen et al., 2017); 0–1 = Most sensitive, 2–16 = Sensitive, 17–29 = Intermediary, 30–100 = Tolerant.

The API can be measured by taking into account the APTI, socio-economic status, and biological factors (Javanmard et al., 2020). API shows the performance of trees affected by the structure of trees, socio-economic principles, and biochemical characteristics (Pandey et al., 2015). API is computed using the relation:

$$API(\%) = \frac{Total\ Positives}{16} \times 100 \tag{4.2}$$

API (%) is categorized as follows: Up to 30 = Not recommended, 31–40 = Very poor, 41–50 = Poor, 51–60 = Moderate, 61–70 = Good, 71–80 = Very good, 81–90 = Exellent, 91–100 = Best (Rai, 2019). The appropriate tree for the construction of urban green infrastructures and biomonitoring of air pollution can be evaluated by acquiring their unique API value. Typically, a maximum of 16 positives plants is assigned as proposed by Mondal et al. (2011).

## 4.6 Impacts of traffic size and road distance on the concentrations of ATEs

Traffic congestion is generally regarded as an important source of airborne and soil contamination along the urban roadsides (Khalid et al., 2018). Traffic is one of the major sources of ATEs pollution and plays a fundamental role in their biogeochemical cycle (Yan et al., 2012; Khan and Kathi, 2014; Olutona et al., 2017). Numerous research works have established that the concentrations of ATEs are usually declining when moving away from roadside and traffic intensity (Zheng et al., 2010; Zhang et al., 2015; Aricak et al., 2019). This was confirmed by Chen et al. (2010) who established that the levels of Cd, Cu, Pb, and Zn showed a decreasing trend with increasing road distance. Khalid et al., (2018) also reported a similar case when they investigated the effect of closeness to the roadside on the concentrations of cadmium, nickel, lead, and zinc in Artemisia. Californica and Hirschfeldia incana. Brown et al. (2017) and Öztürk et al. (2018) also confirmed a direct link between ATEs deposition, proximity to the road, and traffic intensity. Enche-Constantinescu et al. (2015) recorded the highest Pb concentration in *Tillia Tomentosa*. The findings demonstrated the significant link between the traffic volume and the Pb accumulation in the *T. Tomentosa* leaves. Sgrigna et al. (2015) have observed a heavy load of dust on trees in urban areas. They concluded that roadside trees in highly traffic-congested areas could spill the unburnt oil residues on the leaf resulting in high particulate matter (containing ATEs) accumulation. A similar pattern has been recorded by Kord and Kord (2011) when they analyzed concentrations of ATEs in the city of Tehran, Iran. Gunawardena et al. (2013) observed that Cd, Cu, Ni, Pb, and Zn were found to be associated with traffic density.

## 4.7 Drawbacks of using urban roadside trees for ATEs biomonitoring

It is well established that several factors, such as tree species, tree morphology, types of ATEs and their speciation pattern, etc., are involved in influencing the response of urban trees to atmospheric pollution monitoring; however, understanding these factors requires adequate time and in-depth knowledge of atmospheric-plant-metal interactions. Foliar tree organs have been considered to be extremely useful in atmospheric monitoring of ATEs. However, as the large proportion of the ATEs accumulated on the leaves of the tree do not penetrate inside the tissues of the leaves, as described by Tomašević et al. (2011) and Serbula et al. (2012), and the ATEs deposited on the surface of the leaves may be removed easily by winds or washed away by rain. As such, the monitoring of air pollution through the use of tree leaves does not yield satisfactory outcomes and can also produce inaccurate results (Sawidis et al., 2011; Catinon et al., 2012; Minganti et al., 2016). Leaves of some species like the deciduous tree are not available all year round, this will make it difficult to account for seasonal variation of ATEs in tree leaves (Ukpebor et al., 2010; Chrabąszcz and Mróz, 2016). Tree bark can only give spatial data on the distribution of pollutants in a particular area; however, it cannot provide information on the temporal changes (Marć et al., 2015). Moreover, there is a deficiency of a standardized procedure for the preparation of tree bark samples for analyses, which subsequently affects the accuracy and precision of the methods (Świsłowski et al., 2020).

## 4.8 Conclusion and future prospects

Accelerating population growth, urbanization and increased vehicular counts in urban areas have led to a considerable rise in ATEs contamination across the world. ATEs are among of the most important airborne pollutants due to their toxicities and bioaccumulation. As such, there is a pressing need for appropriate monitoring and control approaches for ATEs pollution. Urban roadside trees, due to their simplicity in sampling, cost-effectiveness and ubiquity have nowadays replaced conventional instrumental techniques. Trees act as sink and the first line of defense against ATEs pollution-induced principally by vehicular emissions. However, they are vulnerable to ATEs pollution depending on their tolerance mechanisms. Trees develop a wide range of defense and tolerance mechanisms to counter the effects of ATEs toxicity, and a suitable tree for biomonitoring is the ones that tolerate the metal toxicity. Unlike root metal transfer, which has widely been explored, the mechanism of ATEs transfer from the atmosphere-plant system is poorly studied, and the potential to fight atmospheric pollution by several tree species and their response under different environmental conditions is not clearly understood. So there exist considerable research gaps in the ATEs deposition in urban roadside trees. In view of this, comprehensive research in this area is needed to fill this wide gap.

Moreover, there is insufficient information regarding the movement of metal inside tree's commonly used organs (leaves and barks). Therefore, knowledge of the quantifiable information on the deposition of ATEs should be revamped.

## References

Abhishek, S., 2018. Impact of automobile induced air pollution on roadside vegetation: a review. ESSENCE Int. J. Env. Rehab. Conserv. 9 (1), 101—116.

Achakzai, K., Khalid, S., Adrees, M., Bibi, A., Ali, S., Nawaz, R., Rizwan, M., 2017. Air pollution tolerance index of plants around brick kilnsin Rawalpindi, Pakistan. J Exp. Mang. 190, 252–258.

Adamiec, E., Jarosz-Krzemi ´nska, E., Wieszała, R., 2016. Heavy metals from non-exhaust vehicle emissions in urban and motorway road dusts. Environ. Monit. Assess. 188, 369.

Alahabadi, A., Ehrampoush, M.H., Miri, M., Ebrahimi Aval, H., Yousefzadeh, S., Ghaffari, H.R., Ahmadi, E., Talebi, P., Abaszadeh, F.Z., Babai, F., 2017. A comparative study on capability of different tree species in accumulating heavy metals from soil and ambient air. Chemosphere 172, 459–467.

Alatou, H., Sahli, L., 2019. Using tree leaves and barks collected from contaminated and uncontaminated areas as indicators of air metallic pollution. Int. J. Phytoremediation. 21 (10), 985–997.

Alexandrinoa, K., Viterib, F., Rybarczykc, Y., Andinoa, J.E.G., Zalakeviciutea, R., 2020. Biomonitoring of metal levels in urban areas with different vehicular traffic intensity by using Araucaria *heterophylla needles*. Ecol. Indic. 117, 106701.

Ali, H., Khan, E., Ilahi, I., 2019. Environmental chemistry and ecotoxicology of hazardous Heavy Metals: Environmental persistence, toxicity, and bioaccumulation. J. Chem. 2019, 6730305.

Allahnouri, M., Aghbash, F.G., Pazhouhan, I., 2018. Traffic effects on leaf macro and micro-morphological traits. Folia Oecol. 45 (2), 91–101.

Amodio, M., Catino, S., Dambruoso, P.R., de Gennaro, G., Di Gilio, A., Giungato, P., Laiola, E., Marzocca, A., Mazzone, A., Sardaro, A., Tutino, M., 2014. Atmospheric deposition: sampling procedures, analytical methods, and main recent findings from the scientific literature. Adv. Meteorol. 2014, 1–27.

Aničić, M., Spasic, T., Tomasevic, M., Rajsic, S., Tasic, M., 2011. Trace elements accumulation and temporal trends in leaves of urban deciduous trees (*Aesculus hippocastanum and Tilia ssp.*). Ecol. Indic. 11, 824–830.

Apeagyei, E., Bank, M.S., Spengler, J.D., 2011. Distribution of heavy metals in road dust along an urban-ruralgradient in Massachusetts. Atmos. Environ. 45, 2310–2323.

Aricak, B., Cetin, M., Erdem, R., Sevik, H., Cometen, H., 2019. The Change of some heavy metal concentrations in Scotch Pine (*Pinus Sylvestris*) depending on traffic density, organelle and washing. Appl. Ecol. Environ. Res. 17 (3), 6723–6734.

Badamasi, H., 2017. Biomonitoring of air pollution using plants. MAYFEB J. Environ. Sci. 2, 27–39.

Barbes, L., Barbulescu, A., Radulescu, C., Stihi, C., Chelarescu, E.D., 2014. Determination of heavy metals in leaves and bark of *Populus nigra L.* by atomic absorption spectrometry. Rom. Rep. Phys. 66 (3), 877–886.

Barwise, Y., Kumar, P., 2020. Designing vegetation barriers for urban air pollution abatement: a practical review for appropriate plant species selection. NPJ Clim. Atmos. Sci. 3 (12), 1–19.

Bhandarkar, S., 2013. Vehicular pollution, their effect on human health and mitigation measures. Vehicle Eng. 1 (2), 3340.

Bozdoğan, E., 2016. Heavy metal concentration in leaves of *Melia Azedarach* as a biomonitor of traffic-related air pollution. Oxid. Commun. 39 (1-2), 756–764.

Brown, G., Lu, I., O'Sullivan, G., 2017. Trace metal concentrations in pine needles at varying elevation in proximity to roadways in an urban environment. J. Environ. Prot. 8, 733–743.

Catinon, M., Ayrault, S., Boudouma, O., Asta, J., Tissut, M., Ravanel, P., 2012. Atmospheric element deposit on tree barks: the opposite effects of rain and transpiration. Ecol. Indic. 14 (1), 170.

Chabbi, I., Bahloul, M., Dammak, R., Azri, C., 2018. Dust particle deposition quality assessment in rural areas located not far from a congested highway and several sebkhas: case of Monastir region, Eastern Tunisia. J. Environ. Eng. Landsc. Manag. 26 (2), 141–157.
Chabbi, I., Bahloul, M., Dammak, R., Baati, H., Azri, C., 2020. Wet deposition of major ions and trace elements in rural and coastal-urban sites in Monastir Region. Eastern Tunisia. Arab. J. Geosci. 13, 51.
Charron, A., Polo-Rehn, L., Besombes, J.-L., Golly, B., Buisson, C., Chanut, H., Marchand, N., Guillaud, G., Jaffrezo, J.L., 2018. Identification and quantification of particulate tracers of exhaust and non-exhaust vehicle emissions. Atmos. Chem. Phys. Discuss. 19, 5187–5207.
Chaudhary, I.J., Rathore, D., 2018. Suspended particulate matter deposition and its impact on urban trees. Atmos. Pollut. Res. 9, 1072–1082.
Chaudhary, I.J., Rathore, D., 2019. Dust pollution: Its removal and effect on foliage physiology of urban trees. Sustain. Cities Soc. 51, 101696.
Chen, L., Liu, C., Zhang, L., Zou, R., Zhang, Z., 2017. Variation in tree species ability to capture and retain airborne fine particulate matter ($PM_{2.5}$). Sci. Rep. 7, 1–11.
Chen, X., Xia, X., Zhao, Y., Zhang, P., 2010. Heavy metal concentrations in roadside soils and correlation with urban traffic in Beijing, China. J. Hazard. Mater. 181, 640–646.
Chen, Y., Schleicher, N., Chen, Y., Chai, F., Norra, S., 2014. The influence of governmental mitigation measures on contamination characteristics of PM2:5 in Beijing. Sci. Total. Environ. 490, 647–658.
Chiarantini, L., Rimondi, V., Benvenuti, M., Beutel, M.W., Costagliola, P., Gonnelli, C., Paolieri, M., 2016. Black pine (*Pinus nigra*) barks as biomonitors of airborne mercury pollution. Sci. Total Environ. 569, 105.
Chrabąszcz, M., Mróz, L., 2016. Tree bark, a valuable source of information on air quality. Pol. J. Environ. Stud. 26 (2), 453–466.
Chrzan, A., Marko-Worlowska., M., 2012. Content of heavy metals in soil and in Pine bark in Skalki Twardowskiego Landscape Park in Krakow. Ecol. Chem. Eng. A. 19 (10), 1255–1263.
Chukwuka, K.S., Uka, U.N., 2013. Heavy metals levels in *Carica Papaya L.* tree bark as indicator of atmospheric pollution in Abakaliki municipality. Poll Res 32 (4), 723–726.
Costagliola, P., Benvenuti, M., Chiarantini, L., Lattanzi, P., Paolieri, M., Rimondi, V., 2017. Tree barks as environmental biomonitors of metals—the example of mercury. J. Environ. Sci. Allied Res. 1 (1), 11–18.
Cocozza, C., Ravera, S., Lombardi, C., Maechetti, F., Tognetti, M., 2016. Integrated biomonitoring of airborne pollutants over space and time using tree rings, bark, leaves and epiphytic litchens. Urban For. Urban Green 17, 177–191.
Crilley, L.R., Lucarelli, F., Bloss, W.J., Harrison, R.M., Beddows, D.C., Calzolai, G., Nava, S., Valli, G., Bernardoni, V., Vecchi, R., 2017. Source apportionment of fine and coarse particles at a roadside and urban background site in London during the 2012 summer ClearfLo campaign. Environ. Pollut. 220, 766–778.
de Souza, M.S.P.A., dos Santos, F.S., Magalhães, L.M.S., de Freitas, W.K., de Gois, G., de Oliveira Júnior, J.F., 2019. Poincianella pluviosa as biomonitor of heavy metals in the municipality ofVolta Redonda, RJ, Brazil. Rev. Bras. Eng. Agríc. Ambient. 23 (1), 71–76.
Deepalakshmi, A.P., Ramakrishnaich, H., Ramachandra, Y.L., Kumar, N.N., 2014. Leaves of higher plants as indicators of heavy metal pollution along the urban roadways. Int. J. Sci. Technol. 3 (6), 340.
Doganlara, Z.B., Doganlara, O., Er Arriaga, b, S., Onalc, Y., 2012. Heavy metal pollution and physiological changes in the leaves of some shrub, palm and tree species in urban areas of Adana, Turkey. Chem. Speciat. Bioavailab. 24 (2), 65–78.
Drava, G., Anselmo, M., Brignole, D., Giordani, P., Minganti, V., 2017a. Branch bark of holm oak (*Quercus ilex L.*) for reconstructing the temporal variations of atmospheric deposition of hexavalent chromium. Chemosphere 170, 141–145.

Drava, G., Brignole, D., Giordani, P., Minganti, V., 2017b. The bark of the branches of holm oak (*Quercus ilex L.*) for a retrospective study of trace elements in the atmosphere. Environ. Res. 154, 291–295.

Duong, T.T.T., Lee, B.K., 2011. Determining contamination level of heavy metals in road dust from busy traffic areas with different characteristics. J. Environ. Manag. 92, 554–562.

Ejidike, I.P., Onianwa, P.C., 2015. Assessment of trace metals concentration in tree barks as indictor of atmospheric pollution within Ibadan City, South-West, Nigeria. J. Anal. Methods Chem. 2015, 243601.

El-Khatib, A.A., Barakat, N.A., Youssef, N.A., Samir, N.A., 2020a. Bioaccumulation of heavy metals air pollutants by urban trees. Int. J. Phytoremediation. 22 (2), 210–222.

El-Khatib, A.A., Youssef, N.A., Barakat, N.A., Samir, N.A., 2020b. Responses of *Eucalyptus globulus* and *Ficus nitida* to different potential of heavy metal air pollution. Int. J. Phytoremediation. 22 (10), 986–999.

Enche-Constantinescu, A.M., Madosa, E., Chira, D., Hernea, C., Enche-Constantinescu, R.V., Lalescu, D., Borlea, G.F., 2015. *Tilia spp.* - urban trees for future. Not. Bot. Horti. Agrobo. 43 (1), 259–264.

Fang, G.C., Huang, C.S., 2011. Atmospheric particulate and metallic elements (Zn, Ni, Cu, Cd and Pb) size distribution at three characteristic sampling sites. Environ. Forensics. 12, 191–199.

Fujiwara, F.G., Gómez, D.R., Dawidowski, L., Perelman, P., Faggie, Ana., 2011. Metals associated with airborne particulate matter in road dust and tree bark collected in a megacity (Buenos Aires, Argentina). Ecol. Ind. 11, 240–247.

Gajbhiye, T., Pandey, S.K., Kim, K., Szulejko, J.E., Prasad, S., 2016. Airborne foliar transfer of PM bound heavy metals in *Cassia siamea*: a less common route of heavy metal accumulation. Sci. Total Environ. 573, 123–130.

Gajbhiye, T., Pandeya, S.K., Leeb, S.S., Kimc, K., 2019. Size fractionated phytomonitoring of airborne particulate matter (PM) and speciation of PM bound toxic metals pollution through *Calotropis procera* in an urban environment. Ecol. Indic. 104, 32–40.

Guéguen, F., Stille, P., Millet, M., 2011. Air quality assessment by tree bark biomonitoring in urban, industrial and rural environments of the Rhine Valley: PCDD/Fs, PCBs and trace metal evidence. Chemosphere 85 (2), 195.

Gunawardena, J., Egodawatta, P., Ayoko, G.A., Goonetilleke, A., 2013. Atmospheric deposition as a source of heavy metals in urban stormwater. Atmos. Environ. 68, 235–242.

Hajizadeha, Y., Mokhtarib, M., Farajic, M., Abdolahnejad, A., Mohammadi, A., 2019. Biomonitoring of airborne metals using tree leaves: protocol for biomonitor selection and spatial trend. Methods X 6, 1694–1700.

Harguinteguy, C.A., Cofré, M.N., Fernández, A.C., Pignata, M.L., 2016. The macrophytes *Potamogeton pusillus L.* and *Myriophyllum aquaticum (Vell.) Verdc.* as potential bioindicators of a river contaminated by heavy metals. Microchem. J. 124, 228–234.

Harrison, R.M., Jones, A.M., Gietl, J., Yin, J., Green, D.C., 2012. Estimation of the contributions of brake dust, tire wear, and resuspension to nonexhaust traffic particles derived from atmospheric measurements. Environ. Sci. Technol. 46 (12), 6523–6529.

Hetem, I.G., Andrade, M.F., 2016. Characterization of fine particulate matter emitted from the resuspension of road and pavement dust in the metropolitan area of São Paulo, Brazil. Atmos. Basel. 7, 31.

Janta, R., Chantara, S., 2017. Tree bark as bioindicator of metal accumulation from road traffic and air quality map: a case study of Chiang Mai, Thailand. Atmos. Pollut. Res. 8, 956–967.

Janta, R., Chantara, S., Inta, A., Kawashima, M., Satake, K., 2016. Levels of road traffic heavy metals in tree bark layers of *Cassia fistula* tree. Int. J. Environ. Sci. Dev. 7 (5), 385–388.

Javanmard, Z., Kouchaksaraei, M.T., Hosseini, S.M., Pandey, A.K., 2020. Assessment of anticipated performance index of some deciduous plant species under dust air pollution. Environ. Sci. Pollut. Res. 27, 38987–38994.

Kardel, F., Wuyts, K., Babanezhad, M., Wuytack, T., Potters, G., Samson, R., 2010. Assessing urban habitat quality based on specific leaf area and stomatal characteristics of *Plantago lanceolata L.* Environ. Pollut. 158, 788–794.

Khalid, N., Hussain, M., Young, H.S., Boyce, B., Aqeel, M., Noman, A., 2018. Effects of road proximity on heavy metal concentrations in soils and common roadside plants in Southern California. Environ. Sci. Pollut. R. 25, 35257–35265.

Khan, A.B., Kathi, S., 2014. Evaluation of heavy metal and total petroleum hydrocarbon contamination of roadside surface soil. Int. J. Environ. Sci. Technol. 11, 2259–2270.

Khan, S., Khan, M.A., Rehman, S., 2011. Lead and cadmium contamination of different roadside soil and plants in Peshawar City, Pakistan. Pedosphere. 21 (3), 351.

Kiliç, D.D., 2019. Investigation of heavy metal accumulation and biomonitoring of Calepina irregularis species growing in Amasya (Turkey) province. Anatolian J. Bot. 3 (2), 44–50.

Kim, J-E., Han, Y-J., Kim, P-R., Holsen, T.M., 2012. Factors influencing atmospheric wet deposition of trace elements in rural Korea. Atmos. Res. 116, 185–194.

Kocić, K., Spasić, T., Urŏsević, M.A., Tomăsević, M., 2014. Trees as natural barriers against heavy metal pollution and their role in the protection of cultural heritage. J. Cult. Herit. 15 (3), 227–233.

Kord, B., Kord, B., 2011. Heavy metals levels in pine (*Pinus eldarica* Medw.) tree barks as indicators of atmospheric pollution. Bio Res 6 (2), 927–935.

Kord, B., Tabibian, S., Pourabbasi, S., 2015. Biomonitoring of Pb, Zn, Cu, Ni, and Cr levels in barks of cypress (*Cupressus arizonica* Green.) tree in the atmosphere of Tehran City, Iran. Adv. Bio Res. 6 (1), 133–137.

Krupnova, T.G., Rakova, O.V., Gavrilkina, S.V., Antoshkina, E.G., Baranov, E.O., Dmitrieva, A.P., Somova, A.V., 2020. Extremely high concentrations of zinc in birch tree leaves collected in Chelyabinsk, Russia. Environ. Geochem. Health. 43, 2551–2570.

Kumar, S.R., Arumugam, T., Anandakumar, C.R., Balakrishnan, S., Rajave, D.S., 2013. Use of plant species in controlling environmental pollution—a view. Bull. Environ. Pharmacol. Life Sci. 2 (2), 52–63.

Langner, M., Draheim, T., Endlicher, W., 2011. Particulate matter in the urban atmosphere: concentration, distribution, reduction—results of studies in the Berlin metropolitan area. In: Endlicher, W. (Ed.), Advances in Meteorology in Perspective in Urban Ecology, Ecosystems and Interactions Between Humans and Nature in the Metropolis of Berlin. Springer, Berlin, Germany, pp. 15–41.

Lawal, A.O., Batagarawa, S.M, Oyeyinka, O.D, Lawal, M.O., 2011. Estimation of heavy metals in neem tree leaves along Katsina – Dutsinma – Funtua. J. Appl. Sci. Environ. Manage. 15 (2), 327–330.

Lu, S., Yang, X., Li, S., Chen, B., Jiang, Y., Wang, D., Xu, L., 2018. Effects of plant leaf surface and different pollution levels on PM2.5 adsorption capacity. Urban For. Urban Green 34, 64–70.

Luo, C., Liu, C., Wang, Y., Liu, X., Li, F., Zhang, G., Li, X., 2011. Heavy metal contamination in soils and vegetables near an e-waste processing site, south China. J. Hazard. Mater. 186, 481–490.

Lynam, M.M., Dvonch, J.T., Hall, N.L., Morishita, M., Barres, J.A., 2015. Trace elements and major ions in atmospheric wet and dry deposition across central Illinois, USA. Air Qual. Atmos. Health. 8, 135–147.

Marć, M., Tobiszewski, M., Zabiegała, B., de la Guardia, M., Namie´snik, J., 2015. Current air quality analytics and monitoring: a review. Anal. Chim. Acta. 853, 116–126.

Matin, G., Kargar, N., Buyukisik, H.B., 2016. Bio-monitoring of cadmium, lead, arsenic and mercury in industrial districts of Izmir, Turkey by using honey bees, propolis and pine tree leaves. Ecol. Eng. 90, 331–335.

Mazari, K., Filippelli, G.M., 2020. Using deciduous trees as bioindicators of trace element deposition in a small urban watershed, Indianapolis, IN, USA. J. Environ. Qual. 49 (1), 163–171.

Minganti, V., Drava, G., Giordani, P., Malaspina, P., Modenesi, P., 2016. Human contribution to trace elements in urban areas as measured in Holm oak (*Quercus ilex L.*) bark. Environ. Sci. Pollut. R. 23 (12), 12467.

Miri, M., Ehrampoush, M.H., Ghaffari, H.R., Aval, H.E., Rezai, M., Najafpour, F., Fathabadi, Z.A., Aval, M.Y., Ebrahimi, A., 2016. Atmospheric heavy metals biomonitoring using a *Local Pinus eldarica* tree. Health Scope 6 (1), 1–19.

Mohan, S.M., 2016. An overview of particulate dry deposition: measuring methods, deposition velocity and controlling factors. Int. J. Environ. Sci. Technol. 13, 387–402.

Mondal, D., Gupta, S., Datta, J.K., 2011. Anticipated performance index of some tree species considered for green belt development in urban area. Int. Res. J. Plant Sci. 2 (4), 099–106.

Moreira, T.C.L., Amato-Lourenço, L.F., da Silva, G.T., de André, C.D.S., de André, P.A., Barrozo, L.V., Singer, J.M., Saldiva, P.H.N., Saiki, M., Locosselli, G.M., 2018. The use of tree barks to monitor traffic related air pollution: a case study in São Paulo–Brazil. Front. Environ. Sci. 6, 72.

Moreira, T.C.L., de Oliveira, R.C., Amato, L.F.L., Kang, C.M., Saldiva, P.H.N., Saiki, M., 2016. Intra- urban biomonitoring: source apportionment using tree barks to identify air pollution sources. Environ. Int. 91, 271.

Mori, J., Saebo, A., Hanslin, H.M., Teani, A., Ferrini, F., Fini, A., Burchi, G., 2015. Deposition of traffic-related air pollutants on leaves of six evergreen shrub species during a Mediterranean summer season. Urban For. Urban Green. 14 (2), 264–273.

Muhammad, S., Wuyts, K., Samson, R., 2019. Atmospheric net particle accumulation on 96 plant species with contrasting morphological and anatomical leaf characteristics in a common garden experiment. Atmos. Environ. 202, 328–344.

Mutlu, S.S., Selim, C., Ün, G., 2017. Plant biodiversity of urban roadside trees in Antalya, Turkey. Kastamonu Univ. J. For. Fac. 17 (1), 80–87.

Nie, X., Wang, Y., Li, Y., Sun, L., Li, T., Yang, M., Yang, X., Wang, W., 2017. Characteristics and impacts of trace elements in atmospheric deposition at a high-elevation site, southern China. Environ Sci. Pollut. Res. 24, 22839–22851.

Norouzi, S., Khademi., H., 2015. Source identification of heavy metals in atmospheric dust using *Platanus orientalis L.* leaves as bioindicator. Eurasian J. Soil Sci. 4 (3), 144–152.

Norouzi, I., Khademi, H., Cano, A.F., Acosta, J.A., 2015. Using plane tree leaves for biomonitoring of dust borne heavy metals: A case study from Isfahan, Central Iran. Ecol. Indic. 57, 64–73.

Olajumoke, E.T., Ojo, E.P., 2020. Quantifying vehicular heavy metal deposits on roadside soil and vegetation along Akure-Ilesa express road, South-Western Nigeria. GSC Biol. Pharma. Sci. 2 (1), 054–061.

Olowoyo, J.O., van Heerden, E., Fischer, J.L., 2010. Investigating *Jacaranda mimosifolia* tree as biomonitor of atmospheric trace metals. Environ. Monit. Assess. 164, 435–443.

Olutona, G.O., Oyekunle, J.A.O., Dawodu, M.O., Ogunwale, T.O., Kehinde, P., 2017. Physicochemical characteristics of soil and health risk assessment of potentially toxic metals in soil and vegetables from roadside farmlands in Iwo, Southwestern Nigeria. J. Environ. Sci. Pollut. Res. 3, 213–218.

Ouro-Sama, K., Solitoke, H.D., Tanouayi, G., 2020. Spatial and seasonal variation of trace elements contamination level of waters from the hydrosystem Lake Togo-Lagoon of Aneho (South of Togo). SN Appl. Sci. 2, 811.

Öztürk, A., Yarci, C., Ozyigit, I.I., 2018. Assessment of heavy metal pollution in Istanbul using plant (*Celtis australis L.*) and soil assays. Biotechnol. Biotechnol. Equip. 31 (5), 948–954.

Pace, R., Grote, R., 2020. Deposition and resuspension mechanisms into and from tree canopies: a study modeling particle removal of conifers and broadleaves in different cities. Front. For. Glob. Change. 3, 26.

Pan, Y.P., Wang, Y.S., 2015. Atmospheric wet and dry deposition of trace elements at 10 sites in Northern China. Atmos. Chem. Phys. 15, 951–972.

Pandey, A.K., Pandey, M., Mishra, A., Tiwary, S.M., Tripathi, B.D., 2015. Air pollution tolerance index and anticipated performance index of some plant species for development of urban forest. Urban For. Urban Green. 14 (4), 866–871.

Patel, K., Sharma, R., Dahariya, N., Yadav, A., Blazhev, B., Matini, L., Hoinkis, J., 2015. Heavy metal contamination of tree leaves. Am. J. Anayt. Chem. 6, 687–693.

Patel, K.S., Yadav, A., Sahu, Y.K., Lata, L., Milosh, H., Corns, W.T., Martín-Ramos, P., 2020. Tree bark as a bioindicator for arsenic and heavy metal air pollution in Rajnandgaon district, India. J. Hazard. Toxic Radioact. Waste 24 (1), 05019006.

Petrova, S., Yurukova, L., Velcheva, I., 2014. Possibilities of using deciduous tree species in trace element biomonitoring in an urban area (Plovdiv, Bulgaria). Atmos. Pollut. Res. 5 (2), 196–202.

Popoola, LT., Adebanjo, S.A., Adeoye, B.K., 2018. Assessment of atmospheric particulate matter and heavy metals: a critical review. Int. J. Environ. Sci. Technol. 15, 935–948.

Pulles, T., Denier van der Gon, H., Appelmana, W., Verheul, M., 2012. Emission factors for heavy metals from diesel and petrol used in European vehicles. Atmos. Environ. 61, 641–651.

Rai, P.K., 2013. Environmental magnetic studies of particulates with special reference to bio magnetic monitoring using roadside plant leaves. Atmos. Environ. 72, 113–129.

Rai, P.K., 2016a. Biodiversity of roadside plants and their response to air pollution in an Indo-Burma hotspot region: implications for urban ecosystem restoration. J. Asia Pac. Biodivers. 9 (1), 47–55.

Rai, P.K., 2016b. Impacts of particulate matter pollution on plants: implications for environmental biomonitoring. Ecotoxicol. Environ. Saf. 129, 120–136.

Rai, P.K., 2019. Particulate matter tolerance of plants (APTI and API) in a biodiversity hotspot located in a tropical region: Implications for eco-control. Part. Sci. Technol. 38 (2), 1–10.

Rai, P.K., Lee, S.S., Zhang, M., Tsang, Y.F., Kim, K., 2019. Heavy metals in food crops: health risks, fate, mechanism, and management. Environ. Int. 125, 365–385.

Rehman, K., Fatima, F., Waheed, I., Akash, M.S.H., 2017. Prevalence of exposure of heavy metals and their impact on health consequences. J. Cell. Biochem. 119, 157–184.

Roy, A., Bhattacharya, T., Kumari, M., 2020. Air pollution tolerance, metal accumulation and dust capturing capacity of common tropical trees in commercial and industrial sites. Sci. Total Environ. 722, 137622.

Sæbø, A., Popek, R., Nawrot, B., Hanslin, H.M., Gawronska, H., Gawronski, S.W., 2015. Plant species differences in particulate matter accumulation on leaf surfaces. Sci. Total Environ. 427–428, 347–354.

Sahu, C., Basti, S., Sahu, S.K., 2020. Air pollution tolerance index (APTI) and expected performance index (EPI) of trees in Sambalpur town of India. SN Appl. Sci. 2, 1327.

Sahu, M., Hu, S., Ryan, P.H., Le Masters, G., Grinshpun, S.A., Chow, J.C., Biswas, P., 2011. Chemical compositions and source identification of PM2. 5 aerosols for estimation of a diesel source surrogate. Sci. Total Environ. 409 (13), 2642–2651.

Sakata, M., Asakura, K., 2011. Atmospheric dry deposition of trace elements at a site on Asian-continent side of Japan. Atmos. Environ. 45, 1075–1083.

Sawidis, T., Breuste, J., Mitrovic, M., Pavlovic, P., Tsigaridas, K., 2011. Trees as bioindicator of heavy metal pollution in three European cities. Environ. Pollut. 159, 3560–3570.

Schaubroeck, T., Deckmyn, G., Neirynck, J., Staelens, J., Adriaenssens, S., Dewulf, J., Muys, B., Verheyen, K., 2014. Multilayered modeling of particulate matter removal by a growing forest over time, from plant surface deposition to wash off via rainfall. Environ. Sci. Technol. 48, 10785–10794.

Selmi, W., Weber, C., Rivière, E., Blond, N., Mehdi, L., Nowak, D., 2016. Air pollution removal by trees in public green spaces in Strasbourg city, France. Urban For. Urban Green. 17, 192–201.

Sen, A., Khan, I., Kundu, D., Das, K., Datta, J.K., 2017. Ecophysiological evaluation of tree species for biomonitoring of air quality and identification of air pollution-tolerant species. Environ Monit. Assess. 189, 262.

Serbula, S.M., Miljkovich, D.Dj., Kovacevic, R.M., Ilic, A.A., 2012. Assessment of airborne heavy metal pollution using plant parts and topsoil. Ecotoxicol. Environ. Saf. 76, 209–214.

Sett, R., 2017. Responses in plants exposed to dust pollution. Hort. Int. J. Â 1 (2), 53–56.

Sgrigna, G., Sæbø, A., Gawronski, S., Popek, R., Calfapietra, C., 2015. Particulate matter deposition on Quercus ilex leaves in an industrial city of central Italy. Environ. Pollut. 197, 187–194.

Shafiq, M., Iqbal, M.Z., Arayne, M.S., Athar, M., 2012. Biomonitoring of heavy metal contamination in *Pongamia pinnata* and *Peltophorum pterocarpum* growing in the polluted environment of Karachi, Pakistan. J. Appl. Bot. Food Qual. 85, 120–125.

Shahid, M., Dumat, C., Khalid, S., Niazi, N.K., Antunes, P.M.C., 2016. Cadmium bioavailability, uptake, toxicity and detoxification in soil-plant systemRev. Environ. Contam. Toxicol. Springer, New York, pp. 1–65.

Shahida, M., Dumat, C., Khalida, S., Schreckb, E., Xiong, T., Niazi, N.K., 2016. Foliar heavy metal uptake, toxicity and detoxification in plants: a comparison of foliar and root metal uptake. J. Hazard. Mater. 325, 36–58.

Sharma, P., Yadav, P., Ghosh, C., Singh, B., 2020. Heavy metal capture from the suspended particulate matter by *Morus alba* and evidence of foliar uptake and translocation of PM associated zinc using radiotracer ($^{65}Zn$). Chemosphere 254, 126863.

Sharma, S.K., Mandal, T.K., 2017. Chemical composition of fine mode particulate matter ($PM_{2.5}$) in an urban area of Delhi, India and its source apportionment. Urban Clim. 21 (106), e122.

Simon, E., Baranyai, E., Braun, M., Cserháti, C., Fábián, I., Tóthmérész, B., 2014. Elemental concentrations in deposited dust on leaves along an urbanization gradient. Sci. Total Environ. 490, 514–520.

Simon, E., Braun, M., Vidic, A., Bogyó, D., Fábián, I., Tóthmérész, B., 2011. Air pollution assessment based on elemental concentration of leaves tissue and foliage dust along an urbanization gradient in Vienna. Environ. Pollut. 159, 1229–1233.

Singh, H., Yadav, M., Kumar, N., Kumar, A., Kumar, M., 2020. Assessing adaptation and mitigation potential of roadside trees under the influence of vehicular emissions: a case study of *Grevillea robusta* and *Mangifera indica* planted in an urban city of India. PLoS One 15 (1), e0227380.

Škrbić, B., Milovac, S., Matavulj, M., 2012. Multielement profiles of soil, road dust, tree bark and wood- rotten fungi collected at various distances from high-frequency road in urban area. Ecol. Indic. 13 (1), 168.

Solgi, E., Keramaty, M., Solgi, M., 2020. Biomonitoring of airborne Cu, Pb, and Zn in an urban area employing a broad leaved and a conifer tree species. J. Geochem. Explor. 208, 106400.

Song, F., Gao, Y., 2011. Size distributions of trace elements associated with ambient particular matter in the affinity of a major highway in the New Jersey-New York metropolitan area. Atmos. Environ. 45 (37), 6714–6723.

Sulaiman, F.R., Hamzah, H.A., 2018. Heavy metals accumulation in suburban roadside plants of a tropical area (Jengka, Malaysia). Ecol. Process. 7, 28.

Świsłowski, P., Kříž, J., Rajfur, M., 2020. The use of bark in biomonitoring heavy metal pollution of forest areas on the example of selected areas in Poland. Ecol. Chem. Eng. S. 27 (2), 195–210.

Tak, A.A., Kakde, U.B., 2019. Evaluation of trace elements and particulate matter deposition on plant foliage exposed to vehicular pollution. Acta Bot. Croat. 78 (2), 164–168.

Terekhina, N.V., Ufimtseva, M.D., 2020. Leaves of trees and shrubs as bioindicators of air pollution by particulate matter in Saint Petersburg. Geogr. Environ. Sustain. 13 (1), 224–232.

Tibbetts, J.H., 2015. Air quality and climate change: a delicate balance. Environ. Health Perspect. 123, 148–153.

Tomaševića, M., Anićića, M., Jovanovićb, L.J., Perić-Grujićc, A., Ristićc, M., 2011. Deciduous tree leaves in trace elements biomonitoring: a contribution to methodology. Ecol. Ind. 11, 1689–1695.

Tomašević, M., Aničić, M., 2010. Trace element content in urban tree leaves and SEM-EDAX characterization of deposited particles. Phys. Chem. Technol. 8 (1), 1–13.
Turkyilmaz, A., Sevik, H., Cetin, M., 2018. The use of perennial needles as biomonitors for recently accumulated heavy metals. Landscape Ecol. Eng. 14, 115–120.
Ugolini, F., Tognetti, R., Rasch, A., Bacci, L., 2013. *Quercus ilex L.* as bioacumulator for heavy metals in urban areas: effectiveness of leaf washing with distilled water and considerations on the trees distance from traffic. Urban For. Urban Green. 12 (4), 576–584.
Ukpebor, E.E., Ukpebor, J.E., Aigbokhan, E., Goji, I., Onojeghuo, A.O., Okonkwo, A.C., 2010. *Delonix regia* and *Casuarina equisetifolia* as passive biomonitors and as bioaccumulators of atmospheric trace metals. J. Environ. Sci. 22 (7), 1073–1079.
Venter, A.D., van Zyl, P.G., Beukes, J.P., Josipovic, M., Hendriks, J., Vakkari, V., Laakso, L., 2017. Atmospheric trace metals measured at a regional background site (Welgegund) in South Africa. Atmos. Chem. Phys. 17, 4251–4263.
Winther, M., Slentø, E., 2010. Heavy Metal Emissions for Danish Road Transport. NERI Technical Report No. 780. Aarhus Universitet, Roskilde, Denmark.
World Health Organization (WHO), 2016. Ambient Air Pollution: A Global Assessment of Exposure and Burden of Disease. World Health Organization. World Meteorological Organization, accessed on 20th October, 2020 from. https://public.wmo.int/en.
Wu, Y., Zhanga, J., Nic, Z., Liua, S., Jianga, Z., Huang, X., 2018. Atmospheric deposition of trace elements to Daya Bay, South China Sea: fluxes and sources. Mar. Pollut. Bull. 127, 672–683.
Yang, F., Tan, J., Shi, Z.B., Cai, Y., He, K., Ma, Y., Duan, F., Okuda, T., Tanaka, S., Chen, G., 2012. Five-year record of atmospheric precipitation chemistry in urban Beijing, China. Atmos. Chem. Phys. 12, 2025–2035.
Yan, X., Zhang, F., Zeng, C., Zhang, M., Devkota, L.P., Yao, T., 2012. Relationship between heavy metal concentrations in soils and grasses of roadside farmland in Nepal. Int. J. Environ. Res. Public Health 9, 3209–3226.
Yousaf, M., Mandiwana, K.L., Baig, K.S., Lu, J., 2020. Evaluation of *Acer rubrum* tree bark as a bioindicator of atmospheric heavy metal pollution in Toronto, Canada. Water Air Soil Pollut. 231, 382.
Zare, M., Sanatgar, M., Fatemitalab, R., 2016. Cypress tree (Cupressus arizonica Greene.) as a biomonitor of heavy metal pollution in the atmosphere of Isfahan. J. Ornam. Plants 6 (2), 101–106.
Zhang, F., Yan, X., Zeng, C., Zhang, M., Shrestha, S., Devkota, L.P., Yao, T., 2012. Influence of traffic activity on heavy metal concentrations of roadside farmland soil in mountainous areas. Int. J. Environ. Res. Public Health 9, 1715–1731.
Zhang, H., Wang, Z., Zhang, Y., Ding, M., Li, L., 2015. Identification of traffic-related metals and the effects of different environments on their enrichment in roadside soils along the Qinghai-Tibet highway. Sci. Total. Environ. 521–522, 160–172.
Zhang, L., Gao, M., Cui, J., Yang, F., Wang, H., Fu, C., Huang, Y., 2018. Wet deposition of trace metals at a typical urban site in southwestern China: fluxes, sources and contributions to aquatic environments. Sustainability 10, 69.
Zhang, Q., Ye, J., Chen, J., Xu, H., Wang, C., Zhao, M., 2014. Risk assessment of polychlorinated biphenyls and heavy metals in soils of an abandoned e-waste site in China. Environ. Pollut. 185, 258–265.
Zhang, T., Bai, Y., Hong, X., Sun, L., Liu, Y., 2017. Particulate matter and heavy metal deposition on the leaves of *Euonymus japonicas* during the East Asian monsoon in Beijing, China. PLoS One 12 (6), e0179840.
Zheng, N., Liu, J., Wang, Q., Liang, Z., 2010. Health risk assessment of heavy metal exposure to street dust in the zinc smelting district, Northeast of China. Sci. Total Environ. 408, 726–733.
Zhu, J., Wang, Q., Yu, H., Li, M., He, N., 2016. Heavy metal deposition through rainfall in Chinese natural terrestrial ecosystems, evidences from national-scale network monitoring. Chemosphere 164, 128–133.

# CHAPTER 5

# Proteomics and genomics as an efficient tool for biomonitoring

**Hethesh Chellapandian, Jeyachandran Sivakamavalli**
Department of Biotechnology and Microbiology, National College, Tiruchirappalli, Tamil Nadu, India

## 5.1 Introduction

Proteomics is used in aquaculture research worldwide which is considered as a complementary technique with transcriptomics and genomics (Trapp et al., 2014; Suarez-Ulloa et al., 2015). A group of expressed proteins in the experimental type cells or organism under optimum condition and in particular time period is said to be a proteome. A proteome has the ability to sense and instantaneous responding to environmental signals when compared to the cell's genetic information and its viable regulatory mechanisms. Through proteomics approach the significant role of global protein metabolic changes and post-translational changes and functions can be exclusively characterized (Steinberg et al., 2008; Tedesco et al., 2014). So therefore, proteomics is the novel study of structure and function of various proteins, and the biological mechanisms are also understood as a result in all living organisms. Nowadays, the scientists have covered almost all the aspects of proteomics in the aquaculture proteomics studies. As a result, these achievements have hugely initiated and helped in the development of international aquaculture proteomics for several reasons. First of all, both hosts and pathogens are highly concerned, where the hosts include different varieties of fish, shrimp, crab and shellfish as well as two types of model zebra fish and puffer fish, and the pathogens represent dominant prevailing microbes including pathogenic bacteria and viruses, the applied methodologies are efficient and varied in their own aspects (Bonvallot et al., 2013). Both (2-DE-MS) two-dimensional polyacrylamide gel electrophoresis and mass spectrometry and (LC-MS) Liquid chromatography–tandem mass spectrometry are effectively used in the analytical part of protein isolated from various sources. Validation of differential proteins is done by using rtPCR and qPCR or through Western blotting. Further information is investigated by using an expressed sequence tag (EST). Mechanism studies and transitional biology of potential drug targets of key proteins are

*New Paradigms in Environmental Biomonitoring Using Plants*
DOI: https://doi.org/10.1016/B978-0-12-824351-0.00005-5

identified by functional assays. Third, new novel methods are initiated or the assays are updated from explored research. For example, to characterize a protective immunogen, cross immunoproteomics is used as an approach to identify broad cross-immunogens with a combination of immunoproteomics with bacterial challenge postimmunization. A Blue Native–PAGE (BN-PAGE) method and yeast two hybridization analysis (Fig. 5.1) are used to investigate viral protein–protein interaction and complexes, and a heterogeneous protein pull-down assay to determine host–microbe interaction network. Apart from these two techniques, the protein–protein interactions also envisaged through several servers, algorithms, and computational tools. Combining these all *in vitro*, *in vivo*, and *in silico* approaches, the proteomics will bring the enlightened outputs for the existing problems, which will provides the new insights into aquaculture avenues.

A broad range of proteomics methods has been applied to the research area and a large number of proteome studies focusing on cultured aquaculture

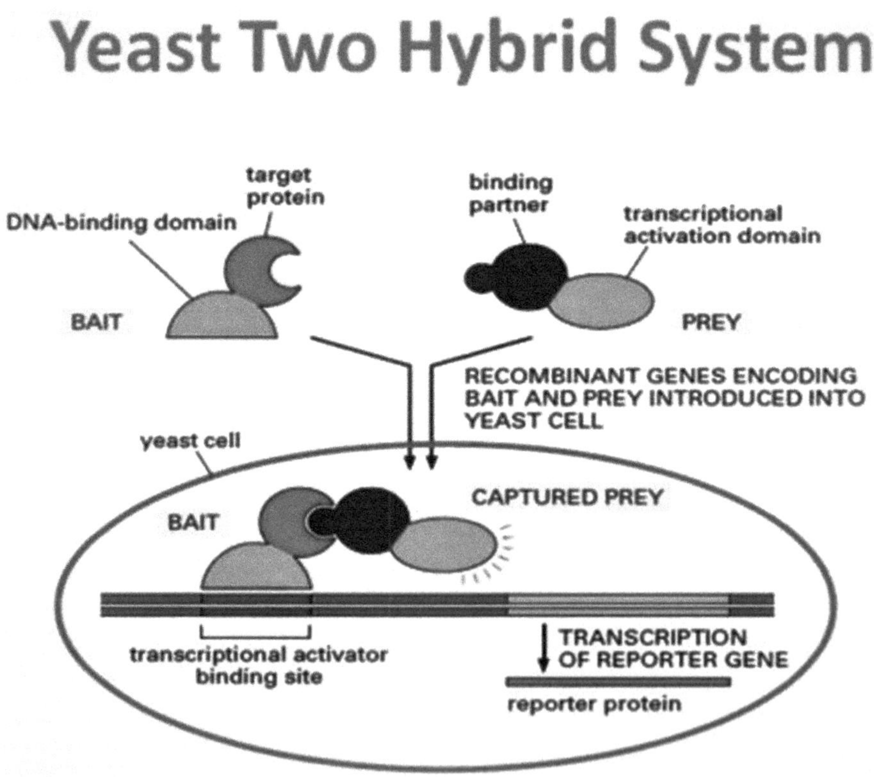


**Fig. 5.1** ***Yeast two hybridization.*** Established methods for proteomics of aquaculture animals and new vaccine development.

animals have been published in recent times. In proteomics-oriented studies, 2-dimensional electrophoresis & mass spectroscopy (2-DE-MS), multidimensional protein identification technology (MudPIT), isotope-coded affinity tag (ICAT), and gas chromatography-mass spectrometry (GC-MS) are widely used. When it comes to aquaculture animals in Asia, 2-DE-MS is the first method to be applied, since it is the first method historically developed for high-resolution proteome separation. It has also been the workhorse in the field of research for various reasons. First of all, the biological samples such as cells or tissues extracted complex mixtures are analyzed using this method. 2-DE begins with 1-DE but lately protein molecules are separated by a 90 degree second property from the first. This causes more effectively separated molecules in 2-DE than in 1-DE because of the two different and distinct properties of the two molecules. By this methodology, numerous proteins of different kind can be resolved and determination of relative amount of each protein can be done. Then, the immobilized pH gradients which are quality controlled with a multitude of pH ranges are made commercial available. The first dimension (1D) separation will be flexible and reproducible which are provided by the lengths. The devices used for casting and running parallel or multiple gels improve the reproducibility of the second dimension (2D) separation. Identification of protein spots are found by employing the MALDI-TOF/MS technique which can be done by Peptide Mass Fingerprinting technique. Acute phase response relative proteome of trauma skin and loach serum can be analyzed by the above-discussed 2-DE–based MALDI-TOF/MS. This can be also used in the identification of proteome species of puffer fish. By the abovementioned methods, certain aquatic animals are used as experiment organisms which are incorporated by modern proteomic technologies. For example, analyzing small proteins or peptides poses a challenge in accessing molecular mass for classic 2-DE PAGE. Due to lower negative charge of tricine, it allows to migrate faster than the ones with glycine. More ion movement and less protein movement are due to its high ionic strength. Low percent acrylamide gels are used to separate the low molecular weight proteins less than 1 kDa. Using the approach, it is identified that APR-related proteins in loach skin and serum with low molecular weight following injury. Tricine-based separated proteins and peptides are separated by 1-DE, but it is previously reported in 2-DE PAGE (Atanasova and Druzhinina, 2010). Due to the potential to be translational medicine the small molecular biological active compounds are in the spotlight. Small molecular mass bioactive compounds are greatly separated by 2-DE done with tricine buffer. In affinity proteomics, the identification of protein in shrimp serum is parallel with IgG. In Parallel, the goat antihuman serum can react specifically with IgG like protein. The purification of the IgG-like protein of the shrimp serum (*Litopenaeus vannamei)* was done by using affinity chromatography with sepharose 4B as matrix. Using 1-DE and 2-DE immunoblotting, MALDI-TOF and Quadruple Time of Flight and tandem mass spectrometry, the IgG-like proteins are analyzed subsequently. Mostly, hemocyanin is the protein mostly isolated from the shrimp serum. Therefore, serum samples are exposed to the antibodies which are immobile, where these proteins are isolated, captured, identified, enriched and all these separation is done using gel-based approach, this approach helps in the enrichment of proteomics in various fields.

## 5.2 Applications

### 5.2.1 Proteomics analyses of immune responses to noninfectious agents

#### *5.2.1.1 Considerable efforts have been made to detect innate immune response in aquaculture animals using comparative proteomics*

The innate immune system is the front-line defense mechanism against the external wounds and injuries. In acute phase reaction (APR), the alteration of cell physiology at the initiation stage of innate immunity occurs during the cell activation. The response is involved in increased production proteins in host, which are produced by acute phase reactions and are generally known as acute phase proteins (APPs). The main function of these APPs is tissue regeneration and repair, protect from infectious and pathogenic microbes, and also indulge in tissue restoration to its normal healthy state. When it comes to APR studies, the researchers mostly focus on the target proteins and the biological samples such as blood or mucus are compared between the control and test groups based amount of the sample present in it. But the abovementioned method is not efficient enough to disclose the novel APPs. The approach is of low efficiency and is incapable of disclosing novel APP members (Suárez-Ulloa et al., 2013). Till date, six different APR-related proteins were identified. In which three were found to be newer APPs. Investigation of APPs with small molecular mass of loach skin serum is still under progress through the abovementioned approach (Bonvallot et al., 2013). Through western blotting technique, the changes in the differential expression proteins can be analyzed and confirmed, and this was performed in the chicken embryo model specifically in the chorioallantonic membrane. Recent study reveals certain new APPs and the role of loach angiostatin-like protein or peptide is demonstrated. And this is applicable not only to loach but also in aquaculture animals which includes both vertebrates and invertebrates. In aquaculture, to protect from various pathogenic diseases and to stimulate innate immunity, various bioactive compounds are used such as immunostimulants, antibiotics, and nutrient rich food. For example, mixture of vitamin C (vC) and Chinese herbs (CH) are used as immunostimulants in shrimps, and protein expression profiles are seen in both test and control groups (Turchini et al., 2019). The use of vitamin C and Chinese herbs mixture showed lower mortality rate against the *Vibrio parahemolyticus* when compared to the control groups.

### 5.2.2 Proteomics characterization of immune response to infectious agents

With the increasing commercialization and intensification of aquaculture production across the globe, disease problems inevitably emerged. When it comes to edible production sector, the bacterial infections are the most important source of problem. In aquaculture, fish and shrimp farms are into loss due to certain bacterial pathogens such as *Vibrio parhemolyticus, Vibrio alginolyticus, Vibrio harveyi,* and various *Vibrio spp.* (Zhao et al., 2011), *Photobacterium damsela, Aeromonas hydrophila, and Edwardsiella tarda.* The infectious diseases caused by the pathogenic bacteria can be controlled by understanding the antiinfective mechanisms of the bacteria. Using the EST library, the serum immune response profile can be investigated combined with differential proteomics (Ulloa et al., 2014). For example, to challenge postimmunization of *A. hydrophila*, the serum of yellow croaker can be used by means of differential proteomics approach. The identification of altered proteins is done by mass spectrometry and, through Western blotting, validation is done. Beta2-microglobulin is one among the altered molecules (protein), which is further investigated since the discovery that it is involved in the pathway of MHC class I antigen. Three clones of beta2 microglobulins are obtained from fish cDNA library, and the polymorphism of the cDNA was revealed. As a result, it indicates the postimmunization of *A. hydrophila* was challenged by the beta2-microglobulin of large yellow croaker. The combination of proteomics and EST library helps in the study of functional proteome in both gene and protein levels, which further helps in understanding host immunity, since in some cases, no gene materials are available for further investigation of targeted molecules.

### 5.2.3 Proteomics analysis of host–microbe interaction

A host–microbe interaction is based diverse variety of organisms involved, where the mode of infection may be obligatory or accidental, but the result in the host is simple: whether health or disease. Therefore, for understanding host immunity and microbial pathogenesis, host–microbiome interaction plays a major role. The causative pathogen of Edwardsiellosis is *E. tarda*, which causes huge loss in the coastal communities across the globe in both marine and freshwater aquaculture. Even though the portal entrance of the pathogen is determined, still there is a void for information concerning the protein–protein interaction of host and pathogen (Xu et al., 2020). The formation of heterogeneous interactome pathogen and host is due to the separation of various different proteins from both pathogen and host. To form a network of heterogeneous interactome, two pairs of molecules are needed, each from both the host and the pathogen. The establishment of host immunity or bacterial infection is controlled by this network. Bacterial outer membrane proteins that are binded to the fish gills as well as bacterial cells interacting with the fish gill proteins can be isolated by the fish gill and bacterial pull-down approaches. For the generation of antibodies, the bacterial protein genes are cloned and expressed, after the generation of antibodies, these antibodies are used to study the interaction fish gills and the bacterial cells like protein–protein interaction. The genes of bacterial proteins were cloned and expressed for preparation of antibodies; the prepared antibodies were used to investigate protein–protein interactions between bacterial cells and fish gills. On the other hand, highly conserved homologues of the fish gill proteins from humans and other animals were known to be involved in host immunity (Liu et al., 2020). This can be used to generate the first heterogeneous interactome between *E. tarda* envelope proteins and fish gill cells, and also reveal the possible role of these interacting bacterial and host proteins in establishing either bacterial infection or host immunity.

Bacterial broad cross-protective immunogens are identified by immunoproteomics approach. When it comes to economic loss in aquaculture practices, bacteria play a major role in causing many infectious bacterial diseases. However, there are vaccines that are capable of resisting from various pathogenic diseases and prevent the animals such as fishes, shrimps, crabs, etc., from these pathogenic infections. But still there is void in the role of vaccines and availability of vaccines and also these vaccines are usually serotype or species specific. Due to the linear or narrow range of bacterial vaccines, the efficiency of these species specific or serotype vaccines are decreased and are epidemically cause a variety of diseases in variety of geographical distribution. However, bacterial pathogens have a variety of serotypes or species which belong to different genera and families. Therefore, development of broadly cross-protective vaccines that can fight against as many species of pathogens as possible is in an urgent need.

### 5.2.4 Proteomics analyses of hosts and pathogens to environmental stresses

A bacterium can be seen in various diverse environments like in a region with higher or lower pH, salinity, iron concentration, temperature, and stress conditions majorly hypoxic stress (oxygen deficiency) which are differed from its host optimum environments. Various reports state that the growth and virulence of a bacterium depends on the environment that they grow on, which must contain desirable conditions such as pH, salinity, $O_2$ concentration, and various mineral components. For example, the proteomes of pacific white shrimp *Litopenaeus vannamei* were studied under various environmental stress along with bacterial pathogen. By means of 2-DE–based proteomics, the proteome profiles of hepatopancreas of *L. vannamei,* which are kept under hypoxic stress are compared. In the hypoxia-stressed hepatopancreas, 33 differential proteins are detected. The detected proteins are isolated and are segregated into various functional proteins such as proteins involved in metabolism, immunity, ungrouped, energy production, cytoskeleton, antioxidant, and chaperones (Campos et al., 2012).

## 5.3 Genomic application in marine biodiversity

The biodiversity assessments are aided by the recent advancements in the research and development such as in molecular sequencing. The previously unknown data of genetic and functional marine biodiversity are revealed by the genomic studies, which also has the ability to change the way we think about the marine biodiversity by implying various genomic methods.

### 5.3.1 Next-generation sequencing

In marine environment, the genome, metabolome, and transcriptome of various marine organisms are analyzed by means of this next-generation sequencing which gives various views on marine biology. The conservation biology and biodiversity assessment in aquatic environment can be explored by studying the variation in the genome of various marine organisms, which can be done in a model fish system based on the relevant data of the survival, adaptation, and voluntary and involuntary mechanisms of the particular organism.

For example, the bottlenose dolphin (Tursiopstruncates) genome shows the evolution data of the nervous system in the dolphin as well as in other huge-brained animals like the humans and other big mammals. The genome isolated from the bottlenose dolphin is compared with many other mammalian animals such as humans, dogs, and even elephants; at the end of the study, almost 10,000 protein coding genome sequences are analyzed which are reported to closely related genomes. Based on the Darwinian relative selection to rest of the mammals, the study has significant changes in the genome which almost discovered 300 new genes in the lineage of dolphin. The slowdown in the molecular change rate in the DNA sequence is the one and only similarity between the humans and the dolphin.

In marine animals, the adaptation, evolution, and diversification of various other marine mammals and their species are studied by the selected genes of dolphin genome by using the whole-genome-wide scans (Yan et al., 2013). For the investigation of the selected 300 genes based on the tissue specificity, tissue annotation database of Uniprot. Further study shows that selected genes are hugely associated in various types of tissues and are expressed in reproductive tissues, nerve cells, and mainly in the immune-related cells and tissues. For the adaptation and to maintain the breathing and body fluid equilibrium, these abovementioned play a major role.

### 5.3.2 Metabolomics

In a biological system, the quantification and identification of whole set of metabolites is done by metabolomics. In a living system, the genetic modification and the physiological stimuli caused by the metabolite response is quantitatively measured. It provides important insights into physiological and disease states, and facilitates in-depth understanding of underlying biochemical pathways. Various insights of auto-immune diseases such as cardiovascular diseases, cancer, Alzheimer's diseases, etc., are obtained by means of metabolomics.

Through metabolomic applications, the changing biotic and abiotic environment of various marine organisms is identified by their physiological responses. The differential effect of various mechanisms of same species as well as closely related species of an individual can be obtained through metabolic applications. Through metabolomics along with proteomics and genomics technologies, various valuable information about the microbial ecosystems and the diversification of ocean can be obtained for further research activities (Zou et al., 2011).

### 5.3.3 Barcoding

It is difficult to find the new species of an organism and prehistoric life using the classic taxonomic method, so instead of that genomic methods are implied and these applications are still in the budding stage but still there are various innovative and exciting technologies to identify the variation in the same species of an individual. DNA barcoding is one such technique through which the closely related species of an individual as well as the closely related species of different organisms are identified phenotypic characters like the major morphological differences. By using cytochrome c oxidase of mitochondria, DNA barcoding can identify any species in the ecosystem. The species of many taxonomic groups that are hard to describe and differentiate and to group under a category can be found by using the DNA barcoding along with the known data of phenotypical or morphological characters of the respective organism, and this also includes the molecular-level identification of specific homologous gene regions like 16S and 18S rRNA compounds and other intercellular spacer region. This kind of molecular-level barcoding is more efficient to find the taxonomical groups of various species especially organism in the marine environment. In developing countries, various practical applications of DNA barcodes are used due to their easily accessible and user friendly genome data of an organism but still there are lot of obstacles like research proficiency, financial support, and skilled personnel to conduct their own research and surveillance in marine biodiversity, fisheries management, and marine bio-conservation.

## 5.4 Genomic applications in aquaculture

Modern biotechnology and genomics have been used to reveal various features of biodiversity in the oceans. The application of these technologies unveils the socio-economic benefits in marine aquaculture production all over world. It can be applied to fisheries and aquaculture; numerous publications of research are published over the years with the help of genomics and also paves a valuable data in the field of marine biology, aquaculture biotechnology, and genomics applications related to the animal feeding, various forms of stress, reproduction abilities, and various others fields. In this chapter, we are discussing about the genomic evolution by understanding the impact of genomics in the field of aquaculture and the methods to treat the pathogenic and infectious diseases in aquaculture organisms (Rodríguez et al., 2020).

In marine ecosystem, lots of information such as adaptation of fishes to different regions, its response to the change in climate and factors involved in the selected harvesting in fisheries around the world can be gained by studying the adaptive evolution of marine organisms and its population genetics. For example, Salmon *(Salmosalar)* is one of the most extensively studied fish and an important model species for fish genetics and evolution. This is due to its easily adaptable capability as a molecular marker which can be used to identify many numbers of different taxonomic groups such as trout, char, and the whitefish (salmonids) and also has the ability to monitor various populations located at different geographical regions. For developing genome-wide markers and cost-efficient production of Single Nucleotide Polymorphisms (SNPs) along with the gene duplication, various genomic tools have been developed. The discovery of genotyping and large-scale marker is done using the RAd-Seq (Restruction site Associated DNA Sequencing). Therefore, these RAD markers are valuable for association as well as quantitative trait loci (QTL) mapping, also in population genetics and evolutionary research. There are certain advantages which cannot be done by any other genomic methods and also discovered the SNPs which majorly contribute in the population genetics and in genotypical studies of organisms which are nonmodel in nature. Through this mapping of genotype–phenotype association can be done along with population genetics and the use of linkage mapping for the scaffold genome assembly (Fig. 5.2).

In the assessment of salmon and various other aquatic animal population genetic structures, SNP arrays have played a major role and the results are successful. These results help in the molecular-level understanding of adaptation of salmon and other aquatic animals to different geographical regions. The sustainability of wild fish populations as well as the farmed or reared aquatic animals such as shrimps, fishes, crabs, etc., can be maintained in their prehistoric forms by the abovementioned applications and also these applications help in the development of aquaculture industries. Without another question, SNPs are considered to be the key markers in research, especially in the field of genetics and epigenetics. Because of their practical use in the analysis of trait–genotype associations, their application in aquaculture is becoming central.

Over the past decade, genome-wide association studies (GWAS) contributed a lot of data significantly in the field of aquaculture genome research. By employing the GWAS and QTL mapping, the association of genotype to phenotypic traits can be used to be investigated. GWAS examine the

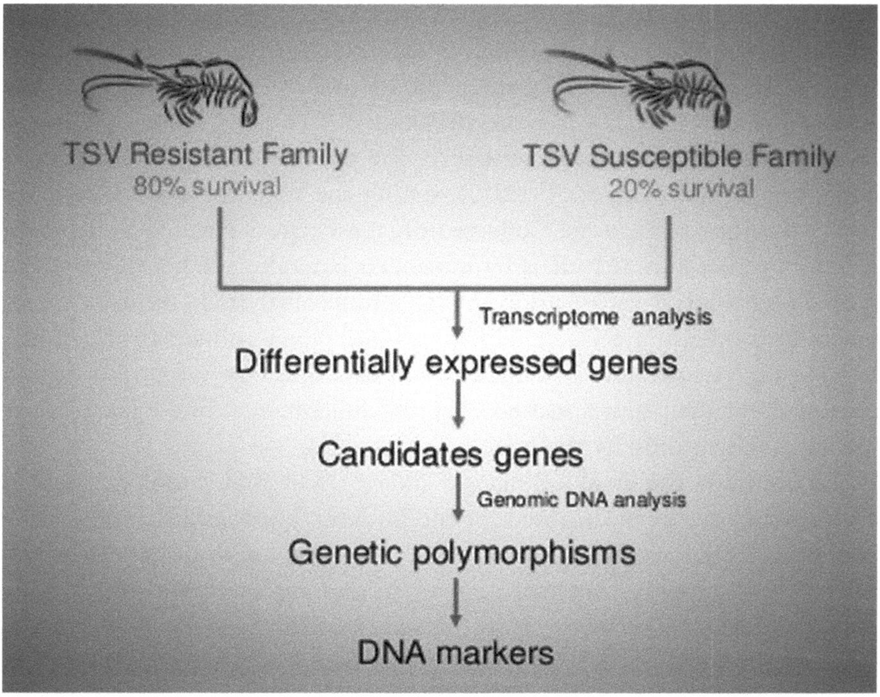


**Fig. 5.2** ***Molecular markers for pathogen resistance.***

association between the specific traits (phenotypes) and widespread genetic variants, usually quantitative, in the genome, transcriptome or proteome. Currently, SNPs are frequently the first choice because of their considerable genome coverage, throughput of assay, and assured quality. In this juncture, the GWAS tests are used to determine the position of marker, whereas the QTL mapping helps in the differentiation of various markers.

## 5.5 Policy issues involved in genomics and marine biotechnology

In developing countries, there is a scarcity of policies for the development of research in the field of postgenomics and marine genomics. To prevent and study the biodiversity of marine organisms present at several regions of Central America, an emerging research program has been introduced for the regional biodiversity studies, the program is named as the "Central American Marine Genomic Initiative" (Huete-Pérez and Quezada, 2013). This program involves in the identification, collection, preservation, and documentation of the collected genomic samples from regional area of

Central America along with pinned coordinates of the sample and its virtual demonstration its morphology. All these information are compiled into database and these data are open for educational purposes as well as the research activities. By the development of the genome archive, a baseline registry of the current state of the region's marine biodiversity is being established (Yan et al., 2013), setting the stage for new discoveries and conservation practices. Collected organisms are identified by standard taxonomic methods as well as by molecular barcoding. When it comes to find the ecological information, transcriptome analysis is the easy to go technique and also a cost-efficient technique when compared to genome sequencing. Various new information and data of a newly found organism usually about their functional ecology and fundamental biological features can be obtained only by the genome sequencing.

To improve the field of marine biology and biotechnology, various applications have been designed for the in-depth knowledge of the marine organisms in different geographical regions. The policies of genomics and marine biotechnology are like majorly promoting the education for the public concern and research bases in areas like institutes, universities, and government-aided laboratories. Assured parameters such as permit for utilizing the genomic resources, good management practices, safety of the workers along with environmental protection are also included in the policy of regulation and the public safety (Huete-Pérez and Quezada, 2013).

## 5.6 Conclusion

In recent days, various research activities are taking place in broad spectrum of fields such as biochemistry, biotechnology, phylogenetics, and ecology or environmental biology. These researches are carried out using the resources from proteomics, genomics, and transcriptomics of many organisms but exclusively from marine organisms and crustacean species. In this chapter, we are mainly focusing on the recent advancements in certain areas of aquaculture and marine biology including from the genotype and phenotypical variation, adaptation to the environment, diversification of species, and some data from the recent reports of proteomic and genomic studies. The proteomic and genomic applications immensely help in the restoration of biodiversity or biorestoration and ecology through the various aspects of research.

In aquaculture and fisheries management, better selective breeding schemes can be obtained by the utilization of molecular-level techniques and genomics and proteomics (omics) techniques-derived tools in industrial level as well as in academic level. Genomics paves a path for the study of genomes based on its structure and function and its evolution over the period of time. In near future, proteomics and genomics help in the study of protein metabolism, expression of genes, and their adaptive traits in large scale in the field of aquaculture. Some of the examples which are mentioned in this chapter demonstrate the potential of genomics and proteomics that helps to improve the biodiversity of marine ecosystem.

## References

Atanasova, L., Druzhinina, I.S., 2010. Global nutrient profiling by Phenotype MicroArrays: a tool complementing genomic and proteomic studies in conidial fungi. J. Zhejiang Univ. Sci. B 11 (3), 151–168.

Bonvallot, N., Tremblay-Franco, M., Chevrier, C., Canlet, C., Warembourg, C., Cravedi, J.P., Cordier, S., 2013. Metabolomics tools for describing complex pesticide exposure in pregnant women in Brittany (France). PLoS One 8 (5), e64433.

Campos, A., Tedesco, S., Vasconcelos, V., Cristobal, S., 2012. Proteomic research in bivalves: towards the identification of molecular markers of aquatic pollution. J. Proteomics 75 (14), 4346–4359.

Huete-Pérez, J.A., Quezada, F., 2013. Genomic approaches in marine biodiversity and aquaculture. Biol. Res. 46 (4), 353–361.

Liu, S., Li, N., Lin, Q., Liu, L., Niu, Y., Liang, H., Fu, X., 2020. Glutaminase 1 in mandarin fish Siniperca chuatsi: molecular characterization, expression pattern and function involving in virus replication. Aquaculture 519, 734924.

Rodríguez, A., Castrejón-Godínez, M.L., Salazar-Bustamante, E., Gama-Martínez, Y., Sánchez-Salinas, E., Mussali-Galante, P., Ortiz-Hernández, M.L., 2020. Omics approaches to pesticide biodegradation. Curr. Microbiol. 77 (4), 545–563.

Steinberg, C.E., Stürzenbaum, S.R., Menzel, R., 2008. Genes and environment—striking the fine balance between sophisticated biomonitoring and true functional environmental genomics. Sci. Total Environ. 400 (1-3), 142–161.

Suárez-Ulloa, V., Fernández-Tajes, J., Manfrin, C., Gerdol, M., Venier, P., Eirín-López, J.M., 2013. Bivalve omics: state of the art and potential applications for the biomonitoring of harmful marine compounds. Marine Drugs 11 (11), 4370–4389.

Suarez-Ulloa, V., Gonzalez-Romero, R., Eirin-Lopez, J.M., 2015. Environmental epigenetics: a promising venue for developing next-generation pollution biomonitoring tools in marine invertebrates. Mar. Pollut. Bull. 98 (1-2), 5–13.

Tedesco, S., Mullen, W., Cristobal, S., 2014. High-throughput proteomics: a new tool for quality and safety in fishery products. Curr. Protein Pept. Sci. 15 (2), 118–133.

Trapp, J., Armengaud, J., Salvador, A., Chaumot, A., Geffard, O., 2014. Next-generation proteomics: toward customized biomarkers for environmental biomonitoring. Environ. Sci. Technol. 48 (23), 13560–13572.

Turchini, G.M., Trushenski, J.T., Glencross, B.D., 2019. Thoughts for the future of aquaculture nutrition: realigning perspectives to reflect contemporary issues related to judicious use of marine resources in aquafeeds. North Am. J. Aquac. 81 (1), 13–39.

Ulloa, P.E., Medrano, J.F., Feijoo, C.G., 2014. Zebrafish as animal model for aquaculture nutrition research. Front. Genet. 5, 313.

Xu, L., Shao, C., Li, G., Shan, A., Chou, S., Wang, J., & Dong, N., 2020. Conversion of broad-spectrum antimicrobial peptides into species-specific antimicrobials capable of precisely targeting pathogenic bacteria. Sci. Rep. 10 (1), 1–9.

Yan, B., Moses, S.A., Gadway, B., Covey, J.P., Hazzard, K.R., Rey, A.M., Ye, J., 2013. Observation of dipolar spin-exchange interactions with lattice-confined polar molecules. Nature 501 (7468), 521–525.

Zhao, Y., Ma, H., Zhang, W., Ai, Q., Mai, K., Xu, W., Liufu, Z., 2011. Effects of dietary β-glucan on the growth, immune responses and resistance of sea cucumber, Apostichopus japonicus against Vibrio splendidus infection. Aquaculture 315 (3-4), 269–274.

Zou, S., Xu, W., Zhang, R., Tang, J., Chen, Y., Zhang, G., 2011. Occurrence and distribution of antibiotics in coastal water of the Bohai Bay, China: impacts of river discharge and aquaculture activities. Environ. Pollut. 159 (10), 2913–2920.

# CHAPTER 6

# Environment biomonitoring with eDNA—A new perspective to identify biodiversity

**Jeyachandran Sivakamavalli**
Department of Biotechnology and Microbiology, National College, Tiruchirappalli, Tamil Nadu, India

## 6.1 Introduction

Environmental DNA (eDNA) terminology denotes the genetic material DNA located in the environmental sources such as sediment, water, and air, including whole cells, extracellular DNA, and potentially whole organisms; these eDNA samples could be helpful for the versatile applications, particularly in biodiversity studies to identify and distinguish the new species (O'Donnell et al., 2017). Such a kind of eDNA sources are available in enormous resources, easy to extract from sources, can be preserved for long years for anthropogenic studies, amplication, sequencing, and categorized based on the identification studies (O'Donnell et al., 2017). This eDNA source could be derived from any material or part of organism skin, mucous, saliva, sperm, secretions, eggs, feces, urine, blood, roots, leaves, fruit, pollen, and rotting bodies of larger organisms, while microorganisms may be obtained in their entirety (Deiner et al., 2015; Bohmann et al., 2014; Barnes and Turner, 2016). eDNA production is dependent on biomass, age, and feeding activity of the organism as well as physiology, life history, and space use (Bohmann et al., 2014; Taberlet et al., 2012; Goldberg et al., 2016).

This eDNA technology is emerging science in the field of science to identify, survey, and monitoring the organismal activity and which act as a better biomonitoring tool in biological applicants. To identify the new species, taxanomic classification inadequate amount of DNA is not sufficient in conventional methods; in eDNA approach, the very little quantity of eDNA can be used for the identification of new species and relatively better, time-consuming process, thus making estimates for entire communities impossible. When compare to other availed technologies, this eDNA-based approach provides the better and enhanced results in targeting different

*New Paradigms in Environmental Biomonitoring Using Plants*
DOI: https://doi.org/10.1016/B978-0-12-824351-0.00002-X

species, sampling greater diversity, and increasing taxonomic resolution (O'Donnell et al., 2017). In addition, eDNA also helps to identify the detection of rare species, but not of determining population quality information such as sex ratios and body conditions, so it is ideal for supplementing traditional studies (O'Donnell et al., 2017; Taberlet et al., 2012).

Through the usage of eDNA sample, we could able to detect the invasive species across the different parts of the location; rarely lower amount of eDNA quantity brings the negative result in identifying the new organism, sometime this information could be derived from extinct organisms. Additionally, some of the drawbacks also exist in this platform while collecting the eDNA, degraded samples also come under this collection vial, such a kind of degraded eDNA limits the genetic information available in the source material. This could be due to the environmental changes warm, tropical regions, and extreme cold environment, along with this water, water flow can affect the quality of eDNA as well as the fine-scale spatiotemporal trends of species and communities (O'Donnell et al., 2017; Barnes and Turner, 2016; Taberlet et al., 2012; Goldberg et al., 2016; Hering et al., 2018).

Despite these drawbacks, eDNA still has the potential to determine relative or rank abundance, as some studies have found it to correspond with biomass, though the variation inherent in environmental samples makes it difficult to quantify (Deiner et al., 2015; Goldberg et al., 2016). While eDNA has numerous applications in conservation, monitoring, and ecosystem assessment, as well as others yet to be described, the highly variable concentrations of eDNA and potential heterogeneity through the water body make it essential that the procedure is optimized, ideally with a pilot study for each new application to ensure that the sampling design is appropriate to detect the target (O'Donnell et al., 2017; Taberlet et al., 2012; Eichmiller et al., 2016). While the definition of eDNA seems straightforward, the lines between different forms of DNA become blurred, particularly in comparison to community DNA, which is described as bulk organismal samples (O'Donnell et al., 2017). A question arises regarding whole microorganisms captured in eDNA samples: do these organisms alter the classification of the sample to a community DNA sample? Additionally, the classification of genetic material from feces is problematic and often referred to as eDNA (O'Donnell et al., 2017). Differentiation between the two is important as community DNA indicates organismal presence at a particular time and place, while eDNA may have come from a different location, from predator feces, or from past presence; however, this differentiation is often impossible (O'Donnell et al., 2017; Carew et al., 2013). For the sake of simplicity, this chapter will consider eDNA loosely classified as including many sectors of DNA biodiversity research, including fecal analysis and bulk samples when they are applicable to biodiversity research and ecosystem analysis. Methods in eDNA research have recently expanded to be able to assess whole communities from a single sample. This process involves metabarcoding, which can be precisely defined as the use of general or universal polymerase chain reaction (PCR) primers on mixed DNA samples from any origin followed by high-throughput next-generation sequencing (NGS) to determine the species composition of the sample. This method has been common in microbiology for years, but is only just finding its footing in assessment of macroorganisms (O'Donnell et al., 2017; Carew et al., 2013).

Ecosystem wide applications of eDNA metabarcoding offer the ability to discover relationships and functional ecology over broad geographical scales; however, erroneous readings due to contamination or other mistakes may restrict this. Altogether, eDNA metabarcoding improves speed, accuracy, and identification while decreasing costs over traditional barcoding, but it must be standardized and unified, combining integrating taxonomy and molecular techniques for comprehensive ecological analysis. The purpose of this review is to offer an overview of methodological and bioinformatics concerns for eDNA metabarcoding research, as well as to highlight published eDNA metabarcoding studies so far, outlining some of the various applications that have resulted from them. Critical concern, opportunities for areas improvement and future goals will be also addressed.

## 6.2 Methods and technical considerations

Within published eDNA metabarcoding studies, the methodologies frequently differ significantly, making cross-study comparisons unfeasible, although the uniqueness of each research location and target taxa likely make standardization a difficult aim. Regardless, a worldwide effort to research and evaluate biodiversity via metabarcoding offers the potential to standardize procedures to some extent. In this section, we will go cover the eDNA metabarcoding procedure in detail, from sample collection to data processing. Because eDNA can be applied to variety of ecosystems, collecting procedures must be tailored to the various sample types. It is possible to simply collect Airborne Microbes and pollen, swab, or scrape the biofilms, precipitate or foist water and treat sediments. However, the choice between precipitation and filtration (and, consequently, the filter sizes) is depending on the target taxa as different taxa can be more effectively separated in other techniques. For water samples high water volumes and numerous field replicates should be utilized (Creer et al., 2016). For soil and sediment samples, higher volumes of sample from larger size classes of species are required, and samples should be obtained from multiple places to avoid sample heterogeneity and to better characterize the region's biodiversity (Pilliod et al., 2013). To maintain independence, field equipment, supplies, and personnel must be segregated from places where PCR, tissue, and organisms are processed or handled in all studies, otherwise decontamination will occur between samples. Among sample types, preservation procedures are straightforward among sample types: drying, freezing at 20°C, or preserving in 100% ethanol or a cell lysis buffer are all suitable ways for sample preservation. Immediate sample preservation is required to avoid deterioration, which can be influenced by environmental factors such as temperature, pH, and exposure to light, as well as target species. Laboratory procedures tend to be more consistent across sample types but vary between studies. eDNA should be processed in a clean lab with nonremovable equipment and supplies; moreover, people and materials must be decontaminated before entering the clean lab. The treatment area of the lab should be cleaned regularly with bleach and UV and filter pippet tips and clean gloves for all operations should be applied during all procedures. In addition to the aforementioned, developing a lab method with positive controls, negative controls, and replicates at all stages is important. Kit-based extraction techniques are probably the most frequent and successful; however, depending on the target taxa, various kits may be more effective, particularly between eukaryotes and prokaryotes. In certain cases, phenol chloroform extractions are also efficient and can isolate more genetic material than other techniques. PCR primer design is one of the most essential factors to consider in eDNA metabarcoding investigations. The coverage, resolution, and bias between taxa of different primers and locations vary. The criteria set by the barcode of life are cytochrome oxidase-I (COI) for metazoans and ribulose bisphosphate carboxylase large chain (rcbL) for plants, although other sections such as 12s or 16s ribosomal RNA may be more suited for species (Ruppert et al., 2019). Primers for eDNA metabarcoding need to be short enough to amplify degraded samples, identical within but variable between species, and flanked by highly conserved regions to amplify a wide range of species without compromising target group specificity. Furthermore, primer selection has the potential to skew finding by preferring amplifying some target sequences more than others, as well as amplifying nontarget groups. The utilization of numerous primer sets, particularly evolutionarily independent primer sets that coincide with standardized barcodes for the target taxonomic groupings, might be one potential answer to this problem. Although this approach can decrease primer bias and enhance taxonomic coverage, it is more expensive and time-consuming. Multiple PCR repetitions improve species detection and reduce the chance of false negatives; however, the number of replicates utilized varies across investigations and is determined by detection probabilities, research objectives, sequencing depth, primer choice, cost constraints, and sequencing platform. It is also worth emphasizing that PCR replicates are not a sufficient alternative for biological replicates; rather, both are necessary for complete results. During NGS PCR, primers can be labeled with short nucleotide sequences to uniquely identify their source, a technique known as multiplexing (Kidziak, 2020). These tags, while beneficial, also have the potential to bias results, particularly when located on the 5' end, thus they must be rigorously tested before implementation. In the sequencing step, a number of platforms are available; however, Illumina sequencing presently beats other NGS systems in terms of depth and accuracy (Shokralla et al., 2012). It is also worth noting that while diversity has been shown to grow with sequencing depth before plateauing, a maximum sequencing depth for particular research may theoretically be identified. Bioinformatics routes are the final technical step to be described. Unique bioinformatics pathways must be established for eDNA metabarcoding and must be altered for each study, previously requiring consultation with a computational specialist. As technology has evolved, however, programs have been designed particularly for analyzing of information from NGS data, making the approach more accessible and user-friendly. NGS leads to in millions of reads which describing the genetic code of each sequenced strand of DNA. These readings are usually categorized into operational taxonomic units (OTUs) which are not consistent between sites, research, or species and cannot be used generally. OTUs are used to distinguish species or taxa based on genetic code similarity as established by sequencing, but they can also be linked to ecological and physiological characteristics for appropriate identification. Various programs are available to assist; however, intragenetic variation impedes this process (Fouhy et al., 2016). OTU clustering is based on sequence similarity and subsequent classification according to similarity cutoffs, with 97%–99% similarity being a common cutoff range; however, ranges for studies rely on intraspecific diversity estimated from existing reference databases, such as GenBank. Species identification from OTUs is based on BLAST results, phylogeny, character (diagnostic regions), classification algorithms, and coalescence (evaluation of quality using likelihood/Bayesian), or even species delimitation, in which sequences are partitioned based on differences rather than similarity to a known sequence, but these should be combined with other ecological characteristics (Pilliod et al., 2013). Unassigned reads are another concern and their impact on analyses should to be kept to a minimum, however, because so many species have yet to be described, analyses may incorrectly exclude target species; poor species representation in databases may lead to incorrect identification and affect assessments of ecological quality. Metabarcoding data with direct observation and widespread occupancy models may best be avoided by mistakes with false positive or negative. Occupancies should be used to evaluate the likelihood of detection, rate of false presences, and if the degree of replication for false negatives is adequate for control. Diversity indices are often employed in order to influence the modeling of diversity and interpretation of findings (Porter and Hajibabaei, 2018).

Quantitative examination of eDNA metabarcoding data is controversial and although certain data show that biomass is linked to number of copies, many confusing factors are obtained, thus several sampling opportunities and replicates as well as lengthy testing or optimization are required. Furthermore, the study's scope must be evaluated, although broad estimations may be appropriate for some studies. There is a prerequisite need about the programming languages to analyze the metabarcogding data and analysis, which helps to operate from the command line, the capacity to use/manipulate/write programs in Perl or Python for analyzing large volumes of data, and familiarity with R. USEARCH, VSEARCH, CROP, swarm, UNIX, QIIME, mothur, OBITools, and phyloseq are often used software to analyze the datasets (Malizia and Ke, 1993; Sarwar et al., 2016). Most essential, data from eDNA metabarcoding studies must be disseminated and accessible in its raw format with defined metadata, ideally in cloud-based storage or common-storage databases to be used by future researchers. Altogether, workflows are frequently dictated by personal taste, available equipment, and cost. It is important to remember that even if all methods in an eDNA metabarcoding investigations are carried out flawlessly with primers that bind perfectly, some mistakes should be expected. Because eDNA is made up of microscopic pieces that are randomly scattered in the environment, there is always the possibility that these fragments will not be picked up. This supports the requirement for lab-based testing before entering the field and replicates across sampling areas, but also the idea that absence cannot be confirmed; even if a species is not detected that does not mean it is not there. Additionally, a positive eDNA result can be introduced via confounding sources. Many issues with eDNA procedures may be overcome over time as the process matures, but the objective should be to reduce false positives rather than flawless detection. Regardless, the data able to be gleaned from eDNA metabarcoding is invaluable and can be applied to a myriad of projects in a variety of ways (Pilliod et al., 2013; Klymus et al., 2017).

## 6.3 Ecosystem and biodiversity monitoring

### 6.3.1 Advantages and disadvantages

Perhaps the most intriguing and thus far the most explored avenue of eDNA metabarcoding research is the application for ecosystem and biodiversity monitoring, either for specific species groups, or potentially for entire ecosystem studies. Biodiversity is known to be declining worldwide, but it is difficult to track using traditional means. In addition, some conventional survey methods rely on damaging procedures and need considerable taxonomic known-how to make them unsatisfactory detection approaches of .eDNA may survive for varying periods of time ranging from weeks to hundreds of thousands of years, enabling for applications in molecular biology, ecology, paleontology, and environmental sciences, particularly when combined with NGS. To further simplify the process, eDNA metabarcoding may be automated. With the further development of the automation system, monitored machine training was utilized to develop predictive models for the monitoring of biodiversity without the imposition of taxonomic tasks, to eliminate the need for interference of people in the development of benchmark data and subsequent assignment of species. eDNA ensures improved detectability of species, reduced labor requirements, does not damage ecosystems, allows species detection without previous information, and may be applied in regions where traditional survey tasks are challenging (Ficetola et al., 2015). Concerns exist with the approach for eDNA metabarcoding, as any new science would expect because each stage of the method has the ability to impact the results, which are then utilized to influence key conservation decisions, it is critical that the procedures be improved and standardized. Many of these concerns are addressed in the preceding section, but some specifically relevant to biodiversity monitoring include: optimization of sample collection and extraction; ensuring samples are free of contamination; designing primers to have minimal bias without sacrificing resolution or spread; development of extensive reference databases; using proper bioinformatics pathways; lack of standardization in classification; and difficulty in estimating DNA degradation rates, thus allowing for the misrepresentation of species presence in areas which they may have vacated (Bush et al., 2019). There is information about populations that simply cannot be obtained via metabarcoding. The population structure and size of population is not possible by utilizing conventional eDNA metabarcoding alone, sex and age ratios, and individual circumstances are frequently essential in developing conservation plans (Deiner et al., 2017; Troth, 2021). Despite this, by using eDNA metabarcoding, it is possible to properly assess biodiversity. The most common solution to this lack of data is to incorporate metabarcoding into the standard monitoring framework. While eDNA metabarcoding cannot determine population structure, it can discover uncommon, cryptic, or elusive species from a single sample, including species that may be considered pests, diseases, or species of conservation concern. The inclusion of both techniques can allow the advantages of each while avoiding both. The integrated approach to biodiversity monitoring is thus suggested for government monitoring initiatives in New Zealand, including biosafety and agriculture. Metabarcoding alone may be able to characterize interactions between species and their consequent effect on ecosystem function, therefore eliminating the requirement for traditional sampling in the identification of ecological networks. Evans et al. (2016) employed the nested tags allowing analysis of samples as individuals, enabling the construction of ecological networks from environmental samples alone (Ruppert, 2020). Traditional methods of defining ecological networks rely on indicator species or bipartite networks, but DNA-based networks have been found to be faster, more efficient, and more comprehensive than traditional methods, especially as the complexity of the region increases or in ecosystems that are poorly understood. Overall, a combination of approaches created the most resolved networks, and they have been recommended for use in conservation, restoration, and analyzing how human-induced climate change affects ecosystem function. Overall, it indicates that combining traditional and innovative approaches is the next step in acquiring information for land management and biodiversity monitoring. The following section describes some of the research that have previously used eDNA metabarcoding for monitoring in marine, estuarine, freshwater, and terrestrial ecosystems (Fig. 6.1).

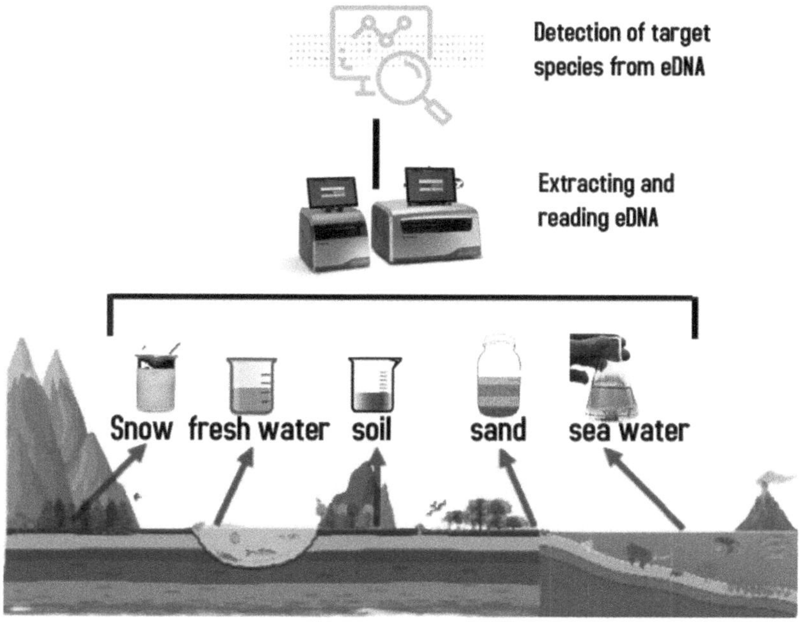


**Fig. 6.1** *Made with PosterMyWall.com.*

### 6.3.2 Marine monitoring

Several collection techniques targeting many different kinds of species have been used to apply sediment eDNA metabarcoding to marine settings. Marine sediments, in particular, are extremely species-rich, play important ecological roles and provide important ecosystem services, and are sensitive to anthropogenic disturbance, but they have been poorly studied using traditional methods and are poorly understood due to the difficulty and rarity of taxonomic work to describe the species found in marine sediments. Although there is a paucity of taxonomic knowledge, eDNA metabarcoding of sand has been utilized in attempts to characterize marine eukaryotic communities. Pawlowski et al. (2011) examined deep-sea eukaryotic richness using metabarcoding on marine sediments and found 942e1756. Pawlowski et al. (2011) taxa per sample, predominantly in dinoflagellates, cercozoans, ciliates, and euglenozoans; even photosynthetic taxa were present. Despite tiny sample quantities, between 31 and 71 metazoan taxa were detected per sample, indicating free, extracellular, residual, or resting DNA/organisms in the sample. They discovered the deep-sea floor to be a worldwide DNA library, defining species living in both the sediment and the water column, making it useful for environmental monitoring. The interpretation of deep-sea bottom DNA data proved difficult owing to a lack of direct observational data, but it nevertheless provides vital insights into live marine creatures and the possibility to research ancient marine life through deep soil core analysis. A similar study by Guardiola et al. (2015) using metabarcoding on deep sea sediments identified over 1600 taxa dominated by Metazoa, Alveolata, Stramenopiles, and Rhizaria, with noticeable differences on a small scale, such as within the same core sample, between core samples, within localities, and between geographic zones (Evans et al., 2016). While composition varied between sites, overall biodiversity was consistent across the sample, with some variation owing to variations in metazoan habits. Guardiola et al. (2016) also published a similar study the following year, finding over 5500 taxa dominated by Metazoa, Alveolata, Rhizaria, Nematoda, Arthropoda, and Annelida and discovering substantial variations across sediment layers, zone (slope or canyon), depth, and season (Guardiola et al., 2015).

Finally, Lanzen et al. (2017) tested the use of metabarcoding genomic DNA obtained from soil samples for determining diversity and community structure of meio- and microbenthic eukaryote communities, finding that sequencing depth and extraction replicates are critical for optimizing diversity estimates (Guardiola et al., 2016). However, the actual number of repetitions or sequencing depth will vary depending on the study's goal. Looking at marine metazoan biodiversity, Fonseca et al. (2010) used metabarcoding to analyze eDNA from sediment in Scotland (Lanzén et al., 2017). From this study, they aimed to determine taxonomic richness based on identifiable taxa, to describe the breadth of undiscovered microbial metazoans, and to assess the ecological importance of various groups. They argued that biodiversity assessments are biased toward large creatures, resulting in a lack of descriptions for millions of tiny metazoans, a group that is believed to account for up to 60% of the animal phyla that undertake critical activities in the marine ecosystem. Their study confirmed this immense lack of knowledge, for which they suggest further exploration of microbial metazoan phylogeny by use of metabarcoding in order to better understand this important and highly diverse group. Invertebrate communities have also been identified using eDNA metabarcoding of sea sand. Cowart et al. (2015) tested metabarcoding by comparing it to morphological methods of identifying invertebrates associated with seagrass meadows (Fonseca et al., 2010). They found metabarcoding to enable them to identify greater diversity, in comparison with morphological surveys, and similar spatial differences; however, 36% of the metabarcoding species were not available in the database and indicated the need for library and procedure optimization prior to future studies. Aylagas et al. (2016) compared taxonomic inferences of benthic macroinvertebrates of known taxonomy between different DNA sources, different COI barcodes, and different amplification conditions (Cowart et al., 2015). Metabarcoding is similar to morphology surveys; however, more benthic macroinvertebrate species have been consistently identified. The distribution of fungal diversity and possible variables influencing that distribution has also been investigated using eDNA metabarcoding of marine sediments. This study detected 816 taxa, 130 recognized genera, 36 orders, 14 classes, and 5 phyla, with Ascomycota making up 72e79% of all units, it was shown that species compositions varied across regions, with temperature determined to be the most important factor in fungal community dispersion. eDNA metabarcoding of sediments has the potential to quickly and accurately assess community structure, benthic health, and responsiveness to environmental changes, making it a promising methodology for biodiversity monitoring. However, it currently continues to require integration or comparison with traditional methods until it is fully optimized.

#### 6.3.2.1 Bulk specimens and biofilms

In addition to sediment, marine eDNA metabarcoding used to identify collected items, particularly tiny species that would otherwise require extensive taxonomic knowledge to identify. This approach has been used in plankton biodiversity research to see if metabarcoding can be used to assess large-scale changes. From bulk samples of zooplankton, Chain et al. (2016) reported approximately 400 metazoan families in ports, including over 30 orders not previously reported (Aylagas et al., 2016). Zooplankton communities were also found to vary temporally and to be different across environment. More recently, Djurhuus et al. (2018) explored the metabarcoding of plankton samples against morphological identification, as well as the possible differences between direct seawater filtration, prefiltering, and plankton tissue metabarcoding (Heather and Chain, 2016). They discovered that several metabarcoding approaches revealed substantial variations in taxonomic makeup, with prefiltering finding less differences than other techniques. Deagle et al. (2018) compared the plankton metabarcoding to morphological identification; however, their analysis depended on the use of continuous plankton recorders which have been used to characterize plankton diversity for decades (Djurhuus et al., 2018). When compared to microscopy, they observed that metabarcoding improved the number of species detected, taxonomic resolution, and total detections. These findings suggest that using metabarcoding in addition to continuous plankton recorders already in place has enormous promise, albeit it would need a shift in mindset on biodiversity surveys to incorporate genetic techniques in what is typically a morphological area. Metabarcoding of specimens has also been used to classify eukaryotes as a whole. Leray and Knowlton (2015) described the variety and structure of animals living on oyster reefs using eDNA metabarcoding of biofilms. They found over 2000 taxonomic units; however, due to wide yet nonspecific taxonomic coverage, only 11% could be matched to reference databases and just 8% to species level (Deagle et al., 2018). Despite coverage issues, they were able to identify that smaller species were more varied and that geographical overlap between locations was limited, demonstrating the method's use for presence/absence surveys, relative abundances, and community-level diversity studies. Wangensteen et al. (2018) used two molecular markers to establish a DNA metabarcoding technique for biodiversity assessment of marine hard-bottom ecosystems (Leray and Knowlton, 2015). They discovered that the studied communities had a substantial amount of unexplained eukaryotic biodiversity, and that while gaps in reference databases were prevalent, they mostly related to "rare" species found on a low number of sequencing reads. Specimen metabarcoding has also been applied to focal groups of nematodes and diatoms. Due to low resolution and insufficient reference databases for the metabarcoding approach, Dell'Anno et al. (2015) revealed that eDNA metabarcoding and nematode morphological data only matched at the family level (Wangensteen et al., 2018). As a result, they recommended for additional research on the topic before it is widely used, although they recognized its promise for researching the hyperdiverse nematodes of the deep-sea benthos. Rivera et al. (2018) investigated the application of metabarcoding biofilms from sea turtle shells to infer turtle behavior in diatoms (Carugati et al., 2015; Rivera et al., 2018). They compared metabarcoding and morphological methods and found that they produced distinct diatom assemblages with each. Metabarcoding could differentiate individual sea turtles based on diatom communities but was restricted by reference databases, whereas microscopy gave additional ecological information on sea turtles, making a combination of the two approaches suitable for studying sea turtle behavior utilizing biofilms. This work demonstrates the broad application of ecosystem eDNA metabarcoding, since it may be used not just to monitor ecosystems but also to movements of larger vertebrates. Overall, metabarcoding is a powerful and sensitive approach for large-scale biodiversity research, and the technology will continue to improve as reference databases are increased. It is a reliable, fast, objective, and cost-effective technique for describing complex communities of any size organism, with the ability to be expanded further by using several primer sets for optimal coverage.

Water metabarcoding of water samples has played an important role in the development of biomonitoring methods. Marine microorganismal diversity was assessed by Brannock et al. (2016) by utilizing metabarcoding to measure microeukaryotic diversity over time at several places along the Alabama continental shelf to examine marine microorganismal diversity (Mayerl et al., 2018). They found that community composition varied seasonally and associated with salinity, temperature, and dissolved silicate. This information on diversity and community variance is crucial to the study of marine ecosystems, but without the use of advanced molecular techniques, it would be nearly impossible to gain a comprehensive understanding. Bridging the gap between micro- and macroorganismal metabarcoding studies, shotgun sequencing was found to be inefficient owing to the huge amount of reads needed to provide a good picture of biodiversity, but the tree-of-life technique was able to identify 287 taxa spanning the major eukaryotic divisions (animals, fungi, protozoa, plantae, chromista, bacteria, archaea). To acquire the entire range of biodiversity, however, numerous sets of "universal" primers were required. Ruppert et al. (2019) reported that the able to isolate haplotypes from a specific fish genus from the same water samples, further expanding the biodiversity applications of eDNA metabarcoding (Pilliod et al., 2013). There are several metabarcoding studies concentrating only on the identification of animals from seawater samples. Thomsen et al. (2012) compared the data from metabarcoding eDNA from seawater to traditional fish surveys to see if it could be used to predict marine fish biodiversity (Brannock et al., 2016). They were able to acquire DNA from 15 different fish species, including those that are essential for human consumption and others that are seldom documented, as well as four different bird species, proving that eDNA metabarcoding identified the same or more species than traditional sampling. Additionally, they found that DNA fragments degraded within days in seawater, indicating that positive eDNA signal simply the presence of a recent species. Kelly et al. (2014) investigated the efficacy of eDNA metabarcoding to recognize bony fish species in a large mesocosm tank, finding a link between eDNA rank abundance and species biomass (Thomsen et al., 2012). They were also able to detect nonnative species used as food in the aquarium. These primers have since been used in numerous fish eDNA metabarcoding studies. Yamamoto et al. (2017) published a research in 2017 that assessed and compared the efficacy of eDNA metabarcoding to identify fish community structure in coastal waters at 47 geographically different stations to 14 years of data from visual surveys (Kelly et al., 2014). The metabarcoding study found 128 species, 62.5% of which were seen during visual surveys, and 20 species that were expected to exist but had never been documented, indicating that the method has the potential to improve marine ecosystem analyses and fish community surveys, especially given its higher detection rate than other methods. Water sample eDNA metabarcoding has also been utilized to characterize whole vertebrate populations. Port et al. (2016) compared the results of eDNA metabarcoding for marine vertebrates along a transect gradient of habitats to simultaneous visual surveys by divers, revealing spatial correspondence between metabarcoding and visual surveys as well as significant differences between habitats, indicating local community compositions (Yamamoto et al., 2017). Furthermore, metabarcoding was found to be capable to reliably detect cryptic species missed by visual surveys, though not all species were consistently detected with both methods, suggesting that the current applications of eDNA metabarcoding are best suited to augment traditional data for the monitoring of community dynamics over space and time. Andruszkiewicz et al. (2017) published a study in which they collected water samples in replicates at multiple depths per station and used eDNA metabarcoding to describe marine vertebrate communities at 10 locations (Port et al., 2016). With the exception of one family, they were able to identify 7 families, 3 subfamilies, 10 genera, and 72 species from 33 different marine vertebrate taxa. They also discovered that communities changed depending on sample depth and sampling location, suggesting that eDNA metabarcoding may accurately census vertebrate communities and hence has great promise for biomonitoring at fine scale throughout space and time. The efficiency of marine eDNA metabarcoding has also been evaluated across sample types. Massana et al. (2015) investigated the taxonomic diversity of both planktonic and benthic protist populations through space and time (Andruszkiewicz et al., 2017). They discovered diverse species compositions in water and sediment samples, as well as differences in community composition between locations. They also discovered that seasonality must be taken into account, as well as the requirement for several samples per site to correctly estimate diversity and community composition. Altogether, the use of sediment and water produced distinct findings in terms of community composition and spatiotemporal differences, further indicating the need to ensure that studies are tailored for the desired species groups. Overall, marine eDNA metabarcoding is noninvasive, efficient, cost-effective, and sensitive, with potential to complement other methods of biodiversity monitoring for ecological study and management on large spatial and temporal scales. Despite the fact that it involves duplicate optimization, marker selection, the avoidance of false negatives/positives, and the construction of a reference database, it has enormous potential across sample types and target taxa.

### 6.3.3 Estuarine monitoring

Because of their similarities, sediment estuarine biodiversity monitoring is frequently combined with marine biodiversity monitoring; however, estuaries are a distinct ecosystem that overlaps both marine and freshwater ecosystems, and as a result, some eDNA metabarcoding studies have focused specifically on estuarine biodiversity. The majority of these studies have focused on identifying benthic ecosystems from sediment samples, which typically contain taxa such as nematodes, which are widely regarded as environmental health indicators. Metabarcoding has been inadequately applied to estuarine macrobenthos communities, and morphology-based characterization of benthic populations has been low-throughput and confusing. Furthermore, eukaryotes, which are a valuable resource for measuring ecosystem health, are underrepresented in models that predict estuarine ecosystem diversity. As such, future studies are needed to explore estuarine sedimentary diversity via metabarcoding; nonetheless, several exploratory studies have been published. Chariton et al. (2010) used a metabarcode in estuarine sediments to identify eucaryotes in ecosystems affected by human effects and successfully collected data sequencing thousands of eucaryotes from 54 phyla from microscopic nonmetazoans to large metazoans, rich in micro- and meiofauna often overlooked by traditional surveys (Massana et al., 2015). Chariton et al. (2015) also published a research using metabarcoding for the assessment of five different conditions' benthic eukaryotic composition of estuaries. They discovered that the eukaryotic makeup of a community reflected environmental conditions in the most disturbed environment and was closely linked to nutrients, turbidity, and pH with the highest biological richness (Chariton et al., 2010). In the same year, Lallias et al. (2015) published a study in which metabarcoding was employed in two distinct estuaries with differing histories of use using longer primers for live persons to evaluate the environmental impacts of the microbe metazoan and protist diversity, with the intention of providing an immediate view of the microbial community (Chariton et al., 2015). They discovered that different species responded differently to hydrodynamics, salinity, and granulometry depending on their life cycle, and that they shared estuarine-specific characteristics within salinity ranges. In development and assessment of metabarcoding methodology, Lobo et al. (2017) created several test communities to assess the viability of metabarcoding, and then compared metabarcoding to morphological identification in field tests at four distinct estuaries (Lallias et al., 2015). They found that field morphological methods underestimated species richness, finding just 23 species as compared to the 61 species identified using metabarcoding. They determined that metabarcoding was capable of producing a large number of high-quality species identifications while also being sensitive enough to detect variations in species compositions between communities. Avo et al. (2017) also tested the ability of metabarcoding to accurately describe communities by comparing metabarcoding of sediments from an estuary that had been previously characterized using traditional methodology (Lobo et al., 2017). Metabarcoding generated 25e26 of the traditionally identified genera and identified 34e43 species in total, including 7e16 new genera for the databases, whereas morphological analysis matched about 90% of the metabarcoding genera. They found significant temporal but not geographic changes of the species composition between the two methodologies, with metabarcoding appearing to be a potential option for standard monitoring methods that are labor-intensive and time-consuming. The eDNA metabarcoding of bulk materials has also been utilized for plankton community research. Abad et al. (2016) have assessed estuarine plankton monitoring metabarcoding capacity by comparing it with conventional taxonomic microscopic analysis (Avó et al., 2017). They found similar spatial and temporal patterns in zooplankton taxonomic diversity but not phytoplankton, perhaps due to a lack of a comprehensive phytoplankton database. Furthermore, there was a good consistency between abundances in the two methods, indicating that metabarcoding has the potential to be used to estimate species abundances as well as detect invading species, since two nonindigenous species were detected. The limitations of eDNA metabarcoding in estuaries are analogous to those observed in other environments and applications; nevertheless, the lack of studies utilizing metabarcoding in estuaries demands additional exploratory research. In general, eDNA is fast, scalable, comprehensive, objective, extremely sensitive, high-taxonome precision metabarcoding, and capable of identifying species of any size. This metabarcode is cost-effective and time-efficient in enhancing conventional monitoring techniques and inform environmental policies.

### 6.3.4 Freshwater monitoring

Biodiversity monitoring studies in freshwater utilizing eDNA metabarcoding are quite similar to those in marine and estuarine environments, with the main areas of focus being in sediment, biofilms, bulk specimens, and water samples. Freshwater fauna are sensitive to environmental change, making them excellent indicators of ecosystem health, yet conventional surveys are labor demanding, entail physical capture which may injure individuals, and often result in low detection. eDNA metabarcoding has potential to improve aquatic biodiversity surveys; however, further research of species characteristics and life history traits that may affect detection and refinement of quantification is needed.

Sediments metabarcoding has been used on freshwater sediments to analyze plant communities. Alsos et al. (2018) analyzed metabarcoding and conventional vegetation sampling to examine the composition of lake vegetation and vascular plant eDNA in 11 lakes in order to assess the extent to which contemporary species are detected in sediment eDNA samples. They recorded 489 taxa by conventional surveys, of which 17e49% were detected using eDNA, while 47 plant taxa were identified using eDNA, of which 73% matched those recorded in vegetation surveys. They concluded those species detected in eDNA but not in surveys were most likely overlooked during the surveys or grew beyond the survey area, and those missed in eDNA may have been found with increased sampling effort. In addition, detection rates differed between plant families and terrestrial plant DNA depended on the sample distance. For aquatic vegetation, eDNA may be superior to traditional surveys; nevertheless, additional tuning is required before the procedure is ready for widespread application. Altogether, metabarcoding of sediments appears to be a promising method for biodiversity monitoring, albeit it requires standardization and expansion of databases before it is able to reach its full potential as a staple in biomonitoring methodology.

#### *6.3.4.1 Bulk specimens and biofilms*

Similar to marine environments, specimen collection for metabarcoding is a popular choice for describing freshwater biodiversity, with many of these studies focusing on plankton. Liu et al. (2017) investigated the influence of prefiltering on the detection of microeukaryotic plankton communities and discovered that roughly 25% of readings in both groups were for metazoans. Additionally, they found no significant differences in richness, diversity, and spatiotemporal patterns, suggesting that prefiltering had little effect to differentiate size classes of eukaryotic plankton in freshwater environments. In the same year, Wurzbacher et al. (2017) published a study in which they tested eDNA metabarcoding for entire plankton community dynamics over time, including eukaryotes and prokaryotes in their analysis (Ruppert et al., 2019). They observed changes in community composition over time that followed the same patterns as morphological approaches, although morphological approaches missed a bloom of a rare dominating bacteria that was affecting the planktonic population. These findings underscore the importance of microbial and macrobial community metabarcoding research in comprehending whole ecosystems, species interactions, and community dynamics. A similar study was reported by Ruppert et al. (2019) in which samples obtained from many places in a storage area have been metabarcoded for over 4 months, from which over 1300 taxonomic units, including metazoans, protists, chlorophytes, and fungi, were identified. They discovered that estimates of diversity vary from one location to another and suggested that various plankton groups responded to distinct environmental factors. Before the research, Sohlberg et al. (2015) investigated groundwater plankton fungal communities linked to deep basement cracks and sampled at all profundities, dominated by Ascomycota, with greater than anticipated variety of fungal. In the fracture zones, fungi have been shown to be active, showing a powerful capacity for cell activity through environmental adaptation. Some of the isolated fungus genomes may represent new species which might play a significant role in the cycling of subface nutrients. Metabarcoding for the characterization of zooplankton communities has been utilized and agreement was reached between the two techniques for species composition, the geographical distribution, the variations between kinds of water bodies and the abundance based on biomass. In addition, metabarcoding was found to identify more species than morphological identification. The biases commonly related to metabarcoding studies such as technical biases, bioinformatics differences, the lack of comprehensive databases, and incapacity to detected individuals' life and health were limited. Another study applying metabarcoding to freshwater specimens was published the same year, in which a new method for analysis of larval fish communities using metabarcoding of bulk ichthyo plankton samples was developed and compared to individual larval sequencing. The two techniques revealed that the results were equivalent, showing metabarcoding potential to enhance knowledge of fish recruitment and river conservation, especially in the tropical areas, where fish larval ecology is rarely investigated due to identifying problems. Metabarcoding of plankton communities has the potential to give greater taxonomic coverage and resolution than morphological methods, and comparison to conventional data indicated comparable abundance and diversity data. However, each method included groups not seen in the other, suggesting the use of the two methods concurrency is the best way to attain optimal coverage and the most accurate estimates of diversity. In addition to plankton, freshwater specimens obtained have been metabarcoded to represent biodiversity in bulk macro invertebrate samples. Carew et al. (2013) have investigated the use of metabarcoding to identify Chironomidae species from field specimens, a major aquatic ecosystem health indicator group. They discovered that 96% of the species could be identified using one gene area, while 99% of the species could be recognized when used with two (COI and CytB). They observed a significant connection between sequence number and individual number indicating that metabarcoding may be used to estimate abundance. Metabarcoding was compared to morphological macroinvertebratic metrics on the gradient connected to watershed characteristics, where the two research types have been shown to consistently demonstrate the same variations across stream gradients. Hajibabaei et al. (2012) published a one-of-a-kind study that investigated the use of preservative ethanol as a nondestructive source of DNA from macroinvertebrates to determine biodiversity and compared this data to both morphological identification of species in the ethanol and metabarcoding of all the specimens homogenized together, as is typically done with specimen metabarcoding (Hajibabaei et al., 2019). They discovered that ethanol-based metabarcoding correctly identified 87% of species whereas tissue metabarcoding correctly identified 89%, with the missing taxa in both cases being those with the lowest abundances in the combination. This technique addresses the need for both sample and voucher specimens by not destroying samples in order to gather data, allowing genetic biodiversity surveys to take advantage of current preservation standards. In addition to the bigger macro invertebrates, it has the ability to target pests, diseases, vectors, or other organisms linked with those preserved in ethanol. Overall, specimen metabarcoding has tremendous promise in the field of freshwater biodiversity monitoring since it has been shown to minimize the cost and time of biomonitoring; nevertheless, like with all metabarcoding research to date, it requires refining and further creation of reference databases. The testing of biofilms for diatom diversity has been applied several times in freshwater diversity; diatoms are frequently a focus of freshwater monitoring due to their position as bioindicators. Several times, metabarcoding of biofilm diatoms has been compared to morphological surveys, with consistent findings. Kermarrec et al. (2014) found that species richness was usually higher using microscopy in a mock community, although they cautioned that some species detections from microscopy might have been from dead cells, which cannot always be distinguished visually from living organisms. Several studies have utilized natural freshwater ecosystems, which are commonly used as indicators of water quality, to evaluate the efficacy of metabarcoding. Zimmermann et al. (2015) discovered that metabarcoding consistently identified more taxa and provided more knowledge on taxonomic diversity than morphological methods, whereas Vasselon et al. (2017) discovered that, while the community structures of the two methods were highly correlated and the ecological quality estimates were comparable, only 13% of species were shared by both approaches, most likely due to overlapping taxonomic diversity (Zimmermann et al., 2015). Similarly, Rivera et al. (2018) discovered comparable assemblies using two techniques organized by same variables for the environment; nevertheless, each methodology has had ecological assessments that are different, potentially due to a lack of a large DNA reference library (Pawlowski et al., 2018). Despite this limitation, the authors propose an integrated approach of taxonomy, autoecology, life forms, cell sizes, and ecological guilds from morphology, molecular data, and ecological aspects for determining freshwater diatom biodiversity; metabarcoding allows for the collection of diversity data, which can then be applied on different spatial, temporal, and taxonomic levels for large-scale analysis. In short, rather than being a full replacement, metabarcoding offers enormous promise for supplementing conventional diatom surveys. Freshwater biofilms were also employed for generic microbial metabarcoding. Besemer et al. (2013) used fluvial biofilms using eDNA metabarcoding to assess the distribution of diversity of ecologically important benthic microorganisms. They discovered that microbial richness dropped downstream, particularly near confluences, using just metabarcoding data, suggesting that environmental and biotic factors may influence biofilm connectivity and diversity. They also observed that the makeup of the species in streams not explained by geographical distance was very variable, which suggests the restriction of microbial spread in fluvial networks. The diversity patterns observed in this metabarcoding study suggest the relevance of headwater communities for regional diversity and conservation, as well as ecosystem function management.

The collection of eDNA in freshwater environment is arguably the most frequent approach for metabarcoding, dominated by research focused mostly on fish and other aquatic animals. One of the earliest metabarcoding studies on freshwater vertebrates contrasted eDNA with electrofishing for stream fishing groups. Electronic fishing was discovered for 12 species and estimated 16 species richness, with 16 species being detected by eDNA sampling, of which 3 never were detected with electrofishing in the body of water. Despite the inability to give particular data on population quality, they found that eDNA metabarcoding has the potential to improve fish surveys without harming fish or habitat in order to better inform management choices. The same year, Civade et al. (2016) used eDNA metabarcoding to assess fish biodiversity in a freshwater ecosystem and explored signal variability by testing organization of fish communities along a longitudinal gradient. They showed that eDNA metabarcoding was more efficient and more probable than conventional species to identify species present and that fish species that have been identifiable as metabarcodes were equivalent to those identified by traditional sampling. Hänfling et al. (2016) attempted to characterize the qualitative and quantitative correctness of eDNA metabarcoding by comparing eDNA metabarcoding of a lake fish population to long-term conventional data. They observed that eDNA was detected from one sampling session by 14 out of the 16 recognized species in the lake, while just a 4 out of the last traditional sample attempt had been detected making eDNA considerably more sensitive. They also discovered that the amount of sequence reads was connected to both site occupancy and abundance estimates from prior surveys, implying that eDNA metabarcoding might be used to describe fish populations quantitatively. Shaw et al. (2016) compared eDNA metabarcoding to traditional approaches, focusing on the netting survey comparison. It is observed that 100% of netted water (including conservation species, seldom, and invading species) were detected in water samples and that, in the description of fish populations, 51 L water samples were more efficient than smaller water samples or sediment samples. Evans et al. (2017) evaluated the next year by comparing it to existing approaches to see how spatial distribution and bioinformatics impact efficiency, eDNA metabarcoding estimates for species richness. They discovered that when all markers were considered independently, eDNA metabarcoding could detect all species detected with conventional sampling plus 11 more. Despite this success with a number of markers, they discovered that just seven known, conventionally caught species were found when species were to be detected by all markers. Furthermore, they discovered that eDNA was relatively evenly spread around the reservoir, but they anticipated that moving water bodies, such as rivers and streams, would not do so. Nakagawa et al. (2018) have evaluated the efficacy of the flowing water bodies eDNA metabarcodings, concentrated on stream fish species and compared metabarcoding results with existing observational data. They discovered that eDNA data from the upstream industry are most comparable to observer data and that they can detect 86.4% of the stated species and two other species. Sato et al. (2017) studied differences in the evaluation of biodiversity in freshwater fish using eDNA metabarcoding pooling water. More species, including moderate geographic correlations, have been discovered in separate samples showing that the pooling method is not suited for measuring wealth, but is effective for comparing fisheries ecosystems within sites. Additionally, they found that increased replicates lead to an increase in water volume sensitivity and a sufficient distance from sampling points and hence advised modest amounts for optimum detectability from numerous places. Evans et al. (2017) have built aquatic mesocosms to evaluate the capacity of eDNA metabarcoding in varied densities as well as relatively large quantities for both fish and amphibian species. They discovered that metabarcoded all species in all mesocosms were detected and a slightly positive connection between species wealth and reading abundance was identified. They demonstrated the potential of eDNA metabarcoding as a species wealth indicator and the assessment of diversity. This is a step forward by developing fish and amphibian primer sets and evaluating them for application in environmental studies and aquatic biodiversity monitoring in both laboratory and field circumstances (Valentini et al., 2016). Initially evaluated with isolated tissues for a reference database and then poured tissue, with spiking water and ultimately field-sampling in regions of familiar species, they tested their primers against 53 amphibian species and 64 freshwater fish and 22 marine fish. The occupation data were compared to historical and traditional survey data after these experiments and the results indicated that the use of eDNA metabarcode with these primers may be a tool to study the biodiversity of freshwater vertebrate. Lim et al. (2016) also assessed species diversity across taxa, using universal metazoan primers on water samples from two reservoirs with known species diversity. They uncovered indications for over 500 species, around 100 of which could be recognized to the genus or species level, including an invasive snail and a rare fish. There have also been amphibian-only studies published. Lopes et al. (2017) compared metabarcoding data to traditional visual and acoustic surveys to test eDNA metabarcoding for anurans in tropical streams. With a single eDNA visit, they were able to detect all the species known to live in the streams, which were more widely read from more common species more closely connected with the streams. They also discovered that detection was increased via purifying enormous quantities of water and sampling multiple sites along streams, but quantification was not found to be achievable. The detection of eDNA per sample event was higher than traditional field surveys could detect several life stages missed from traditional surveys, could be employed outside the reproductive peak species and is noninvasive, and in relatively short times can be used on a wide range of taxonomic and spatial scales, making it helpful in tropical anuranic surveys. The main emphasis was herpetofauna and proved to be metabarcoding useful in identifying species distribution (Lacoursiere-Roussel et al., 2016). The sensitivity of eDNA metabarcoded for large-sized stuff has recently been compared by Harper et al. (2018) to eDNA quantitative PCR (qPCR). They discovered that qPCR provided better detection for this species, but when thresholds were removed from metabarcoding, it performed just as well as qPCR, implying that an initial community survey using metabarcoding followed by species-specific surveys using qPCR for species of concern, as indicated by the metabarcoding studies, would be a good potential methodology. eDNA metabarcoding appears to be an extremely promising avenue in freshwater biodiversity studies with the capability to collect monumental amounts of data in short periods of time without disrupting ecosystems. Although it has the predicted drawbacks described in earlier sections, as technology advances, metabarcoding can be expected to become less expensive and more effective. As in other ecosystems, it is probably best suited as a complementary method to traditional surveys.

### 6.3.5 Surveillance on the ground

Terrestrial eDNA metabarcoding for biodiversity is not an obvious use of the approach; however, there are numerous significant sources of terrestrial eDNA that may be tracked, including soil, saltlicks, water sources, and bulk specimens. It is doubtful that nonsoil dwelling species could be reliably detected using soil samples alone. Indeed, the bulk of terrestrial eDNA metabarcoding investigations have focused on soil samples for invertebrates that come into regular, direct contact with the soil, although additional connections have been identified and investigated. These inventive solutions are one-of-a-kind and appear to be unachievable without the advent of metabarcoding. Rapid survey approaches for monitoring biodiversity might aid in assessing and understanding threats to ecosystems throughout the world, and metabarcoding opens up new avenues for large-scale biodiversity studies as well as being an efficient and economical approach to studying ecosystem structure.

Soil terrestrial metabarcoding began with soil sampling, with a study by Buee et al. (2009) in which universal fungal primers were used on 6 distinct forest soil samples, yielding over 1000 distinct taxonomic units, primarily of Ascomycota and Basidiomycota, with Agaricomycetes as the dominant taxonomic unit. Diversity studies based on this data revealed significant geographical variability and unexpectedly great richness (Buee et al., 2009). Furthermore, they discovered that examined sequences correlated 73% of the time with just 26 recognized species, underlining the critical need for more broad sequencing databases. Soil fungi also play an important ecological role but little is known about the structure, richness, and distribution of different fungal communities and ecological groups, making metabarcoding an interesting potential for soil fungal studies in the future. Other studies have focused on determining plant diversity from soil sample metabarcoding. The composition of the plant community can help define the ecological quality of the sites but above-ground visual surveys are incomplete; eDNA is capable of identifying active and dormant seeds, contaminants and waste of species at the same time as metabarcoding, thus providing a broader view of the plant diversity. Yoccoz et al. (2012) used taxonomic diversity on eight plant communities from two vegetation types using metabarcoding of a short DNA fragment with plant-specific primers and examined DNA persistence by metabarcoding historically farmed soils. They discovered that plant diversity using metabarcoding was consistent with estimates from previous surveys, but identified species that were not detectable in past surveys. Furthermore, soil samples were spatially distinct in species composition and diversity, with crop DNA sequence detection from the formerly cultivated fields dependent on the amount of time since cultivation, dropping off almost completely after approximately 50 years. Fahner et al. (2016) also assessed soil metabarcoding for quantifying plant diversity, evaluating sequence recovery, annotation, and resolution for four plant DNA markers first with database sequences, then with soil samples from a remote wetland. Internal transcribed spacer (ITS)-2 and ribulose bisphosphate carboxylase (rbc)-L were discovered to be the most efficient areas for metabarcoding vascular plants for generalized databases, whereas full-length DNA barcode regions outperformed shorter sequences in taxonomic resolution and detection. This discovery is especially interesting in light of Yoccoz et al. (2012)'s observation of the short barcoding region's limits in taxonomic precision. Metabarcoding of soil and sediment, as in aquatic biodiversity studies, has been used to characterize eukaryotic populations. Drummond et al. (2015) compared a large number of markers to traditional terrestrial eukaryotes control tools, examined in the16S, 18S, trnL, ITS, and COI regions whether eDNA could be an alternative to traditional biodiversity sampling, which combinations of markers are optimal and how sensitive measures are to markers/parameters. They found that alpha and gamma diversities were sensitive to number of reads, but beta diversity remained steady after a few thousand reads. They also observed that the COI area had at least two markers for covering all phylas, notably for plants and invertebrates, using traditional sampling data. Metabarcoding of invertebrate species has also been investigated. Yang et al. (2014) showed that the findings of soil and leaf litter metabarcoding differed from standard arthropod surveys in terms of species composition. However, they described similar ecological information in regards to relative richness and diversity, with metabarcoding perhaps being more sensitive to habitat changes than traditional surveys. Soil metabarcoding may also be used to examine the response of eukaryotic communities to environmental changes.

Štursová et al. (2014) used metabarcoding to examine the response of fungal soil and leaf litter communities during a bark beetle-induced tree dieback, finding that the biomass of the fungal communities decreased, and that community composition changed with a loss of root-associated fungi and an increase in saprotrophic species. Andersen et al. (2012) attempted to evaluate vertebrate taxonomic richness and structural diversity using soil samples from locations with known species compositions such as zoological gardens, safari parks, and farms using metabarcoding. DNA from the soil has been found to represent general taxonomic wealth and relative biomass but may not be responsive to recent community changes. Moreover, they found no impact of soil pH on DNA amplification, but that high organic content inhibited amplification, and that DNA may move across strat a depending on soil texture and structure. They determined that geographical replicates are required to optimize detection, since DNA was not evenly distributed throughout the region, with greater quantities detected around toilet or den sites. Despite this, they determined metabarcoding to be a quick and reliable alternative to traditional surveys. Epp et al. (2012) reported perhaps the most complete study on terrestrial biodiversity from soil samples, in which they produced metabarcodes for fungi, bryophytes, enchytraeids, beetles, and birds, and assessed in silico PCR and in vitro PCR of surface soil and frozen sediment. They discovered that in silico primer design was effective in producing functional primers for the field, and that the most promising universal metabarcoding areas were located on the 12s and 16s, resulting in more coverage than other regions. They also observed substantial variation in the components of current and ancient species.

#### 6.3.5.1 Root associations

In a similar environment to soil samples, sampling of communities associated with roots has been used for metabarcoding biodiversity studies. Blaalid et al. (2012) investigated changes in fungal communities associated with the root systems of Bistorta vivipara, a perennial herbaceous flowering plant, along a primary succession gradient using metabarcoding to detect variation in fungal richness and community composition as ecosystems changed. They discovered 470 taxonomic units that correspond to 8–41 basidiomycetes-dominated fungal sequences. Total taxa grew as the ecosystem evolved but the root system average did not change, even if fungal distribution across root systems showed a significant degree of patchiness. Martínez-García et al. (2015) also looked at root-associated fungal populations by using metabarcoding. They looked at a lengthy soil sequence, including evolution of the ecosystems and backwardness, to determine the relevance of host plants as a factor determining the makeup of the fungal community. They also coupled metabarcoding methodologies (for determining community per site) with terminal fragment length polymorphism (T-RFLP) approaches (for detecting community within sites) to provide a comprehensive picture of community composition, diversity, and change through time. They found fungal communities are strongly organized by ecosystem age and have substantial niche differentiation. They also discovered that as the ecosystem aged, diversity declined, and 39% of fungal variation could be attributed to the host plant, implying that fungal communities are intimately linked to plant communities.

In terrestrial environments, bulk specimen metabarcoding has also been investigated. Conventional ecosystem evaluations based on morphology or barcodes are costly and time-consuming, making them unsuitable for large numbers of samples, which is why metabarcoding will be used in the future. Yu et al. (2012) have established methods for the identification of bulk arthropod sample taxonomy by creating seven communities of arthropod and comparing the composition and diversity between the two datasets. They discovered that it is possible to precisely estimate community differences and diversities by metabarcoding despite the loss of some taxonomic information, and that metabarcoding of bulk samples is effective for measuring biodiversity. In the following year, Ji et al. (2013) validated metabarcoding of bulk arthropod samples by testing it against 3 high quality standard datasets from Malaysia, China and the United Kingdom. For all sites, metabarcoding provided linked estimates of species richness, identified treatments and responses, and returned statistical models in the same taxon. Gibson et al. (2014) performed a similar study in which they tested the capacity of metabarcoding to characterize diversity using several universal primers on bulk arthropod samples. They found that metabarcoding was able to identify 91% of the arthropods and identified microorganisms linked with the arthropods. Furthermore, they determined that metabarcoding was more successful than other methods and greatly reduced the time and cost of biodiversity studies, making it ideal for a variety of further applications in ecology, including interactions between the macro- and microbiome. Metabarcoding of bulk specimens may also have applications in the assessment of host–parasitoid and community interactions within biodiversity studies, an area of study essential to biodiversity monitoring, but difficult to discern using traditional methods.

Šigut et al. (2017) tested the possibility of metabarcoding for identification using fake samples of insect larvae and parasitoids, and they examined the completeness of the barcode database by comparing it to a known host–parasitoid database for the research area. Metabarcoding recognized 92.8% of all taxa present in bogus samples in this investigation and was able to identify taxa consistently across all taxonomic levels. Furthermore, they discovered that 39.4% of parasitoid and 90.7% of host taxa could be recognized using the reference database, indicating a clear need for parasitoid database growth. This study's metabarcoding data show greater parasitoid diversity than traditional surveys, identifying the potential of metabarcoding to obtain more precise identifications from reliable libraries and to better characterize diversity. Miller et al. (2016) established a protocol for metabarcoding fungal communities directly from the DNA extracted from homogenized beetles and found that metabarcoding produced viable information on fungal species abundance associated with beetles particularly when large numbers of PCR repetitions were employed. The use of carrion fly samples to assess vertebrate diversity is a novel use of metabarcoding bulk arthropod collections. Carrion flies feeds on dead and rotting animals all around and no distinctive host preferences have been observed.

Because they have no preference, they are a good prospective model for determining vertebrate diversity, provided that mortality rates between species are generally stable. Applications of this method may even be able to be applied to monitoring efforts of mortality dynamics or identification and monitoring of particular populations. Calvignac-Spencer et al. (2013) used metabarcoding on 201 pooled carrion flies collected in Cote d'Ivoire and Madagascar, retrieving DNA from 22 to 4 species from different forest strata and size ranges, including very rare species, in one of the first studies employing this creative methodology for monitoring terrestrial biodiversity. Though they did not establish a full spectrum of mammalian variety in this study, they anticipated that greater sampling effort may result in a more thorough evaluation of biodiversity. In addition, amphibians and birds were discovered, suggesting that additional applications in biodiversity monitoring are possible. Schubert et al. (2015) tested carrion flies with generic and specific primers to see whether there were any variations in detection rates. They discovered that, while specialized primers increased detection of target species, generic primers could still identify these species. In addition to the planned application of estimating population abundance and assessing overall mammalian diversity, they discovered that the DNA obtained from carrion flies was sufficient to attempt identifying individuals. Lee et al. (2016) have undertaken the calibration of the metabarcoding of carryon flies by using metabarcoding biodiversity estimations in two forests with recognized megafaunas, one without cage trapping, mist trapping, hair trapping, and scat collection techniques conventional. They discovered metabarcoding to detect more species consistently than other approaches and projected that they would continue to exceed standard methods as the time for studies extended. Finally, Rodgers et al. (2017) carried out vertebrate-specific metabarcoding on carrion flies in the region of a renowned mammal population to test and monitor vertebrate detection methods by comparison of diuretic transect count and camera capture findings metabarcoding. All surveys were carried out over the same time period (including fly collection), and metabarcoding was shown to detect more species than other techniques during the same time period, but less than those reported throughout the previous 7 years of surveys. Their data suggest that metabarcoding is powerful and efficient for biodiversity studies, particularly when used in conjunction with current monitoring methods. However, more complete reference databases, PCR optimization, many markers, more sample work, higher sequence depth, feelings shared across other research are necessary. This unique application of metabarcoding may be a rapid, unbiased, and cost- and time-efficient wildlife monitoring methodology with capabilities to detect species and individuals, and the potential to join the standard mammal monitoring field methods. Additional study into fly ecology is necessary before it can adequately be included into the surveillance efforts for terrestrial biodiversity.

#### 6.3.5.2 *Water*

Though they do not reside in the water, terrestrial animals still require water and other nutrient sources to subsist. As such, eDNA can be collected from the sources and analyzed to determine species use. Assessing species presence using eDNA in forest habitats, which are often difficult to survey, particularly when containing elusive or rare species, could be a viable alternative or augmentation to traditional study methods. In 2017, a research in Japan employed metabarcoding to describe the mammalian population using forest ponds, and they were able to detect numerous common animals from the area. However, not all common animals have been discovered because the usage of ponds depends on individual requirement and may vary significantly over time and space. A similar study using universal bird primers has also been published; however, variation in species and a lack of reference sequences reduced detection. Saltlicks or other nutrition centers are another terrestrial place that may be visited more regularly. Environmental metabarcoding of DNA has been used to locate mammalian saltlicks in Borneo; numerous endangered species have been found in this study but many smaller species have been overlooked owing to lack of reference sequences. Regardless, this is a promising field of research given the advancement of methods and reference databases.

## 6.4 Further applications

While the prospects for ecosystem and biodiversity monitoring using eDNA metabarcoding are thrilling and in need of further research, it is not the only area in which the method can be applied. Metabarcoding may be used to reconstruct ancient DNA ecosystems worldwide, to study the interconnection between plants and contaminants, to evaluate diets without food monitoring or stomach flushing, to detect invasive species prior to detecting them with traditional means, to identify Community responses to pollution, and even to assess air quality and its implications for human health. Though these are the topics covered in this review paper, more applications of eDNA metabarcoding are surely yet to come (Fig. 6.2).

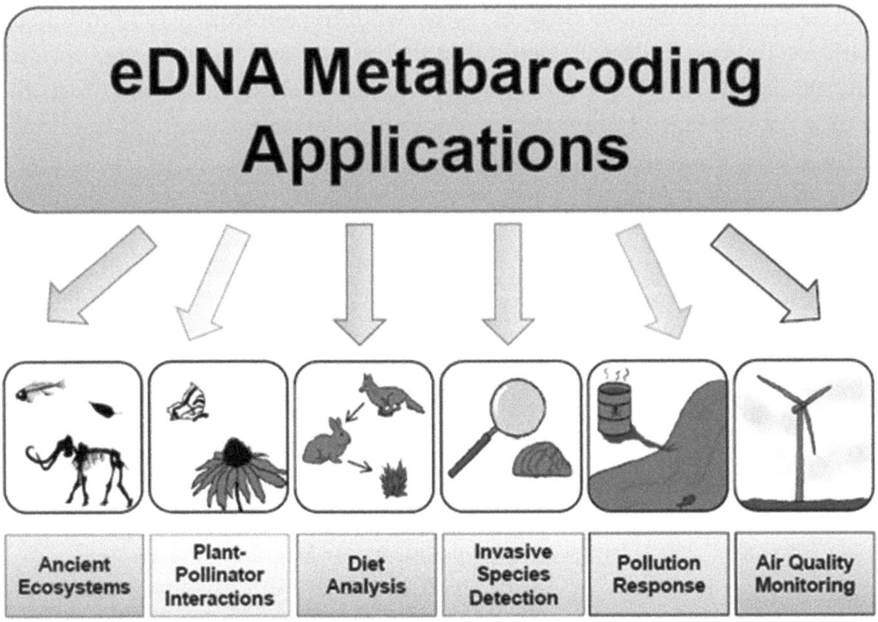


**Fig. 6.2** ***Applications of environmental DNA metabarcoding in aquatic and terrestrial ecosystems.***

### 6.4.1 Paleology and ancient ecosystems

Environmental metabarcoding is not only applicable for monitoring the health and status of modern ecosystems but has also been used to rebuild ancient ecosystems from carbon-dated deep sediment samples that have been carbon dated. As traditional methods of ancient ecosystem reconstruction are time-consuming and prone to human error since they rely on pollen and microfossil identification; metabarcoding investigations, when combined with traditional data, may provide more complete view of ancient ecosystems.These reconstructed ecosystems can in turn be used for future conservation planning, including backwards testing of climate change models, invasive species emergence tracking, and assessment of anthropogenic influences to past biodiversity and landscape. One of the first studies using both metabarcoding and NGS on ancient DNA was published in 2010, wherein Sønstebø et al. (2010) constructed a library of 842 widespread/important arctic flora species and identified chloroplast region P6 as a successful metabarcoding region for these plants.They found that using metabarcoding primers with NGS resulted in greater taxonomic resolution than traditional methods. However, using both methods together was the most successful in identifying the greatest number of taxa to the smallest resolution. Also studying arctic plant composition, Jørgensen et al. (2012) compared sedimentary DNA retrieved from nunataks to data from detailed surveys of the past 130 years. Based on traditional surveys, biodiversity had drastically decreased since 1967 from 60 species to just eight in 2009. They determined that metabarcoding plant DNA was more successful than traditional surveys, as it was able to identify twice as many species as the 2009 survey, including 3 taxa that had never been documented on the nunatak. As there is no animal life on these nunataks, changes in biodiversity can be assumed to directly correlate with climate, making data from this location especially valuable for climate change research. Pedersen et al. (2013) compared metabarcoding of sediment from a lake in Greenland to a previously determined biodiversity record to identify ancient plant biodiversity. Although sedimentary DNA was able to identify two previously undiscovered families at the study site; however, others were missed owing to the incomplete nature of genetic libraries. As such, they determined that DNA metabarcoding has the potential to reconstruct local vegetation in greater detail than traditional means, but that further work is needed on resources such as GenBank to ensure that no species are missed during the process. Despite higher deterioration rates owing to climate variations, even tropical locations can produce meaningful metabarcoding data from sediment. Boessenkool et al. (2014) were able to identify species from tropical sedimentary DNA to a greater taxonomic resolution than was found with pollen sampling; however, more species were identified with pollen. This discrepancy could be due to DNA degradation, or it could be a product of the transient nature of pollen; where sedimentary DNA is relatively local, pollen is able to disperse before settling in sediment, and therefore may not provide an accurate depiction of species composition in a given place. To achieve a more complete look at ancient ecosystems, Willerslev et al. (2014) built upon a previous study from 2003 by studying not only plant diversity from sediment DNA, but also nematode diversity as a proxy for vegetation cover and the diets of ancient herbaceous megafauna as a method of confirming plant species presence and ecosystem-level interactions. By combining all results of ancient eDNA sequencing, the study was able to determine the circumpolar vegetation cover and the changes therein over time. Additionally, metabarcoding of ancient megafauna stomach contents offers information on species' habitat preferences, behavior, and biology of species and their influence on the plant communities. Wood et al. (2012) also used metabarcoding to analyze the diet of extinct herbivores, focusing on the moa of New Zealand. They were able to identify at least 67 plant taxa, revealing a generalist feeding ecology throughout accessible environments. Furthermore, intact seeds indicate the potential importance of moa for plant dispersal. Microorganisms such as fungi also play an important role in ancient ecosystems and can even be used to determine plant and animal diversity based on associated fungal presence. Metabarcoding of permafrost has been successfully used to identify fungal communities with almost double the number of taxonomic units identified than with Sanger sequencing. While this is undoubtedly useful, differences in stability during cryopreservation and bias in PCR amplification and sequencing make it impossible to fully describe ancient fungal communities, though overall changes in diversity over time may be observed. Ancient DNA metabarcoding can also be used to determine anthropogenic effects on the landscape, as explored by Giguet-Covex et al. (2014). Lake sediments were used to reconstruct plant cover and livestock farming history to determine human causes of erosive events as compared to naturally predicted erosion (Giguet-Covex et al., 2014). The presence of high levels of cow and sheep DNA in the sediment suggested intensive grazing, and the extinction of tree species due to deforestation would impair soil stability and lead to erosion. Pansu et al. (2015) also assessed ancient human impacts on plant communities by metabarcoding lake sediments; they were able to identify anthropogenic changes to vegetation communities indicating the establishment of an agropastoral landscape. Although most ancient metabarcoding investigations have employed sediment to assess biodiversity, bone fragments may be sequenced to determine their species of origin. Bone fragments that are unidentifiable based on morphology can be combined and sequenced together to determine the species at a certain site where bones were found. In particular, bulk bone metabarcoding "has applications for the detection of cryptic or morphologically undiscoverable taxa, historical faunal turnover and population fragmentation" and could be a powerful tool for species conservation as well as "identifying taxonomic assemblages, estimating genetic diversity, and assessing general ancient DNA preservation". Bulk bone metabarcoding can be combined with sediment analysis as indicated by Haouchar et al. (2014). They used a variety of general and taxa-specific primers on bone and sediment samples to assess the genetic biodiversity of a cave in Australia, from which at least 15 plant and animal taxa were identified, providing valuable insight to the ancient biodiversity of the area (Haouchar et al., 2014). The possible applications of ancient

DNA metabarcoding are staggering, but it is important to acknowledge potential flaws and limitations of current technology. Limitations of genetic libraries, differing rates of degradation between species and geographic location, and PCR bias are all major concerns. Another major concern is contamination; the presence of a modern species replicated by the primers can not only give false positives, but also false negatives by overwhelming the concentration of taxa to the point where low-concentration DNA is not able to replicate. Human DNA is one of the most difficult contaminants to eliminate, and it might cause issues when using mammal primers. To combat this, Boessenkool et al. (2014) developed human-blocking primers to reduce the amplification of human DNA and found that when the human DNA was repressed, species diversity rose and new taxa appeared. Other issues to consider include climatic/temporal constraints, biological origin/state, stratigraphic dependability, and sterile sampling procedures. The temperature of the location, in particular, limits ancient DNA; most research to far have used the procedure in arctic or temperate climates; tropical regions are generally both warmer and more humid, which accelerates the rate of DNA destruction with time. As such, micro climates such as caves may be essential to the application of ancient eDNA metabarcoding in more tropical regions.

### 6.4.2 Plant–pollinator interactions

eDNA metabarcoding can be used to characterize interactions between organisms and the environment. One of the most promising areas where this is being applied is in analyzing pollinator interactions with plants. Pollinators are essential to human survival and to general ecosystem health, but many populations are declining. The worldwide decline of bee species and colony collapse has the potential to be detrimental to global food production. In determining plant species most used by pollinators, conservation plans can be constructed to ensure these species are available for pollinators to use. Previously, determining which plants pollinators visited required either seeing watching which plants individuals visited or collecting pollen to be identified under a microscope. Both approaches are time-consuming and prone to bias, with the latter needing the identification of a taxonomic pollen specialist. Valentini et al. (2010) used universal barcodes and NGS to determine the geographical origin of honey based on the plant DNA present in the honey samples. They found the method to be simple, fast, robust, and successful in identifying plant presence to pinpoint the origin of the samples. Furthermore, because honey does not degrade DNA, several, longer primer sets might be employed if specific areas are indistinguishable with the universal primers. Hawkins et al. (2015) compared the efficacy of the metabarcoding approach to traditional microscopic pollen particle identification in honey. They found that the two methods had a correspondence of 92% when only high abundance taxa were considered, but that when all taxa were considered, the correspondence dropped to below 50%, with both methods finding taxa that were missed by the other. While metabarcoding is highly sensitive and able to provide greater taxonomic resolution, it is less likely to detect pollen present in low levels than traditional methods due to inherent bias in amplification and sequencing. De Vere et al. (2017) applied metabarcoding of honey samples directly to determination of bee plant preference. They discovered that bees preferentially utilized the same plant taxonomy consisting of native or near-native plants found in forests and hedgerows by evaluating honey in a very varied botanical garden, implying that keeping these natural settings free of invasive species is critical for bees to survive. In addition to honey samples, pollen can be taken directly from pollinators for identification. Keller et al. (2015) compared metabarcoding to microscopy for mixed pollen samples collected from pollinators, evaluating if primers were truly universal while distinguishing species. They discovered that when sequencing findings were 85% matched to genome libraries, 98.9% of samples were classified to genus and 61.6% to species, with metabarcoding uncovering 13 taxa overlooked by microscopy. Microscopy found 4 taxa missed by metabarcoding, but overall richness and abundance estimates were similar. Richardson et al. (2015a,b) published two studies on metabarcoding pollen from pollinators. The first tested a multilocus approach to metabarcoding to characterize pollen from honey bees. This study found that using multiple loci was more reliable for identifying pollen as it helps to eliminate false positives and to avoid primer amplification bias. They also compared traditional pollen analysis to metabarcoding of honey bee pollen samples. The result of this study shows that metabarcoding is sensitive enough for plant identification and is able to identify taxa missed by traditional analysis. However, because of replication bias, it cannot quantify pollen analyses, and they propose that both techniques be used jointly when quantification is required. In the same year, Sickel et al. (2015) reported a dual-index multiplexing method adapted from bacterial studies and applied it to the sequencing of pollen samples from multiple bee species simultaneously. They were able to adapt the method so that labelling and PCR could be done in a single step and determined that 2000–3000 quality reads were needed in order to identify all taxa in the sample. In addition to general plant species usage, metabarcoding of pollen from pollinators can be applied to look at land scape level plant diversity of taxa most used by pollinators. Networking and quantifying plant–pollinator interactions may also be accomplished by metabarcoding. Pornon et al. (2016) tested whether metabarcoding could be used to overcome the limitations of traditional sampling in detecting and quantifying plant–pollinator interactions and found that metabarcoding of insect pollen loads revealed 2.5× more plant–pollinator interactions. There was also a correlation between the number of sequences and the pollen-carrying burden and the number of visits per plant type. These interactions can be used to determine pollen transport networks and plant species choice among groups, as studied by Lucas et al. (2018). They used metabarcoding of pollen from hoverflies to determine that there appears to be individual seasonal specialization in plant choice, but there is redundancy within groups. Using eDNA metabarcoding on the pollen from migratory pollinators, researchers may determine migratory distances and geographic ranges of migratory species. Suchan et al. (2019) used this method on a model butterfly species along the European Mediterranean coast, finding pollen from taxa native to Africa, implying the possibility of insect-mediated intercontinental pollination. Pollen metabarcoding is faster and more effective than direct observation for interactions and permits the investigation of plant–pollinator networks across ecological gradients. Additionally, it is possible to detect numerous interactions that would otherwise be lost by traditional methods, thus better describing pollinator behavior and plant choice. Bell et al. (2017) tested the effectiveness of metabarcoding for creating networks by sampling pollen from 38 bee species across varying environments. They were able to quantify the interactions by matching bee species to pollen presence and the number of interacting individuals based on this data. Though metabarcoding of pollen to determine plant–pollinator interactions has advantages such as faster turnaround time, hands-off work that allows researchers to focus on other activities, the ability to include all pollen grains, more standardized and known techniques, greater taxonomic sensitivity, identification of novel taxa, and more predictable costs that may sometimes be lower than traditional pollen ID, there are some downsides to the technique; namely, lack of complete libraries, the possibility of contamination and misidentification, and possibility of decreased amplification of rare species owing to primer bias.

### 6.4.3 Analyze your diet

One of the early applications of metabarcoding may have been in non-invasive food analysis from feces or stomach contents. Traditional diet analysis is time-consuming and requires skilled morphological identification of partially or totally digested meal items, which is especially challenging for uncommon, tiny, aquatic, nocturnal, fast-moving, elusive, or soft-bodied organism-consuming animals. As a result, a novel approach was required to thoroughly describe the diets of wild animals. Metabarcoding has been proven to be a feasible alternative and to result in more in-depth characterization of food than other techniques for both herbivores and predators, particularly when conventional forward and reverse primers are coupled with a blocking primer to limit host species amplification. Beyond diet characterization, metabarcoding of feces and stomach contents has further uses. It has been used to study resource partitioning, dietary overlap and competition, the effects on forestry and fisheries, comparative conservation strategies, range management, habitat use, species interactions, large-scale monitoring, the ecological impacts of changing populations, and human dimensions. The metabarcoding approach to diet analysis is fast, simple, robust, broadly applicable, typically cheaper than traditional sampling and identification, often has high taxonomic resolution, and can identify DNA from digested materials even for soft-bodied or gelatinous prey. It has even been described as "one of the most accurate approaches available for increasing the understanding on the feeding habits of animals." Despite these undeniable benefits, there are several disadvantages to utilizing metabarcoding for diet research. Poor database quality and imprecise taxonomic resolution are still concerns that must be addressed. Despite this, it can be expected that as technology advances, metabarcoding will render these concerns obsolete. One of the most important specific methodologies for characterization of herbivorous diet using metabarcoding was developed by Valentini et al. (2009): the trnL approach. This approach has been found to be able to identify about 50% of plant taxa to species from degraded DNA from feces by amplifying a 100e150 base pair region of the P6 loop of the chloroplast trnL intron, and can be applied to herbivorous birds, mammals, insects, and molluscs. This method has been applied to numerous herbivore diet analysis studies since its development, including an experiment by Kowalczyk et al. (2011) studying European bison restored after extinction in the wild. In this study, the trnL approach of metabarcoding was used on feces from bison of different feeding groups (intensively fed, less intensively fed, non-fed in forest, and non-fed in agriculture surrounding forest), and found that when fed, bison consumed more herbs, grasses, and sedges, but when not fed they consumed more woody materials. Because supplementary feeding has a negative impact on bison ecology and health, when human-provided food is not available, a conservation plan can be developed based on their dietary choices; notably, the species of trees consumed could be identified and were determined to be insignificant to the forestry industry, further supporting the change to a non-feeding management style. Also using the trnL approach, Baamrane and Abdeljalil (2012) evaluated the dietary patterns of the endangered Moroccan dorcas gazelle to a previously established database of plant species present on the research location by metabarcoding fecal pellets. Plant preferences were found, and the data was compared to the findings of a microhistological research conducted in the 1980s. The current study discovered less taxonomic richness but a far more diverse diet, possibly due to changes in plant nutritional quality through time. Hibert et al. (2013) used the trnL approach to study the diet of tapir, an elusive large rainforest herbivore down to the genus level and discovered strong correlations with plants already known to be eaten by tapirs in the region, but they also discovered new plant families. Furthermore, by using metabarcoding tapir excrement throughout the year, they were able to observe seasonality of food. In the same year, Quem er e et al. (2013) studied the dietary agility of the golden-crowned sifaka lemur in Madagascar by using the trnL approach on fecal material compared to a large DNA reference database. They discovered remarkable dietary diversity among over 130 identified species from varying habitats, experting a high flexibility of foraging strategies, including consumption of cultivated and naturalized plants, and thus confirming the species' ability to adapt to human-induced landscape changes. Metabarcoding has also been used to comprehensively describe lemming diets. Soininen et al. 92013) discovered that lemmings predominantly consumed grass, but that vascular plant intake changed between habitat types, implying regional heterogeneity in lemming diet. The stomach contents of different species of lemmings were metabarcoded, revealing variations in composition and temporal fluctuation. Metabarcoding of lemming feces has been used to assess diet overlap and potential for competition in sympatric lemming species. The species were found to utilize the same plant as their primary food source in proportion with resource availability but differed in other dietary items, indicating a high potential for competition in populations that were not exhibiting any actual competition. Kartzinel et al. (2015) used metabarcoding and NGS of fecal material from seven large mammalian herbivore species in Kenya to evaluate resource partitioning between species. They found diet to be similar within species and strongly divergent between species, even when considering similar groups (i.e., grazing mammals may eat a similar percentage of grass, but different species will consume different grasses), suggesting that food choice is more closely linked to plant species than type. Metabarcoding was recently utilized to assess the diet and seasonal changes of the endangered Italian hare using only 22 fecal pellets, identifying 99 plant species. Metabarcoding applications for herbivorous diets can be coupled with those for carnivorous diets to describe the diverse food items ingested by omnivores such as brown bears, who eat plants, invertebrates, and vertebrates. Metabarcoding of feces and stomach contents can also be used to describe carnivorous invertebrate diets. A study published in 2013 used metabarcoding of feces to analyze the diet of a carnivorous and critically endangered land snail, discovering that the snails mostly consume earthworms that either live in the leaf litter or come to the surface at night to feed on the leaf litter, suggesting that the snail will eat any earthworm it encounters in the leaf litter. This knowledge influences conservation decisions without having to harm or disturb the snails in their natural environment. In the same year, Piñol et al. (2014) published a study which investigated the diet of a generalist arthropod predator (linyphiid spider) using metabarcoding without a predator-blocking oligonucleotide. Of this data, only 6% of usable reads were from prey, but this amounted to over 61,000 reads and was enough to identify many prey taxa and to reach the conclusion that the spider exploits a large range of prey. Additionally, this study reached the conclusion that it is possible to characterize diet without developing a blocking oligonucleotide, which would increase the time, testing volume, and cost. Metabarcoding of vertebrate predator feces emerged around the same time as did that of herbivores. Deagle et al. (2009) compared the technique to traditional food analysis by using four metabarcoding primer sets and a blocking oligonucleotide on the feces of Australian fur seals. They identified 62 species in the seals' diet (including cartilaginous prey missed by traditional identification due to degradation), and were able to identify the most important prey species to the seals as well as location-based variations in prey composition. Also studying aquatic mammalian predators, Berry et al. (2017) applied metabarcoding to Australian sea lion scats to find six classes from three phyla for over eight taxa altogether, some of which verified the consumption known dietary items while others had not previously been reported as prey for these sea lions. These findings suggest that the Australian sea lion is an opportunistic predator primarily consuming demersal and benthic fauna of different species composition between study sites. Metabarcoding of feces has also been used to compare diets of sympatric mammalian predators, such as the study by Bohmann et al. (2011) with African free-tailed bats. They found a wide variety of prey in the samples that were identified to species level at times and were able to determine the differences between the two species' diets. A similar study was recently published by Arrizabalaga-Escudero et al. (2018) in which they used metabarcoding on the feces of two Mediterranean bats in order to determine dietary differences; similar to Bohmann et al. (2011), they were able to characterize the diet of each species and determine the overlap within. Shehzad et al. (2012) developed leopard cat as a model species to create a system for analyzing mammalian carnivore nutrition using feces metabarcoding, and then used the same methodology to snow leopards. To do this, they used a vertebrate primer set on feces samples and tested the methodology both with and without a predator blocking oligonucleotide. In the leopard cat, they found 18 prey taxa including mammals, birds, amphibians, and fish, with majority taxa and reads coming from rodents, supporting prior ideas that the cat has an eclectic diet concentrated on rodents. The methodology was modified for use with snow leopard feces where they observed five large ungulate prey items including domestic livestock. This knowledge of snow leopard diet can help to influence conservation measures concerning livestock predation and retaliatory killing. Deagle et al. (2010) conducted another metabarcoding study in which they pyrosequenced markers from excrement of captive little penguins to see if sequences reflected proportions of species consumed. They found that the same species that were fed were detected, but the proportions were much different, and discovered that DNA can persist in feces for at least 4 days based on identification of species ingested by the penguins before they were taken into captivity for testing. McInnes et al. (2017) investigated seabird nutrition using metabarcoding by establishing scat collecting techniques for detecting food DNA in vertebrate scat using the shy albatross as a case study for approaches that may possibly be extended to a wide range of species. They found that fresh scat yielded highest food sequences whereas dirt collection yielded less food sequences and increased contamination that food DNA detection changed over the mating season, and that older individuals produced scat with more food DNA. They then proceeded to apply this procedure to assess albatross consumption of gelatinous zooplankton, finding that while fish were the primary dietary item, cnidarians were also prevalent in samples and identifiable at all breeding and life stages. They compared this data with fishery statistics from fisheries to establish a link that may imply that jellyfish eating in albatross is selective and provides a significant factor to diet. Because gelatinous zooplankton disintegrate fast if ingested, this investigation would have been very impossible to do before metabarcoding. Finally, metabarcoding has been used to characterize the diets of fish species including catfish, stickleback, and toothfish. Using the feces of a nonnative catfish, Guillerault et al. (2017) compared metabarcoding to traditional prey identification identifying 11 species with DNA and five with stomach content analysis for a total of 14 species, with two species identified using both methods. Even more astonishing, they were able to detect prey indirectly via parasite presence: while eels were not detected, an eel-specific parasite was, indicating that the catfish do in fact consume eels.

Jakubaviciut e et al. (2017) _were able to identify 120 prey taxa consumed by the three-spined stickleback, a fish whose population is currently steeply increasing. They were able to estimate the possible consequences of the sticklebacks' growing numbers based on their extremely high prey variety. Applying metabarcoding to the stomach contents of Antarctic toothfish, Yoon et al. (2017) were able to identify a total of 19 species and to determine the most important prey species for the toothfish. Additionally, they successfully acquired genotypes of prey items via NGS and identified 17 more haplotypes, further extending understanding of toothfish diet. The breadth of dietary metabarcoding studies is already wide and can only be expected to grow as technology improves and techniques are refined.

### 6.4.4 Identifying invasive species

As our globe grows increasingly interconnected, invasive and nonnative species pose some of the most serious challenges to ecosystems. In particular, aquatic vessels have incredible potential to unknowingly transport dangerously prolific species into new habitats where they are capable of outcompeting native species. Current methodologies for aquatic microbiota inventory are diverse and rely on time-consuming sampling and expert taxonomic identification. Additionally, many marine pests cannot be identified morphologically at the larval/juvenile level, so they may go undiscovered during the early phases of infection, when they have the best chances of success. Many factors can impact the transmission of species via vessels, such as activities performed on the boats, types of boats, and the season: motorboats have a higher rate of species translocation, and propagation is more probably during warmer months. Invertebrates may survive in bilge and ballast water for at least 16 days, with protozoan species in particular able to multiply during that time and emerge from the pumps unharmed, possibly long distances away from their native environment. Metabarcoding has the ability to detect these species early and execute management measures, especially for planktonic larval benthic creatures. It is extremely sensitive, able to detect DNA at concentrations from 0.02% to 0.64% of the total biomass analyzed, able to detect a variety of taxa simultaneously, and able to be applied to both freshwater and marine environments. Metabarcoding was initially used to detect nonnative species was first used in marine environments. Ardura et al. (2015) sampled from ballast waters in a ship across its journey from the North Sea to the tropics and found that mollusk species could still be detected, including the invasive European mud snail. In the same year, Zaiko et al. (2015) published two studies comparing results of metabarcoding to traditional identification. In ballast waters, metabarcoding found several noteworthy species that had been missed by traditional methods. In a study with a focus on plankton communities not associated with a vessel, they identified five different non-indigenous species, four of which were not found with traditional monitoring. In 2016, Zaiko et al. (2016) explored the scope of marine eukaryotic metabarcoding by sampling from settlement plates in a busy port. They found more than 400 taxa (several of which were nonindigenous), but even at poor resolution, data were consistent with the 25 taxa found by morphological analysis. In a study testing whether a minimum sampling effort for metabarcoding could compare with previously compiled nonnative invertebrate species, Borrell et al. (2017) found three nonnative species matching the results of traditional studies from only three 1 L bottles of water per port at eight ports. Testing metabarcoding on 30 bilge water samples, Fletcher et al. (2017) were able to identify five different non-indigenous species in 23 water samples, all at high taxonomic resolution. Pochon et al. (2017) took bilge water sampling beyond the status quo, using both eDNA and eRNA metabarcoding to test the potential for invasive identification. They found that eDNA and enhancer RNA (eRNA) gave different results in terms of species presence, with eDNA showing mostly fungi while eRNA showed mostly ciliates and metazoans. They hypothesized that this was due to the increased level of cell activity of ciliates and metazoans, and therefore increased eRNA output. As previously mentioned, metabarcoding is applicable to freshwater ecosystems as well as marine. In 2016, Brown et al. sampled zooplankton in Canadian ports to attempt to identify non-indigenous species via metabarcoding and successfully identified 25 nonindigenous species, 11 of which had never previously been reported in their locations. Blackman et al. (2017) used metabarcoding of freshwater samples from rivers and found it possible to reliably differentiate between native and invasive gammarid species, and that the method could be applied to the assessment of geographic distribution of invasive species. In the following year, Klymus et al. (2017) developed two metabarcoding primers to use with eDNA from the Great Lakes, focusing on the detection of invasive gastropod and bivalve species. Their dual primer design allowed for inference of relative abundance of species, as well as detection of species missed by morphological survey. Most recently, Borrell et al. (2018) used eDNA metabarcoding in two estuaries to test detection of nonnative species, and despite limited sampling, one invasive was identified and later confirmed by classic DNA barcoding studies of the estuaries. Metabarcoding is a strong technique that provides a more thorough means of monitoring and testing for invasive presence than standard identification. Furthermore, the cost of metabarcoding is currently similar to other techniques and is expected to decrease as it becomes more common. As with other applications, it does have drawbacks. For one, it may bias biodiversity assessments by capturing signals from dead organisms, extracellular DNA, or false readings due to taxonomic selectivity and restricted sensitivity of primers. False readings, either positive or negative, can be problematic in invasive species detection; false positives could trigger an unwarranted and potentially expensive response, while false negatives can result in an invasion that could have potentially been prevented. Another drawback to this method is the current lack of comprehensive genetic databases, and the unavoidability of primer and amplification bias make quantification impossible and require careful analysis of results. Regardless, metabarcoding is a unique tool that can only be expected to improve over time from its already wholly applicable and useful state.

### 6.4.5 Pollution response

Uses of eDNA metabarcoding extend to the ability to track ecosystem responses to pollution. While many pollutants do not have DNA and thus cannot be detected, they often cause changes in microbiological communities affected by them. Studies can be done to determine which species are most likely to be present or absent in the case of a specific pollutant to help determine the severity of the pollution. Until that point, responses to pollution can be monitored in areas with known pollutants or in controlled circumstances to determine differences in community composition. Perhaps some of the more high-profile studies of pollution response are those which aim to determine differences in communities surrounding offshore oil drilling platforms or in response to oil pollution events. Smith et al. (2015) used statistical analysis of microbial communities to identify environmental contaminants at a nuclear waste site, and then applied the approach to hydrocarbon contamination from the Deepwater Horizon oil spill. They found that even after the contaminants had been fully degraded, the bacterial communities affected were altered based on the interaction, indicating that bacterial communities have the potential to be used as environmental sensors to evaluate human impacts. Coelho et al. (2016) used microbial activity profiling to evaluate responses to ocean acidification and oil hydrocarbon contamination in lab conditions, determining that the resulting lower relative abundance of sulfate-reducing bacteria may be coupled to changes in sediment archaeal communities. Additionally, they found increased abundance of hydrocarbonoclastic fungi and changes to the sediment chemistry due to the changes in the microbial community, all of which impair the ecosystem's ability to recover from oil contamination events. In terms of community responses surrounding offshore drilling sites, Lanzen et al. (2016) found changes in sediment communities surrounding oil platforms, with metazoans appearing to be the most sensitive eukaryotic taxa to the contaminant effects of drilling, while Laroche et al. (2018) determined that bacteria were highly responsive to drilling activity as well, noting the presence of hydrocarbon-degrading taxa at the petroleum impacted stations, indicating presence of taxa highly specific to site conditions. Both studies identified potential pollution indicator species. Another important sphere of pollution to be considered is that of marine fisheries. Products from these facilities have the potential to affect the marine microbiome, but traditional monitoring of these species is time-consuming and requires extensive specialization. Environmental metabarcoding can be used to estimate community composition, and to identify indicator species. Foraminiferal responses have been studied multiple times, and candidate bioindicator species identified therein. Pawlowski et al. (2014) found high variation between communities collected near the fish farms versus at a distance, with species richness decreasing nearer to the farms, while Pochon et al. (2015) found clear shifts in community structures between farm sites and control sites and differences in structure between high and low flow environments. Bacterial communities may also be important for assessing the impact of fisheries on ecosystems. Dowle et al. (2015) tested the impact of organic sediment enrichment from a salmon farm on bacterial communities within the sediment along a gradient of distances from the farm. They found stark differences between bacterial communities near the farm as compared to control sites, and identified a strong relationship between bacterial communities and total organic matter, suggesting that metabarcoding bacterial communities may be a useful tool in assessing impacts of aquaculture. Stoek et al. (2018) published two studies regarding microbial response to fish farming, one assessing benthic bacterial communities and the other testing protist presence. They found that deltaproteobacteria were predominant in immediate vicinity of cages, while gamma proteobacteria dominated less impacted sites, indicating changes in community structure due to salmon farming. Additionally, alpha and beta diversity correlated with macrofauna diversity metrics, indicating similar responses to farming, thus making bacterial eDNA metabarcoding a potential alternative biomonitoring method. For protist response, they found that ciliate communities were the most powerful indicators of fisheries pollution when compared to traditional macrofaunal surveys, and that the use of ciliates can allow for an even finer scale resolution. Aquatic communities may be subject to pollution and contamination beyond that of oil drilling and commercial fisheries, further extending the potential applications of environmental metabarcoding. Xie et al. (2017) studied the response of eukaryotic communities to various land use types and chemical contaminants, discovering that the structures of eukaryotic communities differed significantly between land use types and identifying key contaminants such as pyrethroids, Mn, Zn, Pb, Cr, and polycyclic aromatic hydrocarbons. Negative associations of eukaryotic classes with these contaminants suggest that there is some degree of toxicity. Other research concentrated on specific contaminant impacts. Chariton et al. (2014) investigated the responses of benthic communities exposed to elevated concentrations of triclosan, a common antimicrobial agent, and found a drop of metazoan diversity, particularly in the taxa Chromadorea, Bivalvia, and Kinorhyncha. Frontalini et al. (2018) tested the response of benthic foraminifera communities exposed to different concentrations of Hg, finding that Hg pollution is detrimental to foraminifera with diversity decreasing as Hg levels increase (Chariton et al., 2014). Potential bioindicators of Hg pollution were also identified. This technique has been used in terrestrial as well as aquatic environments. Bell et al. (2014) compared fungal and bacterial communities of 66 soil samples from the rhizospheres of planted willows and six unplanted control samples at the site of a former petrochemical plant by metabarcoding, finding that fungal communities changed drastically between sites, and bacterial communities changed to a lesser degree between sites with a decrease in diversity in hydrocarbon contaminated sites. Additionally, they found that willow phylogeny influenced fungal diversity in these sites. Heavy metals are also a concern in polluted soils. Azarbad et al. (2015) tested microbial communities associated with forests in metal (Zn and Pb) polluted soils in Poland and found that the metal pollution level impacted bacterial community structure, but not taxon richness or community composition; rather, the relative abundance of specific bacterial taxa changed. In the same year, Zappelini et al. (2015) reported bacterial and fungal communities from both chlor-alkali residue (high in Hg and As) and nearby undisturbed soil to determine community composition. They found a dominance of gammaproteobacteria and Pseudomonas sp. at the study site, and minor differences between genera of fungal communities. Foulon et al. (2016) compared microbial community composition at two sites with varying environmental conditions, but both contaminated with Zn, Cd, and Pb: Pierrelaye, which was a sandier soil type enriched with P, C, and total N, and Leforest, which was a more silt/clay soil type with a lower pH and higher amounts of Cd and Zn. They found that Ascomycota dominated overall poplar soil, root samples at Pierrelaye, and unplanted soil, while Basidiomycota was more dominant in root samples from Leforest (Foulon et al., 2016). The dominance of an ectomycorrhizal community at Leforest was potentially favored due to strong resistance to environmental stress, while at Pierrelaye where soils were more fertile and resources more available, other species are able to thrive. Most recently, fungal communities at an Hg-contaminated phytomanagement site of poplar trees were characterized using environmental metabarcoding. Communities were found to be negatively affected by Hg concentration, and aboveground and belowground fungi communities were found to be almost completely isolated from one another. Even areas specifically denoted to contain pollution are capable of having an effect on wildlife. Klymus et al. (2017) tested eDNA metabarcoding as a survey tool to examine potential use of uranium mine containment ponds as water sources by wildlife by testing ponds near mines and one mine containment pond using two markers for vertebrates. Traditional surveys missed two species, including the presence of tiger salamanders in the confinement pond. Finally, while perhaps not a response to a pollutant in the traditional sense, environmental metabarcoding can be used to test the effects of invasive species on native communities. Wangensteen et al., 2018) used multigene metabarcoding to observe littoral community responses to three invasive, canopy-forming seaweeds. They found that the presence of invasive seaweed had a substantial impact on two of the three species' understory communities. Overall, pollution response studies have found metabarcoding to be more sensitive in revealing community changes than traditional surveys; however, it is still subject to amplification bias, interpretation uncertainties, and other roadblocks previously discussed.

### 6.4.6 Air quality monitoring

DNA metabarcoding can also have applications in human health; determining the microbiome in the air can influence health decisions regarding allergens, pathogenic organisms, and air pollution. Current methods in studying airborne allergens are microscope-based, labor intensive, and cannot identify to the genus level in all relevant groups. It is feasible to assess the makeup of the airborne microbiome and its potential consequences for human and environmental health by collecting air samples and using metabarcoding and NGS technologies. Metabarcoding of pollen was found in 2014 to be more efficient and more sensitive than current pollen monitoring techniques; however, a simple pollen count is needed in addition to metabarcoding data in order to properly quantify airborne pollen. Leontidou et al. (2018) created a standardized procedure for sample preparation and DNA extraction for airborne pollen metabarcoding by comparing metabarcoding results to morphological identification (Banchi et al., 2018). They found that even in complex samples with more than 46 families, plants can be identified efficiently by metabarcoding with greater taxonomic resolution than by morphological identification. Airborne fungal diversity is another important facet of the aerial microbiome. Presence and abundance of taxa in fungal airborne samples can be variable over time and impacted by climatic conditions, and because many mycelium fragments and spores have the potential to be toxic or cause allergies, it is important to describe airborne fungal diversity. Previous studies have relied on a few indicator species and taxa of key importance, but with metabarcoding all taxa can be analyzed. Oh et al. (2014) used metagenomics to study airborne fungal diversity and found over 1000 unique OTUs, with a set of 40 potentially allergy-inducing genera identified accounting for approximately 20% of the airborne fungal diversity. Banchi et al. (2018) specifically compared metabarcoding to traditional airborne fungal analysis and found that metabarcoding provided a 10× more comprehensive determination of taxa compared to microscopy; even so, some taxa were only identified in one technique or the other, making both methods necessary for comprehensive characterization of the fungal community (Cao et al., 2014). While the most frequent use of environmental metabarcoding is in natural environments, the methodology can be applied to indoor locations as well. In particular, airborne microorganisms are a critical concern in hospitals, where some have the potential to cause hospital-associated infections. Tong et al. (2017) compared results of metabarcoding air samples and traditional culture-based methods to describe the microbial diversity. In this study, few bacteria were grown after cultivation, but sequencing revealed the relative abundance of fungi, archaea, bacteria, and viruses (Tong et al., 2017). Knowing the species present can inform infection control plans in the case of an outbreak, as sampling different areas of the hospital can offer a better idea of what control methods are necessary. Cao et al. (2014) applied a similar method of characterizing the full microbiome; however, their study was outdoors in Beijing during a severe smog event. While particulate matter air pollution is well-understood, inhalable microorganisms have not been well studied, and most studies have been purposed for taxonomic identification rather than pathogenic potential (Cao et al., 2014). This study employed metabarcoding to describe the airborne microbial community and was able to identify most taxa to the species level with the results. In addition, they discovered the presence of some respiratory microbial allergens and pathogens.

### 6.4.7 Future directions

For eDNA metabarcoding to truly take off, current assessments of ecological quality would need to be adapted to the eDNA metabarcoding framework. These changes must be feasible on a large scale, particularly when considering thresholds between countries and the differences between traditional and molecular methods. To calibrate, molecular methods would need to be applied simultaneously with existing systems in key environmental gradients which would likely be accomplished during development and testing of molecular methods. As such, a comprehensive eDNA monitoring program could be possible in just a few years across a variety of environments and taxa, thus providing a global framework for ecosystem modeling and function to inform management decisions. To handle this shift, reference databases would need to be expanded to include identifiable sequences for all target biodiversity, and advanced facilities will need to be created to house the growing data libraries. Wide-spread implementation would benefit from optimized bioinformatics pathways with user-friendly interfaces, as well as general improvements to the eDNA collection and extraction methodologies, particularly in terms of accounting for degradation. Metabarcoding has the potential to complement higher cost biodiversity measures and even replace them in some instances, especially as it becomes more widespread and the cost goes down. Ecological indicators are essential for conservation science, but full biodiversity assessments that are rapid, cost-effective, and noninvasive are also important to conservation biology. While eDNA metabarcoding cannot give information as to the health or demographics of a population, it has the potential to assess long-term changes in ecosystems and local extinctions over time, including for modeling the impacts of climate change on populations. eDNA also has applications in citizen science by engagement of citizen collection of eDNA using commercially available sampling kits, involving the public in biodiversity sampling in a way that could be complementary to already established methods. However, it is limited by the difficulty in visualizing and explaining data to laypeople. More user-friendly methods of data exploration are needed. It may even be possible to automate eDNA sampling in the future, but further research is needed into the temporal longevity and spatial dispersal of eDNA in all ecosystems for studies of abundance and richness to be considered fully valid. Metagenomics, or shotgun sequencing of environmental samples without relying on a PCR amplification step, avoids primer bias and could theoretically be used for biodiversity studies. However, because it requires a huge increase in sequencing power to acquire adequate depth, the cost is often prohibitive. As technology advances, this may become a commonplace method of biodiversity monitoring. There is a definite need to let methods mature as eDNA metabarcoding is a developing field, and thus apply the evolution of the method to reliable indices and ensure both forward and backward compatibility.

## 6.5 Conclusion

In conclusion, the potential of eDNA metabarcoding in biological research seems almost limitless, but the technique requires scientific collaboration and coordination. As technology advances and procedures are optimized, metabarcoding will likely emerge as an essential tool for a number of scientific tasks not only in biodiversity monitoring but also in the numerous areas discussed above. eDNA metabarcoding brings together several fields for the assessment of communities for purposes ranging from ecosystem restoration to human health, making it extremely versatile and vital to the future of molecular research.

## References

Abad, D., Albaina, A., Aguirre, M., Laza-Martínez, A., Uriarte, I., Iriarte, A., Villate, F., Estonba, A., 2016. Is metabarcoding suitable for estuarine plankton monitoring? A comparative study with microscopy. Mar. Biol. 163 (7), 1–13.

Alsos, I.G., Lammers, Y., Yoccoz, N.G., Jørgensen, T., Sjögren, P., Gielly, L., Edwards, M.E., 2018. Plant DNA metabarcoding of lake sediments: how does it represent the contemporary vegetation. PLoS One 13 (4), e0195403.

Andersen, K., et al., 2012. Meta-barcoding of 'dirt'DNA from soil reflects vertebrate biodiversity. Mol. Ecol. 21 (8), 1966–1979.

Andruszkiewicz, E.A., et al., 2017. Biomonitoring of marine vertebrates in Monterey Bay using eDNA metabarcoding. PLoS One 12 (4), e0176343.

Ardura, A., et al., 2015. Environmental DNA evidence of transfer of North Sea molluscs across tropical waters through ballast water. J. Molluscan Stud. 81 (4), 495–501.

Avó, A.P., Daniell, T.J., Neilson, R., Oliveira, S., Branco, J., Adão, H., 2017. DNA barcoding and morphological identification of benthic nematodes assemblages of estuarine intertidal sediments: advances in molecular tools for biodiversity assessment. Front. Marine Sci. 4, 66.

Aylagas, E., et al., 2016. Benchmarking DNA metabarcoding for biodiversity-based monitoring and assessment. Front. Marine Sci. 3, 96.

Azarbad, H., et al., 2015. Microbial community composition and functions are resilient to metal pollution along two forest soil gradients. FEMS Microbiol. Ecol. 91 (1), 1–11.

Baamrane, A., Abdeljalil, M., et al., 2012. Assessment of the food habits of the Moroccan dorcas gazelle in M'Sabih Talaa, west central Morocco, using the trn L approach. PLoS One 7 (4), e35643.

Banchi, E., et al., 2018. DNA metabarcoding uncovers fungal diversity of mixed airborne samples in Italy. PLoS One 13 (3), e0194489.

Barnes, M.A., Turner, C.R., 2016. The ecology of environmental DNA and implications for conservation genetics. Conserv. Genet. 17 (1), 1–17.

Bell, K.L., et al., 2017. Applying pollen DNA metabarcoding to the study of plant–pollinator interactions. Appl. Plant Sci. 5 (6), 1600124.

Bell, T.H., et al., 2014. Linkage between bacterial and fungal rhizosphere communities in hydrocarbon-contaminated soils is related to plant phylogeny. ISME J. 8 (2), 331–343.

Berry, T.E., et al., 2017. DNA metabarcoding for diet analysis and biodiversity: a case study using the endangered Australian sea lion (Neophoca cinerea). Ecol. Evol. 7 (14), 5435–5453.

Besemer, K., Singer, G., Quince, C., Bertuzzo, E., Sloan, W., Battin, T.J., 2013. Headwaters are critical reservoirs of microbial diversity for fluvial networks. Proc. R. Soc. B Biol. Sci. 280 (1771), 20131760.

Blaalid, R., et al., 2012. Changes in the root-associated fungal communities along a primary succession gradient analysed by 454 pyrosequencing. Mol. Ecol. 21 (8), 1897–1908.

Blackman, R.C., et al., 2017. Detection of a new non-native freshwater species by DNA metabarcoding of environmental samples—first record of Gammarus fossarum in the UK. Aquat. Invasions 12 (2).

Boessenkool, S., et al., 2014. Use of ancient sedimentary DNA as a novel conservation tool for high-altitude tropical biodiversity. Conserv. Biol. 28 (2), 446–455.

Boessenkool, S., et al., 2014. Use of ancient sedimentary DNA as a novel conservation tool for high-altitude tropical biodiversity. Conserv. Biol. 28 (2), 446–455.

Bohmann, K., et al., 2011. Molecular diet analysis of two African free-tailed bats (Molossidae) using high throughput sequencing. PLoS One 6 (6), e21441.

Bohmann, K., et al., 2014. Environmental DNA for wildlife biology and biodiversity monitoring. Trends Ecol. Evol. 29 (6), 358–367.

Borrell, Y.J., et al., 2017. DNA in a bottle—Rapid metabarcoding survey for early alerts of invasive species in ports. PLoS One 12 (9), e0183347.

Borrell, Y.J., et al., 2018. Metabarcoding and post-sampling strategies to discover non-indigenous species: a case study in the estuaries of the central south Bay of Biscay. J. Nat. Conserv. 42, 67–74.

Brannock, P.M., et al., 2016. Metabarcoding reveals environmental factors influencing spatio-temporal variation in pelagic micro-eukaryotes. Mol. Ecol. 25 (15), 3593–3604.

Buee, M., et al., 2009. 454 Pyrosequencing analyses of forest soils reveal an unexpectedly high fungal diversity. New Phytol. 184 (2), 449–456.

Bush, A., et al., 2019. Studying ecosystems with DNA metabarcoding: lessons from biomonitoring of aquatic macroinvertebrates. Front. Ecol. Evol. 7, 434.

Calvignac-Spencer, S., et al., 2013. Carrion fly-derived DNA as a tool for comprehensive and cost-effective assessment of mammalian biodiversity. Mol. Ecol. 22 (4), 915–924.

Cao, C., et al., 2014. Inhalable microorganisms in Beijing's PM2. 5 and PM10 pollutants during a severe smog event. Environ. Sci. Technol. 48 (3), 1499–1507.

Carew, M.E., et al., 2013. Environmental monitoring using next generation sequencing: rapid identification of macroinvertebrate bioindicator species. Front. Zool. 10 (1), 1–15.

Carugati, L., et al., 2015. Metagenetic tools for the census of marine meiofaunal biodiversity: an overview. Mar. Geonomics 24, 11–20.

Chariton, A.A., Court, L.N., Hartley, D.M., Colloff, M.J., Hardy, C.M., 2010. Ecological assessment of estuarine sediments by pyrosequencing eukaryotic ribosomal DNA. Front. Ecol. Environ. 8 (5), 233–238.

Chariton, A.A., Stephenson, S., Morgan, M.J., Steven, A.D., Colloff, M.J., Court, L.N., Hardy, C.M., 2015. Metabarcoding of benthic eukaryote communities predicts the ecological condition of estuaries. Environ. Pollut. 203, 165–174.

Chariton, A.A., et al., 2014. A molecular-based approach for examining responses of eukaryotes in microcosms to contaminant-spiked estuarine sediments. Environ. Toxicol. Chem. 33 (2), 359–369.

Civade, R., Dejean, T., Valentini, A., Roset, N., Raymond, J.C., Bonin, A., Taberlet, P., Pont, D., 2016. Spatial representativeness of environmental DNA metabarcoding signal for fish biodiversity assessment in a natural freshwater system. PLoS One 11 (6), e0157366.

Coelho, F.J.R.C., et al., 2016. Multitaxon activity profiling reveals differential microbial response to reduced seawater pH and oil pollution. Mol. Ecol. 25 (18), 4645–4659.

Cowart, D.A., et al., 2015. Metabarcoding is powerful yet still blind: a comparative analysis of morphological and molecular surveys of seagrass communities. PLoS One 10 (2), e0117562.

Creer, S., et al., 2016. The ecologist's field guide to sequence-based identification of biodiversity. Methods Ecol. Evol. 7 (9), 1008–1018.

De Vere, N., et al., 2017. Using DNA metabarcoding to investigate honey bee foraging reveals limited flower use despite high floral availability. Sci. Rep. 7 (1), 1–10.

Deagle, B.E., et al., 2018. Genetic monitoring of open ocean biodiversity: An evaluation of DNA metabarcoding for processing continuous plankton recorder samples. Mol. Ecol. Resour. 18 (3), 391–406.

Deagle, B.E., et al., 2010. Pyrosequencing faecal DNA to determine diet of little penguins: is what goes in what comes out? Conserv. Genet. 11 (5), 2039–2048.

Deagle, B.E., Kirkwood, R., Jarman, S.N., 2009. Analysis of Australian fur seal diet by pyrosequencing prey DNA in faeces. Mol. Ecol. 18 (9), 2022–2038.

Deiner, K., et al., 2015. Choice of capture and extraction methods affect detection of freshwater biodiversity from environmental DNA. Biol. Conserv. 183, 53–63.

Deiner, K., et al., 2017. Environmental DNA metabarcoding: Transforming how we survey animal and plant communities. Mol. Ecol. 26 (21), 5872–5895.
Djurhuus, A., et al., 2018. Evaluation of marine zooplankton community structure through environmental DNA metabarcoding. Limnol. Oceanogr. Methods 16 (4), 209–221.
Dowle, E., et al., 2015. Assessing the effects of salmon farming seabed enrichment using bacterial community diversity and high-throughput sequencing. FEMS Microbiol. Ecol. 91 (8), fiv089.
Drummond, A.J., et al., 2015. Evaluating a multigene environmental DNA approach for biodiversity assessment. GigaScience 4 (1) s13742-015.
Eichmiller, J.J., Miller, L.M., Sorensen, P.W., 2016. Optimizing techniques to capture and extract environmental DNA for detection and quantification of fish. Mol. Ecol. Resour. 16 (1), 56–68.
Epp, L.S., et al., 2012. New environmental metabarcodes for analysing soil DNA: potential for studying past and present ecosystems. Mol. Ecol. 21 (8), 1821–1833.
Evans, D.M., et al., 2016. Merging DNA metabarcoding and ecological network analysis to understand and build resilient terrestrial ecosystems. Funct. Ecol. 30 (12), 1904–1916.
Evans, N.T., Li, Y., Renshaw, M.A., Olds, B.P., Deiner, K., Turner, C.R., Jerde, C.L., Lodge, D.M., Lamberti, G.A., Pfrender, M.E., 2017. Fish community assessment with eDNA metabarcoding: effects of sampling design and bioinformatic filtering. Can. J. Fish. Aquat. Sci. 74 (9), 1362–1374.
Fahner, N.A., et al., 2016. Large-scale monitoring of plants through environmental DNA metabarcoding of soil: recovery, resolution, and annotation of four DNA markers. PLoS One 11 (6), e0157505.
Ficetola, G.F., et al., 2015. Replication levels, false presences and the estimation of the presence/absence from eDNA metabarcoding data. Mol. Ecol. Resour. 15 (3), 543–556.
Fletcher, L.M., et al., 2017. Bilge water as a vector for the spread of marine pests: a morphological, metabarcoding and experimental assessment. Biol. Invasions 19 (10), 2851–2867.
Fonseca, V.G., et al., 2010. Second-generation environmental sequencing unmasks marine metazoan biodiversity. Nat. Commun. 1 (1), 1–8.
Fouhy, F., et al., 2016. 16S rRNA gene sequencing of mock microbial populations-impact of DNA extraction method, primer choice and sequencing platform. BMC Microbiol. 16 (1), 1–13.
Foulon, J., et al., 2016. Environmental metabarcoding reveals contrasting microbial communities at two poplar phytomanagement sites. Sci. Total Environ. 571, 1230–1240.
Gibson, J., et al., 2014. Simultaneous assessment of the macrobiome and microbiome in a bulk sample of tropical arthropods through DNA metasystematics. Proc. Natl. Acad. Sci. 111 (22), 8007–8012.
Giguet-Covex, C., et al., 2014. Long livestock farming history and human landscape shaping revealed by lake sediment DNA. Nat. Commun. 5 (1), 1–7.
Goldberg, S.B., et al., 2016. Creating a climate for therapist improvement: a case study of an agency focused on outcomes and deliberate practice. Psychotherapy 53 (3), 367.
Guardiola, F., et al., 2016. Impact of date palm fruits extracts and probiotic enriched diet on antioxidant status, innate immune response and immune-related gene expression of European seabass (Dicentrarchus labrax). Fish Shellfish Immunol. 52, 298–308.
Guardiola, M., et al., 2015. Deep-sea, deep-sequencing: metabarcoding extracellular DNA from sediments of marine canyons. PLoS One 10 (10), e0139633.
Guillerault, N., et al., 2017. Application of DNA metabarcoding on faeces to identify European catfish Silurus glanis diet. J. Fish Biol. 90 (5), 2214–2219.
Hajibabaei, M., Porter, T.M., Robinson, C.V., Baird, D.J., Shokralla, S., Wright, M.T., 2019. Watered-down biodiversity? A comparison of metabarcoding results from DNA extracted from matched water and bulk tissue biomonitoring samples. PLoS One 14 (12), e0225409.

Hänfling, B., Lawson Handley, L., Read, D.S., Hahn, C., Li, J., Nichols, P., Blackman, R.C., Oliver, A., Winfield, I.J., 2016. Environmental DNA metabarcoding of lake fish communities reflects long-term data from established survey methods. Mol. Ecol. 25 (13), 3101–3119.

Haouchar, D., et al., 2014. Thorough assessment of DNA preservation from fossil bone and sediments excavated from a late Pleistocene–Holocene cave deposit on Kangaroo Island, South Australia. Quat. Sci. Rev. 84, 56–64.

Harper, L.R., et al., 2018. Needle in a haystack? A comparison of eDNA metabarcoding and targeted qPCR for detection of the great crested newt (Triturus cristatus). Ecol. Evol. 8 (12), 6330–6341.

Hawkins, J., et al., 2015. Using DNA metabarcoding to identify the floral composition of honey: a new tool for investigating honey bee foraging preferences. PLoS One 10 (8), e0134735.

Heather, J.M., Chain, B., 2016. The sequence of sequencers: the history of sequencing DNA. Genomics 107 (1), 1–8.

Hering, D., et al., 2018. Implementation options for DNA-based identification into ecological status assessment under the European Water Framework Directive. Water Res. 138, 192–205.

Hibert, F., et al., 2013. Unveiling the diet of elusive rainforest herbivores in next generation sequencing era? The tapir as a case study. PLoS One 8 (4), e60799.

Ji, Y., et al., 2013. Reliable, verifiable and efficient monitoring of biodiversity via metabarcoding. Ecol. Lett. 16 (10), 1245–1257.

Jørgensen, T., et al., 2012. Islands in the ice: detecting past vegetation on Greenlandic nunataks using historical records and sedimentary ancient DNA Meta-barcoding. Mol. Ecol. 21 (8), 1980–1988.

Kartzinel, T.R., et al., 2015. DNA metabarcoding illuminates dietary niche partitioning by African large herbivores. Proc. Natl. Acad. Sci. 112 (26), 8019–8024.

Keller, A., et al., 2015. Evaluating multiplexed next-generation sequencing as a method in palynology for mixed pollen samples. Plant Biol. 17 (2), 558–566.

Kelly, L., et al., 2014. Overview of the share/clef ehealth evaluation lab, International Conference of the Cross-Language Evaluation Forum for European Languages, 2014. Springer.

Kidziak, S., 2020. The Ability of PCR-Based Methods and Next Generation Sequencing to Detect Secondary Predation in the Gut Contents of Aquatic Macroinvertebrates. La Trobe.

Klymus, K.E., Marshall, N.T., Stepien, C.A., 2017. Environmental DNA (eDNA) metabarcoding assays to detect invasive invertebrate species in the Great Lakes. PLoS One 12 (5), e0177643.

Klymus, K.E., Marshall, N.T., Stepien, C.A., 2017. Environmental DNA (eDNA) metabarcoding assays to detect invasive invertebrate species in the Great Lakes. PLoS One 12 (5), e0177643.

Kowalczyk, R., et al., 2011. Influence of management practices on large herbivore diet—Case of European bison in Białowieża Primeval Forest (Poland). Forest Ecol. Manag. 261 (4), 821–828.

Lallias, D., Hiddink, J.G., Fonseca, V.G., Gaspar, J.M., Sung, W., Neill, S.P., Barnes, N., Ferrero, T., Hall, N., Lambshead, P.J.D., Packer, M., 2015. Environmental metabarcoding reveals heterogeneous drivers of microbial eukaryote diversity in contrasting estuarine ecosystems. ISME J. 9 (5), 1208–1221.

Lanzén, A., et al., 2017. DNA extraction replicates improve diversity and compositional dissimilarity in metabarcoding of eukaryotes in marine sediments. PLoS One 12 (6), e0179443.

Laroche, O., et al., 2018. A cross-taxa study using environmental DNA/RNA metabarcoding to measure biological impacts of offshore oil and gas drilling and production operations. Mar. Pollut. Bull. 127, 97–107.

Lee, P.-.S., et al., 2016. Field calibration of blowfly-derived DNA against traditional methods for assessing mammal diversity in tropical forests. Genome 59 (11), 1008–1022.

Leontidou, K., et al., 2018. DNA metabarcoding of airborne pollen: new protocols for improved taxonomic identification of environmental samples. Aerobiologia 34 (1), 63–74.

Leray, M., Knowlton, N., 2015. DNA barcoding and metabarcoding of standardized samples reveal patterns of marine benthic diversity. Proc. Natl. Acad. Sci. 112 (7), 2076–2081.

Lim, N.K., Tay, Y.C., Srivathsan, A., Tan, J.W., Kwik, J.T., Baloğlu, B., Meier, R., Yeo, D.C., 2016. Next-generation freshwater bioassessment: eDNA metabarcoding with a conserved metazoan primer reveals species-rich and reservoir-specific communities. R. Soc. Open Sci. 3 (11), 160635.

Lobo, J., Shokralla, S., Costa, M.H., Hajibabaei, M., Costa, F.O., 2017. DNA metabarcoding for high-throughput monitoring of estuarine macrobenthic communities. Sci. Rep. 7 (1), 1–13.

Lopes, C.M., Sasso, T., Valentini, A., Dejean, T., Martins, M., Zamudio, K.R., Haddad, C.F.B., 2017. eDNA metabarcoding: a promising method for anuran surveys in highly diverse tropical forests. Mol. Ecol. Resour. 17 (5), 904–914. doi:10.1111/1755-0998.12643.

Malizia, E.E., Ke, S., 1993. The influence of economic diversity on unemployment and stability. J. Region. Sci. 33 (2), 221–235.

Martínez-García, L.B., et al., 2015. Host identity is a dominant driver of mycorrhizal fungal community composition during ecosystem development. New Phytol. 205 (4), 1565–1576.

Massana, R., et al., 2015. Marine protist diversity in E uropean coastal waters and sediments as revealed by high-throughput sequencing. Environ. Microbiol. 17 (10), 4035–4049.

Mayerl, C.J., et al., 2018. The impact of keels and tails on turtle swimming performance and their potential as models for biomimetic design. Bioinspir. Biomim. 14 (1), 016002.

McInnes, J.C., et al., 2017. Optimised scat collection protocols for dietary DNA metabarcoding in vertebrates. Methods Ecol. Evol. 8 (2), 192–202.

Miller, K.E., et al., 2016. Metabarcoding of fungal communities associated with bark beetles. Ecol. Evol. 6 (6), 1590–1600.

Nakagawa, H., Yamamoto, S., Sato, Y., Sado, T., Minamoto, T., Miya, M., 2018. Comparing local-and regional-scale estimations of the diversity of stream fish using eDNA metabarcoding and conventional observation methods. Freshwater Biol. 63 (6), 569–580.

O'Donnell, J.L., et al., 2017. Spatial distribution of environmental DNA in a nearshore marine habitat. Peer J 5, e3044.

Pansu, J., et al., 2015. Reconstructing long-term human impacts on plant communities: an ecological approach based on lake sediment DNA. Mol. Ecol. 24 (7), 1485–1498.

Pawlowski, J., Kelly-Quinn, M., Altermatt, F., Apothéloz-Perret-Gentil, L., Beja, P., Boggero, A., Borja, A., Bouchez, A., Cordier, T., Domaizon, I., Feio, M.J., 2018. The future of biotic indices in the ecogenomic era: integrating (e) DNA metabarcoding in biological assessment of aquatic ecosystems. Sci. Total Environ. 637, 1295–1310.

Pedersen, M.W., et al., 2013. A comparative study of ancient environmental DNA to pollen and macrofossils from lake sediments reveals taxonomic overlap and additional plant taxa. Quat. Sci. Rev. 75, 161–168.

Pilliod, D.S., et al., 2013. Estimating occupancy and abundance of stream amphibians using environmental DNA from filtered water samples. Can. J. Fish. Aquat. Sci. 70 (8), 1123–1130.

Piñol, J., et al., 2014. A pragmatic approach to the analysis of diets of generalist predators: The use of next-generation sequencing with no blocking probes. Mol. Ecol. Resour. 14 (1), 18–26.

Pochon, X., et al., 2015. Accurate assessment of the impact of salmon farming on benthic sediment enrichment using foraminiferal metabarcoding. Mar. Pollut. Bull. 100 (1), 370–382.
Pochon, X., et al., 2017. Wanted dead or alive? Using metabarcoding of environmental DNA and RNA to distinguish living assemblages for biosecurity applications. PLoS One 12 (11), e0187636.
Pornon, A., et al., 2016. Using metabarcoding to reveal and quantify plant-pollinator interactions. Sci. Rep. 6 (1), 1–12.
Port, J.A., et al., 2016. Assessing vertebrate biodiversity in a kelp forest ecosystem using environmental DNA. Mol. Ecol. 25 (2), 527–541.
Porter, T.M., Hajibabaei, M., 2018. Scaling up: A guide to high-throughput genomic approaches for biodiversity analysis. Mol. Ecol. 27 (2), 313–338.
Rivera, S.F., et al., 2018. DNA metabarcoding and microscopic analyses of sea turtles biofilms: complementary to understand turtle behavior. PLoS One 13 (4), e0195770.
Rodgers, T.W., et al., 2017. Carrion fly-derived DNA metabarcoding is an effective tool for mammal surveys: evidence from a known tropical mammal community. Mol. Ecol. Resour. 17 (6), e133–e145.
Ruppert, K.M., 2020. Development and Assessment of an Environmental DNA (eDNA) Assay for the Rio Grande Siren and Review of eDNA Metabarcoding Applications. The University of Texas, Rio Grande Valley.
Ruppert, K.M., Kline, R.J., Rahman, M.S., 2019. Past, present, and future perspectives of environmental DNA (eDNA) metabarcoding: a systematic review in methods, monitoring, and applications of global eDNA. Global Ecol. Conserv. 17, e00547.
Sarwar, M.A., Rehman, A., Ferzund, J., 2016. Database search, alignment viewer and genomics analysis tools: big data for bioinformatics. Int. J. Comput. Sci. Inf. Secur. 14 (12), 317.
Sato, H., Sogo, Y., Doi, H., Yamanaka, H., 2017. Usefulness and limitations of sample pooling for environmental DNA metabarcoding of freshwater fish communities. Sci. Rep. 7 (1), 1–12.
Schubert, G., et al., 2015. Targeted detection of mammalian species using carrion fly-derived DNA. Mol. Ecol. Resour. 15 (2), 285–294.
Shaw, J.L., Clarke, L.J., Wedderburn, S.D., Barnes, T.C., Weyrich, L.S., Cooper, A., 2016. Comparison of environmental DNA metabarcoding and conventional fish survey methods in a river system. Biol. Conserv. 197, 131–138.
Shehzad, W., et al., 2012. Prey preference of snow leopard (Panthera uncia) in South Gobi, Mongolia. PLoS One 7 (2), e32104.
Shokralla, S., et al., 2012. Next-generation sequencing technologies for environmental DNA research. Mol. Ecol. 21 (8), 1794–1805.
Sickel, W., et al., 2015. Increased efficiency in identifying mixed pollen samples by metabarcoding with a dual-indexing approach. BMC Ecol. 15 (1), 1–9.
Šigut, M., et al., 2017. Performance of DNA metabarcoding, standard barcoding, and morphological approach in the identification of host–parasitoid interactions. PLoS One 12 (12), e0187803.
Smith, M.B., et al., 2015. Natural bacterial communities serve as quantitative geochemical biosensors. MBio 6 (3) e00326-15.
Soininen, E.M., et al., 2013. Shedding new light on the diet of Norwegian lemmings: DNA metabarcoding of stomach content. Polar Biol. 36 (7), 1069–1076.
Sønstebø, J.H., et al., 2010. Using next-generation sequencing for molecular reconstruction of past Arctic vegetation and climate. Mol. Ecol. Resour. 10 (6), 1009–1018.
Štursová, M., et al., 2014. When the forest dies: the response of forest soil fungi to a bark beetle-induced tree dieback. ISME J. 8 (9), 1920–1931.
Suchan, T., et al., 2019. Pollen metabarcoding as a tool for tracking long-distance insect migrations. Mol. Ecol. Resour. 19 (1), 149–162.

Taberlet, P., et al., 2012. Towards next-generation biodiversity assessment using DNA metabarcoding. Mol. Ecol. 21 (8), 2045–2050.

Thomsen, P.F., et al., 2012. Monitoring endangered freshwater biodiversity using environmental DNA. Mol. Ecol. 21 (11), 2565–2573.

Tong, X., et al., 2017. High diversity of airborne fungi in the hospital environment as revealed by meta-sequencing-based microbiome analysis. Sci. Rep. 7 (1), 1–8.

Troth, C., 2021. Transforming a Research Concept Into Commercial Practice: Addressing the 'Hurdles' of Single-Species eDNA-Based Detection. University of Derby, United Kingdom.

Valentini, A., et al., 2009. New perspectives in diet analysis based on DNA barcoding and parallel pyrosequencing: the trnL approach. Mol. Ecol. Resour. 9 (1), 51–60.

Valentini, A., Miquel, C., Taberlet, P., 2010. DNA barcoding for honey biodiversity. Diversity 2 (4), 610–617.

Valentini, A., Taberlet, P., Miaud, C., Civade, R., Herder, J., Thomsen, P.F., Dejean, T., 2016. Next-generation monitoring of aquatic biodiversity using environmental DNA metabarcoding. Mol. Ecol. 25 (4), 929–942. doi:10.1111/mec.13428.

Wangensteen, O.S., et al., 2018. DNA metabarcoding of littoral hard-bottom communities: high diversity and database gaps revealed by two molecular markers. Peer J 6, e4705.

Wangensteen, O.S., et al., 2018. Under the canopy: Community-wide effects of invasive algae in Marine Protected Areas revealed by metabarcoding. Mar. Pollut. Bull. 127, 54–66.

Willerslev, E., et al., 2014. Fifty thousand years of Arctic vegetation and megafaunal diet. Nature 506 (7486), 47–51.

Wood, J.R., et al., 2012. High-resolution coproecology: using coprolites to reconstruct the habits and habitats of New Zealand's extinct upland moa (Megalapteryx didinus). PLoS One 7 (6), e40025.

Xie, Y., et al., 2017. Environmental DNA metabarcoding reveals primary chemical contaminants in freshwater sediments from different land-use types. Chemosphere 172, 201–209.

Yamamoto, S., et al., 2017. Environmental DNA metabarcoding reveals local fish communities in a species-rich coastal sea. Sci. Rep. 7 (1), 1–12.

Yang, C., et al., 2014. Using metabarcoding to ask if easily collected soil and leaf-litter samples can be used as a general biodiversity indicator. Ecol. Indic. 46, 379–389.

Yoccoz, N.G., et al., 2012. DNA from soil mirrors plant taxonomic and growth form diversity. Mol. Ecol. 21 (15), 3647–3655.

Yoon, T-Ho, et al., 2017. Metabarcoding analysis of the stomach contents of the Antarctic Toothfish (Dissostichus mawsoni) collected in the Antarctic Ocean. Peer J 5, e3977.

Yu, D.W., et al., 2012. Biodiversity soup: metabarcoding of arthropods for rapid biodiversity assessment and biomonitoring. Methods Ecol. Evol. 3 (4), 613–623.

Zaiko, A., et al., 2015. Metabarcoding approach for the ballast water surveillance—an advantageous solution or an awkward challenge? Mar. Pollut. Bull. 92 (1-2), 25–34.

Zaiko, A., et al., 2016. Metabarcoding improves detection of eukaryotes from early biofouling communities: implications for pest monitoring and pathway management. Biofouling 32 (6), 671–684.

Zappelini, C., et al., 2015. Diversity and complexity of microbial communities from a chlor-alkali tailings dump. Soil Biol. Biochem. 90, 101–110.

Zimmermann, J., Glöckner, G., Jahn, R., Enke, N., Gemeinholzer, B., 2015. Metabarcoding vs. morphological identification to assess diatom diversity in environmental studies. Mol. Ecol. Resour. 15 (3), 526–542.

# CHAPTER 7

# Pollution biomarkers in environmental biomonitoring: An insight into air pollution

**Nivedita Chaudhary**
Department of Environmental Science, School of Earth Sciences, Central University of Rajasthan, Ajmer, Rajasthan, India

## 7.1 Introduction

Pollutants in the environment are increasing and dispersing all over the environmental matrices due to numerous anthropogenic activities employing stress-causing deleterious effects on organisms. Over the last few decades, the world is experiencing an increase in the levels of environmental pollution due to anthropogenic activities, for instance, an accelerated rise in air pollution (Lelieveld et al., 2019). In 2019, globally declining air quality caused premature deaths, massive crop loss, biodiversity loss and also, responsible for climate change (Flower et al., 2020). Environmental pollutants can be persistent or mobile in all environmental matrices causing harmful effects on vegetation and nearly half of the world's population despite diminishing air pollution (Shaddick et al., 2020). The assessment of pollutant's impact on the organism is usually difficult due to different levels of sensitivities among organisms and alteration in the pollutants' approach to cause-effect and the possibility of their chemical transformation (Kaviraj et al., 2014). Cullen (1990) suggested the importance of biological parameters for estimating the variations in the environment and impact on the organism in responses to pollution and designated them as biomarkers. Therefore, alteration in normal conditions in the biological system due to exposure to pollution quantitatively measured and considered as a biomarker (Lionetto et al., 2019). Assessments on biota provide early warning signs about the changes in the environment due to pollution (Peakall, 1994; Lam and Gray, 2003). Estimation of the effects of environmental pollution on organisms at the molecular, cellular, biochemical and physiological levels used as biomarkers and provides information related to the concentration of the pollutant or substance inside the organism (Wolterbeek, 2002).

*New Paradigms in Environmental Biomonitoring Using Plants*
DOI: https://doi.org/10.1016/B978-0-12-824351-0.00009-2

de Temmerman et al. (2005) suggested biomonitoring of air pollution by using plants at the different biological organizations from molecular to ecosystem level.

The process of acquiring quantitative attributes at the biological level, sequentially from organism to ecosystem-level due to environmental changes termed biomonitoring, can also reduce the adverse effects. Cascade of biological responses stimulated due to the environmental pollution and these responses manifested the changes in the organism designated as biomarkers (Lam and Gray, 2003). Therefore, biomarker represents the changes in the environment due to pollution and organism (plant or animals) conveying responses for biomonitoring (Gorelova and Frontasyeva, 2017). The present chapter provides information about the biomarker's responses with a reliable exposure to environmental pollution, particularly to air pollution and usage of the biomarkers in biological monitoring to assess impacts.

## 7.2 General characteristics of biomarkers and role in biomonitoring

Biomarkers are quantifiable parameters at the sub-organism level, for instance, molecular, cellular, physiological, and biochemical levels of an organism, to serve as the indicators of contaminants in the environment through the changes in the responses of the biomarkers. Generally, at the ecological level, responses at individual and population levels are acknowledged; however, the pollution stress effect is less changeable and more damaging at the ecosystem level than sub-organism levels (Hamza-Chaffai, 2014). Lam and Gray (2003) reported that pollution-induced changes at organism, population, and ecological levels are generally signified as bioindicators, however, with no well-defined consent. For the usage of biomarkers, required to be sensitive enough to depict the environmental change at an early stage, biomarker can be specific to the pollutant, and response of biomarker can also change according to the pollutant levels for biomonitoring. However, due to several pollutants causing numerous effect; therefore, a collection of biomarkers belongs at different biological levels are essential for biomonitoring programs (Handy et al., 2003). Biomarker possibly shows the effect of pollution, although the absence of enough pollutant level and repair mechanism of biological organism possibly reduce the response; therefore, integration with the approach of physio-chemical analysis provide the better biological consequences caused by environmental pollution (Lam and Gray, 2003). Peakall (2012) suggested the quantitative aspect of biomarker determine by the dose of pollutant/contaminants relationship with the organism, the time of evaluation, duration of exposure, and stability of biomarker response over a certain period for biomonitoring. The measurement of biomarker responses in sensitive species (plants or animals) can be used to assess alteration in the environment due to early warning signals, mainly used surveys of polluted environments (Parmar et al., 2016). In the case of aquatic pollution, certain invertebrates showed biomarker responses in the form of cellular and biochemical alterations and showed early warning signs in sensitive species (Hagger et al., 2008).

Similarly, in soil pollution, assessment of biomarkers on organisms living in soil are considered as relevant bioindicators and significantly measure quality of soil such as soil invertebrates (Sanchez-Hernandez et al., 2006). Lionetto et al. (2019) suggested the biomarker responses in terrestrial and aquatic organisms exposed to trace metals as metallothionein, cysteine-rich metal-binding protein and acetylcholinesterase inhibition. For air pollution, notably, ozone pollution Nali et al. (2006) suggested the definition of biomonitoring assessing the biological organism responses and information about the integrated effect of environmental changes, pollutants and biological changes challenging to determine through analytical measurements. For instance, qualitative and quantitative assessment of air pollutants can be done by native organism/plants considered as passive biomonitoring where as those organism or plants consciously required to be introduced in the research area for a particular period of time considered for active biomonitoring (Agathokleous et al., 2020). For instance, in a particular area native and cultivated plants observed for ozone biomonitoring termed as passive monitoring. For instance, *Centaurea jacea L.* a native plant in central Europe consider as a potential ozone biomonitor. Furthermore, response analysis of the selected plant species exposed to $O_3$ under typical experimental conditions referred to as active monitoring such as tobacco (*Nicotiana tabacum* L.) Bel-W3 has been used comprehensively as a consistent bioindicator of ozone (Nali et al., 2006). Biomarker responses in the form of visible foliar injury symptoms and biomass ratio among sensitive and resistant varieties of vascular plants (*Phaseolus vulgaris* L., *Trifolium repens* L. and *Nicotiana tabacum* L.) significantly assess the impact of ozone pollution (Agathokleous et al., 2020). Also, the ultimate purpose of biomonitoring is not to substitute the physicochemical analytical approach to assess the impact of pollutants, preferably the combination of both with the biological and physicochemical process to evaluate environmental pollution impact.

## 7.3 Biomarkers for air pollution biomonitoring

### 7.3.1 Plant physiological assessment as a biomarker for air pollution

Air pollution enters plants through stomata and impeded photosynthesis, transpiration, respiration, carboxylation activity, and electron transport (Fig. 7.1). These physiological effects are extensively within and between plant species (Darrall, 1989). Physiological alterations later turn into observable changes in the plants, such as foliar injury symptoms, plant growth, and productivity. Therefore, these physiological responses are considered biomarkers. It is also of utmost essential to identify physiological biomarker air pollution impact under natural occurring stress conditions. Air pollutant, when uptake by plants through stomata undergoes the formation of reactive oxygen species (ROS) and alteration in the plant's defense mechanism in the substomatal cavity, affects the plant's physiology. For instance, Heath (1994) suggested that $O_3$ pollution induces the stomatal closure when the plant is exposed to $O_3$ lead metabolic alternation internal $CO_2$ utilization, and fixation of $CO_2$ leads to the decline in the photosynthesis.

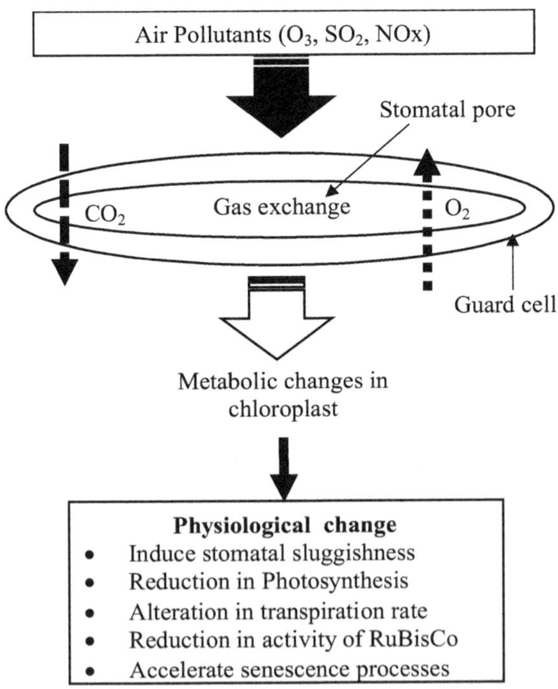


**Fig. 7.1** ***Overview of physiological responses of air pollutants on plants.***

Earlier studies on chloroplast suggested ultrastructural changes in polluted areas mainly due to the thylakoid system and carbohydrate mechanism (Zellnig et al., 1989). Mainly, ROS generated by air pollutants generates oxidative stress in plants, causes effects on the photosynthetic mechanism by changes in the thylakoid structure and performance. Additionally, lipid peroxidation induces disintegration and instability of chloroplast membrane causes chlorophyll reduction. Also, the newly accumulated lipid may lead to thylakoid swelling (Dhir, 2016). Air pollutants also can affect the ultrastructure of cellular membranes by oxidation of unsaturated fatty acids and causing metabolic modifications of membrane lipids (Sakaki, 2002).Various studies suggested air pollution stimulated physiological effects directly by reducing photosynthetic rate or indirectly by accelerating senescence processes and alteration in stomatal conductance (Agathokleous et al., 2017; Paoletti and Grulke, 2005; Heath, 1994).

Similarly, biomonitoring of air pollution such as $O_3$, $SO_2$, and NOx suggested reducing in the physiological characteristics of wheat (Rajput and Agrawal, 2005). Exposure of air pollutants ($O_3$, NOx, and $SO_2$) to various plant species showed a reduction in photosynthesis, deterioration in the electron transport system, reduced activity of carboxylation, direct and indirect effects on stomata (Darrall, 1989). However, elevated levels of $CO_2$, increase in photosynthesis due to increased activity of Rubisco helps in the carboxylation of RuBP (Ribulose-1,5-Bisphosphate). It utilizes $O_2$ for photorespiration, which generally occurs at the extra expense of the plants. Overall, the initial response of high levels of $CO_2$ increases photosynthesis but under long-term $CO_2$ exposure eventually begins the reduction, as observed by several crop plant studies with elevated $CO_2$ exposure (Long et al., 2005).

Apart from gaseous air pollutants, plant encounters with the particulate pollutants are considered the sink due to its leaf surface area. Particulate matter with heavy metals such as Cd, Hg, Pb, Co, Ni, Zn, Fe, As, Cr and Ag directly causes the effect on the photosynthetic system due to ROS production. For instance, the impact on plants is assessed in crop plants and trees, for instance, due to black dust alteration in gaseous exchange and photosynthetic decline in a reduction in *Viburnum tinus*. A similar observation in *Zygophyllum prismatocarpum* due to limestone dust and Eugenia uniflora showed due to iron dust, *Triumfetta rhomboidea* due to Coal-smoke emissions reviewed by Rai (2016). Certain heavy metals in the air permeated inside the leaves, causing a detrimental effect on the photosynthetic system, mainly the functioning of photosystem-I and II and dark reaction in the plant reviewed by Rai et al. (2016).

### 7.3.2 Photosynthetic pigments as a biomarker of air pollution

Chlorophyll content considered a biophysical biomarker in plants under environmental stress. Rai (2016) suggested that chlorophyll pigment endure various phytochemical reactions in plants lead to biochemical and physiological alteration. Reviewed studies related to higher air pollution, causing the reduction in the photosynthetic pigment. Air pollutants significantly affect the conversion of chlorophyll a and pheophytin also damage and loss phytol from chlorophyll b suggested the damage of chloroplast (Ram et al., 2015). For instance, the biomonitoring assessment of chlorophyll and carotenoids suggested being an important biomarker under elevated $O_3$, NOx, and $SO_2$. Table 7.1 showed recent studies suggesting that alteration and reduction in chlorophyll pigments in various plant and tree species are considered a potential biomarker for air pollution. $O_3$ biomonitoring

**Table 7.1** Alteration in photosynthetic pigments as a potential biomarker due to air pollution.

| S. No. | Plants/trees | Pollutants | Effects | References |
|---|---|---|---|---|
| 1. | Morus alb | $NO_2$ | Reduction in chlorophyll contents | Wang et al., 2019 |
| 2. | *Betula pendula* | $O_3$ | Increase in the amount of chloroplasts or vacuole | Hartikainen et al., 2020 |
| 3. | *Mangifera indica* | Unhealthy Air Quality Index | Reduction in chlorophyll content | Mukherjee, et al., 2019 |
| 4. | *Euonymus kiautschovicus, Ligustrum vicaryi*, and *Syringa oblata* | $SO_2$ | Reduction in contents of chlorophyll a and chlorophyll b | Duan et al., 2019 |
| 5. | *Tillandsia usneoides* | $NO_2$ | Carotenoids, flavonoids, pheophytins and lipofuscins are chloroplast metabolites fluorescence emission profile showed alterations. | Giampaoli et al., 2020 |
| 6. | *T. aestivum* and *P. vulgaris* | $O_3$ | Reduction in chlorophyll content | Hayes et al., 2020 |
| 7. | *Phaseolus vulgaris* | $O_3$ | Emergence of foliar damage | Leung et al., 2020 |
| 8. | *Pinus taeda L.* and *Liquidambar styraciflua L.* | $O_3$ | Clear visible discoloration of foliage | Reviewed by Grulke and Heath (2020), |

| S. No. | Plants/trees | Pollutants | Effects | References |
|---|---|---|---|---|
| 9. | *Byrsonima sericea, Cordia verbenacea,* and *Psidium guineense* | $SO_2$ | Leaf showed necrotic regions and various foliar micromorphological changes | da Silva et al., 2020 |
| 10. | Joannesia princeps | Acid precipitation | Early appearance of wilting of epidermal cells even before emergence of visible symptoms | Andrade et al., 2020 |
| 11. | Populus tremuloides and Betula pendula | $O_3$ | Appearance of dark and pigmented dots on upper side of leaf surface | Reviewed by Oksanen and Kontunen-Soppela, 2021 |
| 12. | Ficus religiosa, Ficus benghalensis, Tamarindus indica, Annona squamosa, and Delonix regia | Urban air pollution | Reduced chlorophyll b content | Kamble et al., 2021 |
| 13. | *Phoenix dactylifera L., Punica granatum L., Ficus carica L., and Olea europaea L.* | Industrial air pollution | Decrease in total chlorophyll concentration and appearance of visual symptoms | Amor et al., 2021 |
| 14. | Mangifera indica *L. and* Polyalthia longifolia *(Sonn.),* Saraca asoka *(Roxb.) Willd.*, and Ficus benghalensis *L.* | Urban and industrial air pollution mainly $SO_2$ | Reduction in chlorophyll content | Karmakar et al., 2021 |

different cultivars of clover showed a reduction in both chlorophyll and carotenoids (Chaudhary and Agrawal, 2013); similarly, Rajput and Agrawal (2005) suggested the reduction in photosynthetic pigments of wheat cultivars use of different air pollutant's biomonitoring. The case of lichens, used as the biomonitors, showed an increase in the chlorophyll content when at the place of high vehicular traffic causing high levels of NOx and $SO_2$ pollutants (Conti and Cecchetti, 2001). Formation of acid radicals due to $SO_2$ and NOx in the atmosphere and their affinity to form acid radicals after diffusing inside the leaf affects chlorophyll molecules; similarly, particulate pollution constituents with low and high pH resulted in the alteration of the photosynthetic pigment (Rai 2016).

### 7.3.3 Biochemical responses as a biomarker of air pollution

Biochemical responses as biomarkers considered to be specific to pollutants/contaminants caused by different environmental stress conditions; therefore, their benefit in identifying the bioavailability of pollutants in the environment and effectual biological effect (Peakall et al., 2012). Air pollutants produce ROS eventually causes oxidative stress causing impairment from cellular to molecular levels of the plant (Apel and Hirt, 2004). Oxidative stress induced by ROS, such as $O_2^{\bullet-}$ (superoxide radical), $OH^{\bullet}$ (hydroxyl radical), $^1O_2$ (singlet oxygen), and $H_2O_2$ (hydrogen peroxide), and which cause oxidative stress. The plant consists of an antioxidative defense mechanism having various enzymatic and non-enzymatic activities that work against ROS. However, in the case of improper removal of ROS, an effect designated as "oxidative stress" (Gill and Tuteja, 2010) (Fig. 7.2). Schulz and Härtling (2001) suggested the consideration of biochemical parameters (ascorbic acid, superoxide dismutase, glutathione reductase, glutathione, α-Tocopherol and pigments) as biomarkers to assess the early effect of $O_3$, NOx, and $SO_2$ on *Pinus sylvestris*.

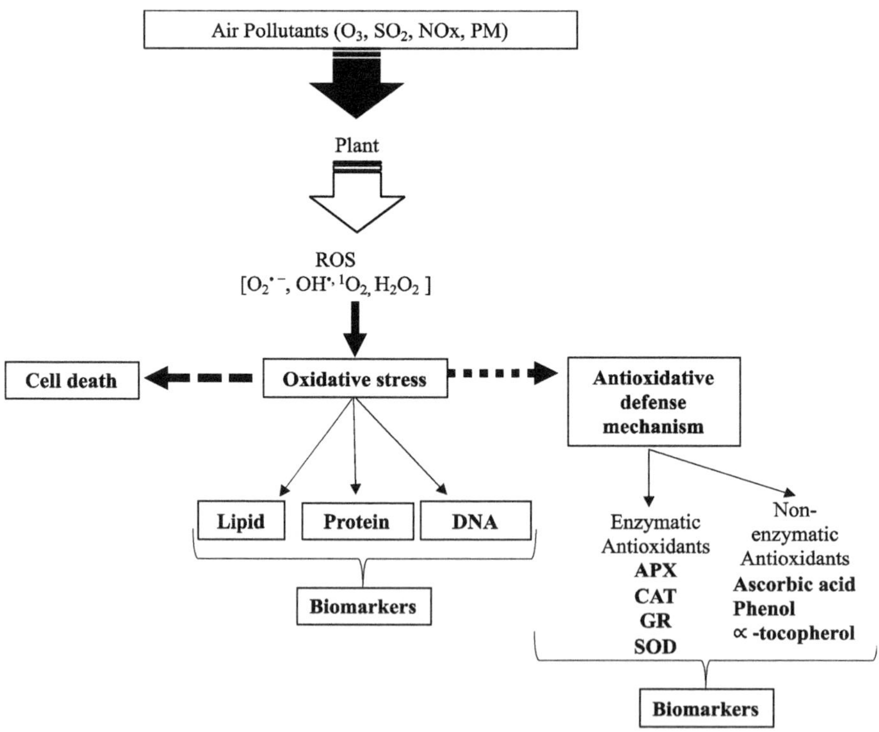


**Fig. 7.2** *Overview of biochemical responses of air pollutants on plants.*

#### 7.3.3.1 ROS, lipid peroxidation, metabolites, and antioxidative enzyme

Different environmental pollutants cause deregulation in the biochemical plant metabolism, such as stimulation in the enzyme activity and alteration in the metabolites due to air pollution exposure; therefore, they can consider as biochemical biomarkers for qualitative assessment of pollution (Peakall et al., 2012). Quantification of these biochemical responses serves as the biochemical biomarker for identifying the effects of environmental pollution. The magnitude of ascorbic acid, malondialdehyde content, antioxidative enzymes in plants altered under the influence of stress generated by the air pollution. Various studies on the effect of air pollutants on plants suggested the magnitude of antioxidative enzyme activities-initiated plants signal and may also consider involving as biomarkers to assess the phytotoxicity reviewed by Rai (2016). Air pollution-induced deleterious ROS causes oxidative stress in the plants, causing oxidative stress affecting the peroxidative damage to typical cellular components (Tiwari et al., 2006). High lipid peroxidation proline content observed in *Mangifera indica* suggested using as biomarkers (Mukherjee et al., 2019).

Similarly, ROS assessment revealed the damage in different cultivars of clover used to assess the oxidative caused by $O_3$ (Chaudhary and Agrawal, 2013). Fangmeier et al. (1994) suggested estimation of malondialdehyde content, ascorbic acid, protein and the activities of antioxidative enzymes (catalase and peroxidase) as biomarkers to assess the $O_3$ stress in wheat plants. Similarly, these biochemical parameters were assessed among different in clover cultivars and their magnitudes varies under ambient and elevated levels of $O_3$ (Chaudhary and Agrawal, 2013). Air pollution-induced free radical production has detrimental effects, along with the antioxidative protection mechanisms considered biomarkers in plants and lichens (Cuny et al., 2002). In an industrial area using assessment of oxidative stress biomarkers such as glutathione (GSH), glutathione S-transferase (GST) activity, chlorophyll, malondialdehyde, and protein contents in *C. libani* (Oztetik 2015).

#### 7.3.3.2 Molecular and genetic biomarker for air pollution

Environmental pollution can stimulate the plant genome and assess the genotoxicity by using the system by the transgenic plant. Integrating the genome of the marker gene of the identified sequence as target mutagenic effects can be used for quantifying the activity mutagenicity of pollutant or contaminant in the environment can be assessed (Kovalchuk et al., 2001). Steinkellner et al. (1998) also suggested that the mutations in plants are established on revealing chromosomal aberrations in *Vicia faba*, *Allium cepa*,

*Tradescantia* and the methodology used for the air quality monitoring affected by a load of metal.

An essential issue of the transgenic approach is using the right transgene to extrapolate the mutation frequency in the transgene used in the whole genome. It reduces the possibility that selection processes influence the frequency of particular mutations. Air pollutants are also considered to affect the DNA and can consider as the biomarker; genotoxicity was estimated by AFLP (Amplified Fragment Length Polymorphism) in *Trifolium repens* suggesting DNA damage (Piraino et al., 2006). To assess the effect of lichen affected by environmental pollution, a molecular biomarker used as the Random Amplified Polymorphic DNA Polymerase Chain Reaction to define the variation in DNA band pattern (Cansaran-Duman et al., 2011). Amato-Lourenco et al. (2017) suggested Tradescantia micronucleus (Trad-MCN) bioassay to assess the genotoxic response of vehicular traffic air pollution and metal accumulation on trees and considered as a useful biomarker to quantify air pollution. Similarly, genotoxic activity caused by air pollution, particularly $SO_2$ assessed by Trad-MCN on wild plants (Mišík et al., 2007). Similarly, air pollution-induced genotoxicity assessed by using Trad-MCN assay in a 5-year study in Brazil for vehicular pollution (Pereira et al., 2013).

The utilization of plants in the biomonitoring of environmental pollution Trad-MCN is advantageous in estimating of environmental pollution, particularly in industrial and urban areas and landfills. The approach is simple and less necessary to allow the setting up of a map of genotoxicity to develop the pollution site monitoring (Sadowska and Pluygers, 2002). Genotoxicity of environmental pollutants also evaluated with a molecular marker of AFLP suggested that environmental pollutants caused changes patterns of a band from AFLP analysis, implicating DNA damage in the lichen species *Pseudevernia furfuracea* and suggesting use of lichen for in biomonitoring (Vardar et al., 2014).

### 7.3.4 Visual biomarker for pollution biomonitoring

Air pollutants, for instance, $SO_2$, NOx, $O_3$ and particulate matter, are recognized to be toxic and deleterious effect observed when the concentration exceeded over the threshold level of pollutants causes effect at the morphological, physiological, biochemical, and molecular level under the prevailing environmental conditions (Fuhrer and Bungener, 1999). Emberson et al. (2001) suggested that observing visible symptoms of injury on plants due to air pollutants under the studies conducted in the field, transect studies and experiments in controlled exposure of pollutants defining the exposure–response relationship. Table 7.2 showed air polution

**Table 7.2** Air pollution biomarker response as foliar injury symptoms.

| S. No. | Plants/trees | Pollutants | Visual injury | References |
|---|---|---|---|---|
| 1. | Nicotiana tabacum | $O_3$ | Leaf injury due to environmental factors and $O_3$ | Käffer et al., 2019 |
| 2. | *Medicago sativa* L. | $SO_2$ | Harmful effects on leaf demography and growth | Cotrozzi et al., 2020 |
| 3. | *Amaranthus hypochondriacus* | $O_3$ | Foliar injury corresponds to higher yield reduction among sensitive cultivars | Yadav et al., 2020 |
| 4. | *Phaseolus vulgaris* L. | $O_3$ | Foliar injury on sensitive genotypes | Leung et al., 2020 |
| 5. | Hybrid poplar, *Picea abies*, *T. aestivum* | $SO_2$, $O_3$, $H_2S$, NOx | Distinct symptoms chlorotic and necrotic at margins and interveinal portions | Reviewed by Papazian and Blande et al., 2020 |
| 6. | *Sida cordifolia L.* | $O_3$ | Development of foliar injury symptoms in the adaxial surface of mature leaves | Ansari et al., 2021 |
| 7. | *Astronium graveolens Jacq.* | $SO_2$, Hydrocarbons, Particulate matters | Emergence of foliar damage correlated with phyt-toxic dose of $O_3$ | Fernandes et al., 2021 |
| 8. | *Pinus taeda L.* and *Liquidambar styraciflua L.* | $O_3$ | Clear visible discoloration of foliage | Reviewed by Grulke and Heath, 2020 |
| 9. | Azardirachta indica, Ceasalpinia sepiaria Albizzia lebbek, and Guaiacum officinale | Vehicular emission ($N_2O$, $SO_2$, CO, and particulates) | Chlorosis on leaves | Reviewed by Muthu et al., 2021 |
| 10. | Byrsonima variabilis, Myrceugenia alpigena, and Eremanthus erythropappus | Fluoride | Chlorosis and necrosis | Gomes et al., 2021 |

visual biomarkers considerded as foliar injury symptoms in plants. Foliar injury assessment for utilization in the biomonitoring program to identify pollution. For instance, $O_3$ biomonitoring for revealing monitoring comprises verifying $O_3$-induced visible injury to well-known sensitive plant species under (Nali 2006). Krupa and Kickert (1993) suggested utilizing

$O_3$ sensitive plant *Nicotiana tabacum* L. Bel-W3 as a bioindicator depicting specific $O_3$ injury symptoms. Similarly, *Trifolium repens* L. Regal depicted to test the potential effect and levels of $O_3$ utilized widely in the USA (Nali 2006; Heagle et al., 1994).

Similarly, different cultivars of clover assessed under the exposure-response experiment, and visual injury symptoms identified for the manifestation of the effect of $O_3$ (Chaudhary and Agrawal, 2013). Manning et al. (2002) also reported the bioindicator plants showed visual injury symptoms in *Centaurea nigra* L. as a bioindicator utilized to appraise the correlation between $O_3$ exposure and effects. The adverse effect of environmental pollutants causes alterations in the metabolism of plant metabolism leads to the identification by visual effects in terms of foliar injury symptoms (Peakall et al., 2012). Falla et al. (2000) reported that polycyclic aromatic hydrocarbons induced foliar injuries identified on tobacco plants in the form of necrosis. Besides, gaseous air pollution, particulate matter also resulted in injury symptoms such as necrosis, patches, spots and, senescence and ultimate leaves death; similarly, injuries observed in the presence of high level of heavy metals in the dust causing numerous physiological changes to lead to visual such as symptoms of injuries (Saha and Padhy, 2011).

## 7.4 Conclusion

Environmental pollution is a significant environmental problem affecting organisms, and the responses of biomarkers can assess the consequences. Biomonitoring is a crucial tool for insight into emerging environmental concerns. The degree of contracting of biomarkers complies with assessing the environmental quality by physiological, biochemical, genetic, and genotoxic responses. Identification of biomarkers and quest to develop and promote biomarkers to assess specific environmental pollution is considered an emerging research area. Further, the usage of biomarkers at the field level and its integrated with the environmental conditions will help to assess environmental impact caused by pollution. Therefore, a significant amount and specific biomarker in the field condition help to demonstrate the stress. Specific to air pollution, a comprehensive analysis prescribed, thus integrating the physicochemical approach with biological analysis, provides an insight into the biomonitoring.

## Acknowledgment

NC is thankful to UGC-BSR startup grant, New Delhi, India.

## References


Agathokleous, E., Saitanis, C.J., Burkey, K.O., Ntatsi, G., Vougeleka, V., Mashaheet, A.M., Pallides, A., 2017. Application and further characterization of the snap bean S156/R123 ozone biomonitoring system in relation to ambient air temperature. Sci. Total Environ. 580, 1046–1055.

Agathokleous, E., Saitanis, C.J., Feng, Z., De Marco, A., Araminiene, V., Domingos, M., Sicard, P., Paoletti, E., 2020. Ozone biomonitoring: a versatile tool for science, education and regulation. Curr. Opin. Environ. Sci. Health 18, 7–13.

Amato-Lourenco, L.F., Lobo, D.J.A., Guimarães, E.T., Moreira, T.C.L., Carvalho-Oliveira, R., Saiki, M., Saldiva, P.H.N., Mauad, T., 2017. Biomonitoring of genotoxic effects and elemental accumulation derived from air pollution in community urban gardens. Sci. Total Environ. 575, 1438–1444.

Amor, A.B., Bagues, M., Elloumi, N., Chaira, N., Rahmani, R., Nagaz, K., 2021. Suitability of four main Mediterranean tree crops for their growth in peri-urban agriculture and restoration (Gabes, Tunisia). Environ. Sci. Pollut. Res. 28 (15), 19034–19045.

Andrade, G.C., Castro, L.N., da Silva, L.C., 2020. Micromorphological alterations induced by simulated acid rain on the leaf surface of *Joannesia princeps* Vell. (Euphorbiaceae). Ecol. Indic. 116, 106526.

Ansari, N., Agrawal, M., Agrawal, S.B., 2021. An assessment of growth, floral morphology, and metabolites of a medicinal plant *Sida cordifolia* L. under the influence of elevated ozone. Environ. Sci. Pollut. Res. 28 (1), 832–845.

Apel, K., Hirt, H., 2004. Reactive oxygen species: metabolism, oxidative stress, and signal transduction. Annu. Rev. Plant Biol. 55, 373–399.

Cansaran-Duman, D., Atakol, O., Aras, S., 2011. Assessment of air pollution genotoxicity by RAPD in *Evernia prunastri* L. Ach. from around iron-steel factory in Karabük, Turkey. J. Environ. Sci. 23 (7), 1171–1178.

Chaudhary, N., Agrawal, S.B., 2013. Intraspecific responses of six Indian clover cultivars under ambient and elevated levels of ozone. Environ. Sci. Pollut. Res. 20 (8), 5318–5329.

Conti, M.E., Cecchetti, G., 2001. Biological monitoring: lichens as bioindicators of air pollution assessment—a review. Environ. Pollut. 114 (3), 471–492.

Cotrozzi, L., 2020. Leaf demography and growth analysis to assess the impact of air pollution on plants: a case study on alfalfa exposed to a gradient of sulphur dioxide concentrations. Atmos. Pollut. Res. 11 (1), 186–192.

Cullen, P., 1990. Biomonitoring and environmental management. Environ. Monit. Assess. 14 (2), 107–114.

Cuny, D., Pignata, M.L., Kranner, I., Beckett, R., 2002. Biomarkers of pollution-induced oxidative stress and membrane damage in lichens Monitoring With Lichens—Monitoring Lichens. Springer, Dordrecht, pp. 97–110.

da Silva, L.C., de Freitas-Silva, L., Rocha, D.I., Pereira, J.D.S.C., de Freitas Assis, D.E., 2020. Leaf morpho-anatomical structure determines differential response among resting a species exposed to emissions from an iron ore pelletizing plant. Water Air Soil Pollut. 231 (4), 1–9.

Darrall, N.M., 1989. The effect of air pollutants on physiological processes in plants. Plant Cell Environ. 12 (1), 1–30.

de Temmerman, L., Bell, N., Garrec, J.P., Klumpp, A., Krause, G.H., Tonneijck, A.E., 2005. Biomonitoring of air pollutants with plants. Environ. News 11 (2), 5–6.

Dhir, B., 2016. Air pollutants and photosynthetic efficiency of plants Plant Responses to Air Pollution. Springer, Singapore, pp. 71–84.

Duan, J., Fu, B., Kang, H., Song, Z., Jia, M., Cao, D., Wei, A., 2019. Response of gas-exchange characteristics and chlorophyll fluorescence to acute sulfur dioxide exposure in landscape plants. Ecotoxicol. Environ. Saf. 171, 122–129.

Emberson, L.D., Ashmore, M.R., Murray, F., Kuylenstierna, J.C.I., Percy, K.E., Izuta, T., Zheng, Y., Shimizu, H., Sheu, B.H., Liu, C.P., Agrawal, M., 2001. Impacts of air pollutants on vegetation in developing countries. Water Air Soil Pollut. 130 (1), 107–118.
Falla, J., Laval-Gilly, P., Henryon, M., Morlot, D., Ferard, J.F., 2000. Biological air quality monitoring: a review. Environ. Monit. Assess. 64 (3), 627–644.
Fangmeier, A., Brunschön, S., Jäger, H.J., 1994. Time course of oxidant stress biomarkers in flag leaves of wheat exposed to ozone and drought stress. New Phytol. 126 (1), 63–69.
Fernandes, F.F., Moura, B.B., 2021. Foliage visible injury in the tropical tree species, Astronium graveolens is strictly related to phytotoxic ozone dose (PODy). Environ. Sci. Pollut. Res., 1–10.
Fowler, D., Pyle, J.A., Sutton, M.A., Williams, M.L., 2020. Global Air Quality, Past, Present, and Future: An Introduction. Phil. Trans. R. Soc. A., 378, 20190323.
Fuhrer, J., Bungener, P., 1999. Effects of air pollutants on plants. Analusis 27 (4), 355–362.
Giampaoli, P., Fernandes, F.F., Tavares, A.R., Domingos, M., Cardoso-Gustavson, P., 2020. Fluorescence emission spectra of target chloroplast metabolites (flavonoids, carotenoids, lipofuscins, pheophytins) as biomarkers of air pollutants and seasonal tropical climate. Environ. Sci. Pollut. Res. 27 (20), 25363–25373.
Gill, S.S., Tuteja, N., 2010. Reactive oxygen species and antioxidant machinery in abiotic stress tolerance in crop plants. Plant Physiol. Biochem. 48 (12), 909–930.
Gomes, T.F., Lima, A.M., Marques, A.P.P., da Silva, L.C., 2021. Effects of fluoride emission on the morphoanatomy of three plant species endemics to Brazil using passive biomonitoring. Environ. Sci. Pollut. Res., 1–13.
Gorelova, S.V., Frontasyeva, M.V., 2017. The use of higher plants in biomonitoring and environmental bioremediation Phytoremediation. Springer, Cham, pp. 103–155.
Grulke, N.E., Heath, R.L., 2020. Ozone effects on plants in natural ecosystems. Plant Biol. 22, 12–37.
Hagger, J.A., Jones, M.B., Lowe, D., Leonard, D.P., Owen, R., Galloway, T.S., 2008. Application of biomarkers for improving risk assessments of chemicals under the Water Framework Directive: a case study. Mar. Pollut. Bull. 56 (6), 1111–1118.
Hamza-Chaffai, A., 2014. Usefulness of bioindicators and biomarkers in pollution biomonitoring. Int. J. Biotechnol. Wellness Ind. 3 (1), 19–26.
Handy, R.D., Galloway, T.S., Depledge, M.H., 2003. A proposal for the use of biomarkers for the assessment of chronic pollution and in regulatory toxicology. Ecotoxicology 12 (1), 331–343.
Hartikainen, K., Kivimäenpää, M., Nerg, A.-M., Mäenpää, M., Oksanen, E., Rousi, M., Holopainen, T., 2020. Elevated temperature and ozone modify structural characteristics of silver birch (*Betula pendula*) leaves. Tree Physiol. 40, 467–483. doi:10.1093/treephys/tpz127.
Hayes, F., Sharps, K., Harmens, H., Roberts, I., Mills, G., 2020. Tropospheric ozone pollution reduces the yield of African crops. J. Agron. Crop Sci. 206 (2), 214–228.
Heagle, A.S., Miller, J.E., Sherrill, D.E., 1994 A White Clover System to Estimate Effects of Tropospheric Ozone on Plants23. American Society of Agronomy, Crop Science Society of America, and Soil Science Society of America, pp. 613–621.
Heath, R.L., 1994. Possible mechanisms for the inhibition of photosynthesis by ozone. Photosynth. Res. 39 (3), 439–451.
Käffer, M.I., Domingos, M., Lieske, I., Vargas, V.M., 2019. Predicting ozone levels from climatic parameters and leaf traits of Bel-W3 tobacco variety. Environ. Pollut. 248, 471–477.
Kamble, P., Bodhane, P.S., Beig, G., Awale, M., Mukkannawar, U., Mane, A.V., Mujumdar, M., Kuchekar, S.R., Patil, V.N., 2021. Impact of transport sector emissions on biochemical characteristics of plants and mitigation strategy in Pune, India. Environ. Chall. 4, 100081.
Karmakar, D., Deb, K., Padhy, P.K., 2021. Ecophysiological responses of tree species due to air pollution for biomonitoring of environmental health in urban area. Urban Climate 35, 100741.

Kaviraj, A., Unlu, E., Gupta, A., El Nemr, A., 2014. Biomarkers of Environmental Pollutants. Biomed Res. Int. 2014, 1–22.
Kovalchuk, I., Kovalchuk, O., Hohn, B., 2001. Biomonitoring the genotoxicity of environmental factors with transgenic plants. Trends Plant Sci. 6 (7), 306–310.
Krupa, S.V., Kickert, R.N., 1993. The greenhouse effect: the impacts of carbon dioxide ($CO_2$), ultraviolet-B (UV-B) radiation and ozone ($O_3$) on vegetation (crops)$CO_2$ and Biosphere. Springer, Dordrecht, pp. 223–238.
Lam, P.K., Gray, J.S., 2003. The use of biomarkers in environmental monitoring programmes. Mar. Pollut. Bull. 46 (2), 182–186.
Lelieveld, J., Klingmüller, K., Pozzer, A., Burnett, R.T., Haines, A., Ramanathan, V., 2019. Effects of fossil fuel and total anthropogenic emission removal on public health and climate. Proc. Natl. Acad. Sci. 116 (15), 7192–7197.
Leung, F., Pang, J., Tai, A.P., Lam, T., Tao, D.K., Sharps, K., 2020. Evidence of ozone-induced visible foliar injury in Hong Kong using *Phaseolus vulgaris* as a bioindicator. Atmosphere 11 (3), 266.
Lionetto, M.G., Caricato, R., Giordano, M.E., 2019. Pollution biomarkers in environmental and human biomonitoring. Open Biomark. J. 9 (1), 1–9.
Long, S.P., Ainsworth, E.A., Leakey, A.D., Morgan, P.B., 2005. Global food insecurity. Treatment of major food crops with elevated carbon dioxide or ozone under large-scale fully open-air conditions suggests recent models may have overestimated future yields. Philos. Trans. R. Soc. B Biol. Sci. 360 (1463), 2011–2020.
Manning, W.J., Godzik, B., Musselman, R., 2002. Potential bioindicator plant species for ambient ozone in forested mountain areas of central Europe. Environ. Pollut. 119 (3), 283–290.
Mišík, M., Mičieta, K., Solenská, M., Mišíková, K., Pisarčíková, H., Knasmüller, S., 2007. In situ biomonitoring of the genotoxic effects of mixed industrial emissions using the Tradescantia micronucleus and pollen abortion tests with wild life plants: Demonstration of the efficacy of emission controls in an eastern European city. Environ. Pollut. 145 (2), 459–466.
Mukherjee, S., Chakraborty, A., Mondal, S., Saha, S., Haque, A., Paul, S., 2019. Assessment of common plant parameters as biomarkers of air pollution. Environ. Monit. Assess. 191 (6), 1–8.
Muthu, M., Gopal, J., Kim, D.H., Sivanesan, I., 2021. Reviewing the impact of vehicular pollution on road-side plants—future perspectives. Sustainability 13 (9), 5114.
Nali, C., Francini, A., Lorenzini, G., 2006. Biological monitoring of ozone: the twenty-year Italian experience. J. Environ. Monit. 8 (1), 25–32.
Oksanen, E., Kontunen-Soppela, S., 2021. Plants have different strategies to defend against air pollutants. Curr. Opin. Environ. Sci. Health 19, 100222.
Oztetik, E., 2015. Biomarkers of ecotoxicological oxidative stress in an urban environment: using evergreen plant in industrial areas. Ecotoxicology 24 (4), 903–914.
Paoletti, E., Grulke, N.E., 2005. Does living in elevated $CO_2$ ameliorate tree response to ozone? A review on stomatal responses. Environ. Pollut. 137 (3), 483–493.
Papazian, S., Blande, J.D., 2020. Dynamics of plant responses to combinations of air pollutants. Plant Biol. 22, 68–83.
Parmar, T.K., Rawtani, D., Agrawal, Y.K., 2016. Bioindicators: the natural indicator of environmental pollution. Front. Life Sci. 9 (2), 110–118.
Peakall, D.B., 1994. The role of biomarkers in environmental assessment (1). Introduction. Ecotoxicology 3 (3), 157–160.
Peakall, D.B., Walker, C.H., Migula, P., 2012Biomarkers: A Pragmatic Basis for Remediation of Severe Pollution in Eastern Europe54. Springer Science & Business Media.
Pereira, B.B., de Campos Júnior, E.O., Morelli, S., 2013. In situ biomonitoring of the genotoxic effects of vehicular pollution in Uberlândia, Brazil, using a Tradescantia micronucleus assay. Ecotoxicol. Environ. Saf. 87, 17–22.

Piraino, F., Aina, R., Palin, L., Prato, N., Sgorbati, S., Santagostino, A., Citterio, S., 2006. Air quality biomonitoring: Assessment of air pollution genotoxicity in the Province of Novara (North Italy) by using *Trifolium repens* L. and molecular markers. Sci. Total Environ. 372 (1), 350–359.
Rai, P.K., 2016. Impacts of particulate matter pollution on plants: Implications for environmental biomonitoring. Ecotoxicol. Environ. Saf. 129, 120–136.
Rai, R., Agrawal, M., Agrawal, S.B., 2016. Impact of heavy metals on physiological processes of plants: with special reference to photosynthetic systemPlant Responses to Xenobiotics. Springer, Singapore, pp. 127–140.
Rajput, M., Agrawal, M., 2005. Biomonitoring of air pollution in a seasonally dry tropical suburban area using wheat transplants. Environ. Monit. Assess. 101 (1), 39–53.
Ram, S.S., Majumder, S., Chaudhuri, P., Chanda, S., Santra, S.C., Chakraborty, A., Sudarshan, M., 2015. A review on air pollution monitoring and management using plants with special reference to foliar dust adsorption and physiological stress responses. Crit. Rev. Environ. Sci. Technol. 45 (23), 2489–2522.
Sadowska, A., Pluygers, E., 2002. Genotoxicity testing using higher plants: an evaluation of the tradescantia micronucleus assay. NATO Science Series Sub Series I Life and Behavioural Sciences 340, 224–239.
Saha, D.C., Padhy, P.K., 2011. Effects of stone crushing industry on *Shorea robusta* and *Madhuca indica* foliage in Lalpahari forest. Atmos. Pollut. Res. 2 (4), 463–476.
Sakaki, T., 2002. Effects of air pollutants on lipid metabolism in plantsAir Pollution and Plant Biotechnology. Springer, Tokyo, pp. 93–109.
Sanchez-Hernandez, J.C., 2006. Earthworm biomarkers in ecological risk assessment. Rev. Environ. Contam. Toxicol. 188, 85–126.
Schulz, H., Härtling, S., 2001. Biochemical parameters as biomarkers for the early recognition of environmental pollution on Scots pine trees. II. The antioxidative metabolites ascorbic acid, glutathione, α-tocopherol and the enzymes superoxide dismutase and glutathione reductase. Zeitschrift für Naturforschung C 56 (9-10), 767–780.
Shaddick, G., Thomas, M.L., Mudu, P., Ruggeri, G., Gumy, S., 2020. Half the world's population are exposed to increasing air pollution. NPJ Clim. Atmos. Sci. 3 (1), 1–5.
Steinkellner, H., Mun-Sik, K., Helma, C., Ecker, S., Ma, T.H., Horak, O., Kundi, M., Knasmüller, S., 1998. Genotoxic effects of heavy metals: comparative investigation with plant bioassays. Environ. Mol. Mutagen. 31 (2), 183–191.
Tiwari, S., Agrawal, M., Marshall, F.M., 2006. Evaluation of ambient air pollution impact on carrot plants at a sub urban site using open top chambers. Environ. Monit. Assess. 119 (1), 15–30.
Vardar, Ç., Basaran, E., Cansaran-Duman, D., Aras, S., 2014. Air-quality biomonitoring: Assessment of genotoxicity of air pollution in the Province of Kayseri (Central Anatolia) by use of the lichen *Pseudevernia furfuracea* (L.) Zopf and amplified fragment-length polymorphism Markers. Mut. Res. Genet. Toxicol. Environ. Mutagenesis 759, 43–50.
Wang, Y., Jin, W., Che, Y., Huang, D., Wang, J., Zhao, M., Sun, G., 2019. Atmospheric nitrogen dioxide improves photosynthesis in mulberry leaves via effective utilization of excess absorbed light energy. Forests 10 (4), 312.
Wolterbeek, B., 2002. Biomonitoring of trace element air pollution: principles, possibilities and perspectives. Environ. Pollut. 120 (1), 11–21.
Yadav, P., Mina, U., Bhatia, A., 2020. Screening of forty Indian Amaranthus hypochondriacus cultivars for tolerance and susceptibility to tropospheric ozone stress. Nucleus 63 (3), 281–291.
Zellnig, G., Gailhofer, M., Pfeifhofer, H.W., Grill, D., 1989. Ultrastructure and Pigment Composition of Chloroplasts of Differently Damaged Green and Yellowed Needles of Picea abies. Phyton (Austria) 29 (2), 213–225.

# CHAPTER 8

# Biomonitoring potential of tropospheric ozone in plants utilizing visible injury and biomarkers

**Aditya Abha Singh**
University Department of Botany, Babasaheb Bhimrao Ambedkar Bihar University, Muzaffarpur, Bihar, India

## 8.1 Introduction

Increases in human activities and escalating economic developments have led to the problem of air pollution which is a growing concern across the globe. Usually, the air pollutants generated from automobiles, industries, or various other sources are the primary pollutants, but they may easily get converted into secondary pollutants undergoing chemical reactions (Pal, 2014). One such secondary pollutant is tropospheric ozone ($O_3$) which is formed by the reactions between oxides of nitrogen ($NO_x$), volatile organic compounds (VOCs), methane ($CH_4$), and carbon monoxide (CO) in the presence of high temperature and bright sunshine (Jacob and Winner, 2009). Globally, the $O_3$ concentrations have increased twice the concentrations during the preindustrial levels (Monks et al., 2015) and are projected to increase further in the future. Tropospheric $O_3$ is very reactive and toxic at appropriate exposures (Felzer et al., 2007) being harmful to humans as well as plants (El-Khatib, 2003; Agathokleous et al., 2020). Ozone causes oxidative stress in plants resulting in visible injuries, substantive reductions in plant growth and biomass, disrupting photosynthetic efficiency, induce early leaf death, peroxidation of lipid membrane, and negatively affecting the yield (Morgan et al., 2003; Ashmore, 2005; Mishra et al., 2013; Singh et al., 2014; Singh et al., 2018; Leung et al., 2020) resulting in economic losses. Ozone is considered as a regional rather than an urban problem due to its intercontinental transport concerns (Wang et al., 2011). Moreover, $O_3$ forming potential is considerably higher in rural than in urban areas. The impact of $O_3$ is regardless of the political boundaries or the geographical

*New Paradigms in Environmental Biomonitoring Using Plants*
DOI: https://doi.org/10.1016/B978-0-12-824351-0.00008-0

barriers (Mahmood et al., 2020). Therefore, $O_3$ is an issue of concern, whose concentrations need to be monitored.

Plants are known to be good responders for particular air pollutants. The technique of air pollutant monitoring through plants is known as biomonitoring (Gupta and Kulshrestha, 2016). Historically, the recurrent incidence of foliar injury symptoms on native and cultivated plant species in Los Angeles observed in the late 40s of the 20th century was the landmark incident that prompted the study of pollutant on plants, and for years, certain plant species have been used as bioindicators (Manning, 1998). Biomonitoring on plant species is a simple and efficient tool to identify bioindicator species (Pal, 2014). Plant biomonitoring includes the changes occurring in morphology, biochemistry, and physiology of plants as the indicators of pollution impact. The morphological changes observed include visible injury symptoms like chlorosis, necrosis, etc. (Tiwari and Agrawal, 2018). Certain plant species manifest typical foliar injury symptoms when exposed to $O_3$, making them useful as bioindicators. There are however plants which have nonvisible effects, but they respond to the presence of air pollutant and are known as biomarkers. Biomarkers offer direct evidence of $O_3$ stress or potential stress. Ozone may often cause functional alterations in plants without the onset of visible foliar symptoms (Gottardini et al., 2014). They have effects at physiological, biochemical, cellular, and molecular levels, which needs to be analyzed by physiological or microscopic techniques, as well as biochemical investigations (Gupta and Kulshrestha, 2016). Various genomics, proteomics, and metabolomics tools and techniques are available for biomarker discovery.

Bioindicator plants offer a simple and effective way to assess the level of air pollution when automatic measurements are not available (Klumpp et al., 2006a) and $O_3$ biomonitoring has been introduced in many parts of the world. Biomonitoring gives comparative information about the $O_3$ prevalence at different sites so that the ambient $O_3$ impact is measured directly (Tiwari and Agrawal, 2018). Hence, they can be used to monitor the spatial and temporal distribution of $O_3$ pollution and different scales are possible which can be local, regional, and international (Temmerman et al., 2001). However, damage rates of $O_3$ may vary with time and space depending on the local environmental and meteorological conditions, therefore biomonitoring responses may be influenced by the prevailing meteorological conditions. In this chapter, $O_3$ formation in the troposphere, its worldwide concentrations and futuristic predictions have been addressed. An overview of biomonitoring of $O_3$ pollution using various bioindicators and biomonitor species is addressed. Moreover, the different types of visible $O_3$ injury in plants are summarized. The molecular and cellular markers which are used to detect the $O_3$ stress in plants have also been emphasized. Apart from this, the climatic impact on $O_3$ biomonitoring is also addressed.

## 8.2 Formation of $O_3$ in troposphere

Tropospheric $O_3$ is a phytotoxic air pollutant distributed worldwide (Cho et al., 2011; Singh et al., 2015) and climatic conditions play a key role in $O_3$ production and destruction (Royal Society, 2008). Ozone is a secondary air pollutant formed by the chemical reactions between nitrogen oxides ($NO_x - NO_2 + NO$), methane and nonmethane VOCs (Kleanthous et al., 2014), and carbon monoxide (CO), which are known commonly as precursors of $O_3$. The past few decades have seen industrial development and urbanization which has increased fossil fuel burning increasing the emission of VOCs and $NO_x$ which are the two most significant $O_3$ precursors. Nonetheless, due to the complex photochemistry involved besides the $O_3$ precursor's abundance, the formation of $O_3$ also depends on meteorological conditions like solar radiation, variations in air temperature, wind speed, wind direction, and other climatic influences (Ahammed et al., 2006; Tiwari et al., 2008; Coates et al., 2016; Monks et al., 2015).

Hydrocarbons such as methane or nonmethane compounds in a complex set of reactions react with oxygen molecules to give formaldehyde (HCHO)/RCHO, water vapor, and $O_3$ (Fig. 8.1). Another pathway for $O_3$ formation involves the CO oxidation that results from the presence of hydroxyl radical ($^{\bullet}OH$). The hydrogen atom thus formed rapidly reacts with oxygen to give a peroxy radical ($^{\bullet}HO_2$). Peroxy radicals then combine with NO to give $NO_2$, which photo dissociates to produce atomic oxygen. Furthermore, the atomic oxygen reacts with the molecular oxygen to form $O_3$. Nevertheless, the shelf life of $O_3$ is very short under these conditions; therefore, $O_3$ again reacts with NO to produce $NO_2$ and $O_2$. The entire process, therefore, represents a chain reaction, in which $O_3$ becomes photo dissociated by ultraviolet radiation to form atomic oxygen. The excited oxygen atom then again reacts with water vapor to regenerate $^{\bullet}OH$ radicals, which drives the chain reaction process. The destruction of $O_3$ also results from the photochemical reactions involving NO, $HO_2$, or $^{\bullet}OH$. A steadiness is, therefore, established between the formation and degradation of $O_3$ (Lorenzini and Saitanis, 2003). The source of NO involved in the above reactions comes from the emission of nitrogen dioxide ($NO_2$) by the reaction

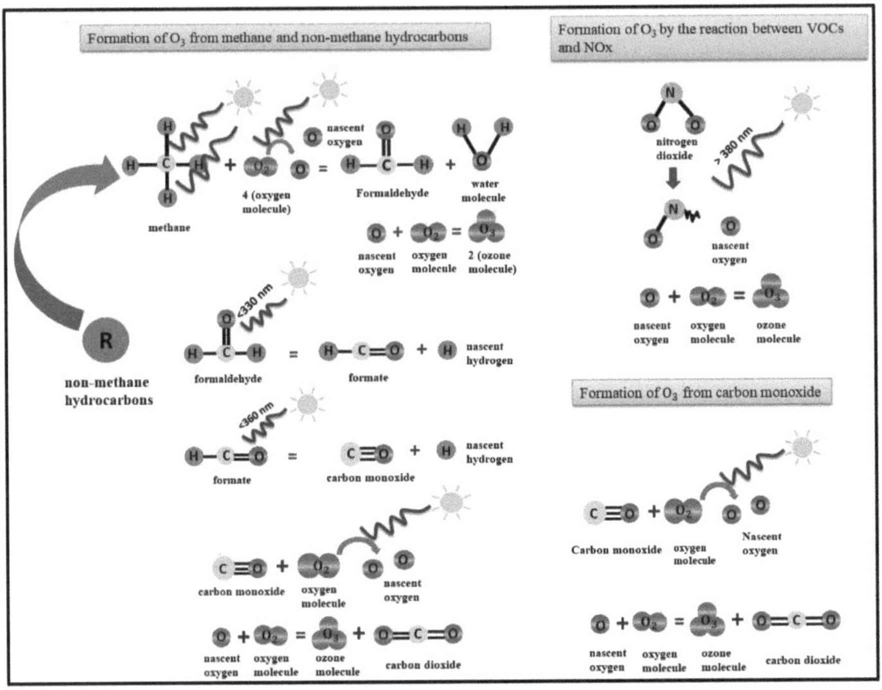


Fig. 8.1 ***Chemical reactions involved in the formation of secondary pollutant $O_3$ in troposphere from its various precursors in presence of sunlight.***

between $N_2$ and $O_2$ present in the atmosphere under high-temperature conditions found in the ignition chambers of the automobile engine. Nitrogen dioxide thus produced is easily broken down by the ultraviolet component of the solar radiation into nitric oxide (NO) and nascent oxygen (O) and the nascent oxygen combines with molecular oxygen ($O_2$) to form $O_3$ (Fig. 8.1). Staehelin and Poberaj (2008) have stated that the $NO_x$ species restrict $O_3$ formation by having their concentration gradually decreased by the formation of $HNO_3$. Carbon monoxide directly reacts with $O_2$ to give $CO_2$ and nascent $O_3$. This nascent $O_3$ then combines with $O_2$ to produce $O_3$. The complex reactions involved in the formation of tropospheric $O_3$ are elaborated in Fig. 8.1. These reactions occur incessantly in polluted air which leads to a prevalence of $O_3$ at distant sites downwind from the initial polluting source which is usually the urban areas. Urban areas are the chief sources of $O_3$ precursors, and increased regional transportation causes an efficient dispersion which leads to an increased $O_3$ concentration in rural areas (Agrawal et al., 2003; Paoletti et al., 2014). Ozone is also generated from the natural phenomenon in low concentrations ranging between 10 and 40 ppb chiefly produced by the high-intensity sunlight driven reactions involving the hydrocarbons like terpenes and isoprenes emitted by plants and nitrogen dioxide usually generated by lighting or may be emitted by the bacteria present in soil (Mauzerall and Wang, 2001). Another natural phenomenon contributing to $O_3$ presence in the troposphere is downward movement from the stratosphere to the lower atmosphere level (Andersen et al., 2013).

## 8.3 Concentrations and trends of worldwide ozone and future projections

Tropospheric $O_3$ concentrations were approximately 10 ppb during the preindustrial times (Van Dingenen et al., 2009) but, in recent times, its concentration has shown a steady increase. Its volume fraction has been increased approximately twice to more than four times with respect to the preindustrial age and continues to increase (Vingarzan, 2004). Assessment Report Five (AR5) of IPCC indicated that if the existing trend continues, $O_3$ probably would rise to 40%–60% by the end of this century (IPCC, 2014). Furthermore, increased demand for energy production, fossil fuels, and transport are expected to further increase the precursor of $O_3$ emissions in near future (Winiwarter and Klimont, 2011). Moreover, long-range transport also contributes significantly to the $O_3$ load, since $O_3$ precursor molecules travel long distances from their site of origin, apart from the contributions from the nearby sources (Paoletti et al., 2014). Ozone is relatively greater in the Northern Hemisphere as compared to the Southern Hemisphere due to considerable enhancement in industrialization and urbanization (Rai et al., 2016). Ozone concentrations of the Northern Hemisphere are high enough in the present times and are predicted to remain elevated over the coming decades. Due to the execution of $O_3$ precursor emission abatement policies, a declining trend in peak $O_3$ concentrations in Japan, the USA, and Europe have been observed, while in Asian countries $O_3$ concentrations have increased and are still at the higher risk due to rising $O_3$ concentrations (Naja and Akimoto, 2004; Royal Society, 2008; Ghosh et al., 2018). The USEPA has reported that reductions in the emission of $O_3$ precursors have been substantial over the past 29 years (USEPA, 2009). However, despite of the reductions in the emissions of anthropogenic $O_3$ precursors, an escalation in background $O_3$ in the lower troposphere of Europe is reported (Chevalier et al., 2007). In South Asia, a steep enhancement in the future scenario is forecast based on the postulation of global and emission pathways (EEA, 2016).

Phytotoxic levels of tropospheric $O_3$ occur in the Mid-Atlantic and North-Eastern areas of United States (Coulston et al., 2003). In agriculture regions of the USA, average $O_3$ concentration was reported between 50 and 60 ppb (USEPA, 2006). A study presented the influence of meteorological conditions and precursor $NO_2$ level on tropospheric $O_3$ concentration over the region of Eastern Texas, the USA. The 8-h average $O_3$ concentration in the area ranged from 32.9 ppb to 59.5 ppb for the duration May 1, 2012 to May 7, 2012 with maximum concentrations in the counties of Denton, Grayson, Harris, Harrison, Liberty, and Tarrant (Gorai et al., 2015). Surface-based measurements from the European Monitoring Evaluation Programme and Airbase networks were used to evaluate the changes in surface $O_3$ levels during 1995–2014 over Europe. Significant enhancement in $O_3$ (0.20–0.59 $\mu gm^{-3}$ $yr^{-1}$) for the annual means over the European suburban and urban stations were observed during the period 1995–2012 based on the Airbase sites (Yan et al., 2018). All the countries of Europe are encountering episodic $O_3$ periodically every year with several days of peak $O_3$ concentrations exceeding 50 ppb (Hayes et al., 2007). In northern Fennoscandia, $O_3$ concentrations generally range between 30 and 40 ppb with acute, short-term exposures up to 50–80 ppb (Manninen et al., 2009). Ground-level $O_3$ is an increasing environmental concern in Italy in the entire Mediterranean basin (Nali et al., 2006). Out of the 214 monitored stations, an increase of 0.64% $year^{-1}$ was recorded in urban and 0.46% $year^{-1}$ in suburban stations while 58% of the rural stations revealed an average reduction of 0.43% $year^{-1}$ in concentrations of $O_3$ at Mediterranean region for the period 2000–2010 (Sicard et al., 2013). The highest 1-hour $O_3$ concentration of 240 ppb was recorded in France (Pellegrini et al., 2011). The hourly $O_3$ data from 332 background monitoring stations in France were examined over the period 1999–2012. It was found that the average annual concentrations reduced by −0.12 ppb $year^{-1}$ at rural stations while the background $O_3$ level was rising at 76% of the urban sites (+0.14 ppb $year^{-1}$), chiefly during the cold period (Sicard et al., 2016). Hourly $O_3$ concentrations frequently exceeded 70 ppb, while the average concentration of $O_3$ during early afternoon hours (12:00–19:00) for the months of June, July, and August was 65.4 ppb at Tripolis plateau located in Greece (Saitanis et al., 2015). An average enhancement of 0.14 ppb $year^{-1}$ in the concentration of $O_3$ was found in 13 rural sites in United Kingdom from 1990 to 2006 (Jenkin, 2008).

Enormous parts of South Asia face up to 50–90 ppb mean 7 h $O_3$ concentration (Emberson et al., 2009). India being a developing economy faces a high level of pollution including the tropospheric $O_3$. Due to anthropogenic activities, chemical transport models have assessed a considerable increase in $O_3$ production over India and prominently in Indo-Gangetic Plain (IGP) (Ojha et al., 2012). North Indian states like Uttar Pradesh, Bihar, Uttaranchal, Rajasthan, Haryana, and Punjab encounter high levels of $O_3$ concentrations during summer and winter (Ohara et al., 2007). Mean 7-hour $O_3$ concentration of 53.0 ppb was recorded from October 2011 to January 2014 at Mohali (Sinha et al., 2015). Ozone monitoring at various urban sites of Delhi during different times shows a tendency of increasing $O_3$ concentrations. Average $O_3$ concentration for the consecutive years for the duration October to March 2009–2010 and 2010–2011 was 28 and 33 ppb, respectively (Singh et al., 2013). Ozone concentrations were recorded constantly at three sites in Delhi-NCR being urban background, urban/traffic and rural during the consecutive years 2013–2014. The diurnal variation of $O_3$ showed the day-time maxima and night-time minima having maximum day time concentrations of 50.2, 46.1, and 56.7 ppb at urban background, urban/traffic, and rural sites, respectively (Kumar et al., 2015). IGP is among the most polluted regions of the world in terms of gases and aerosol loading contributing to $O_3$ formation (Ghude et al., 2011; Singh and Agrawal, 2017). The highest trend of increasing $O_3$ concentration per decade has been observed over the IGP region (Lal et al., 2012). In Varanasi, high $O_3$ concentrations prevalence is there in the range of 46.3–67.9 ppb during December to April of 2011–2012 (Singh et al., 2014). Yadav et al. (2020) observed the mean ambient $O_3$ concentration for the period December 2015 to March 2016 being 51.7 ppb. The average 8-hour $O_3$ concentration was 46.6 $\pm$ 0.7 ppb from November 2018 to March 2019 (Ghosh et al., 2020). Daytime 8 h $d^{-1}$ $O_3$ concentrations ranged from 7 to 83 ppb at Lucknow for the monitoring duration August 5 to November 22, 2011 (Pandey et al., 2015). From southern India, Reddy et al. (2012) in Anantapur have found maximum $O_3$ concentrations in summer (70.2 ppb) in the year 2010. The annual average $O_3$ diurnal profile showed a peak of (30.3 $\pm$ 10.4) ppbv in the late afternoon and the maximum value of $O_3$ mixing ratio was observed in winter months being 44 $\pm$ 3.1 ppbv at Kannur, a tropical site located in the coastal region of India for the monitoring period November 2009 to October 2010 (Nishanth et al., 2012). Another site located in the southern coastal Indian region Nagercoil displayed an average annual $O_3$ concentration of 19.8 ppb (Elampari et al., 2013). Based on modeling studies, $O_3$ monitoring throughout the year showed an increasing pattern during summer and a decreasing trend in the winter season at National Environmental Engineering Research Institute Nagpur in Maharashtra (Kumar et al., 2017). Apart from India, fast urbanization and industrialization rates across China in the last two decades have resulted in a remarkable increase in $O_3$ concentrations due to which tropospheric $O_3$ has become one of the major air pollutants in China (Chen et al., 2020). As compared to the 1990s, $O_3$ concentration in China has increased by approximately 13% by the year 2013 (Brauer et al., 2016). Hong Kong is one of the most heavily populated cities in the world, with a problem of severe air pollution together with rising $O_3$ concentrations (Leung et al., 2020). Monitoring results of Wang et al. (2009) for the period 1994-2007 revealed that $O_3$ concentrations have increased by 0.87 ppb year$^{-1}$ by comparing the average concentrations during 1994–2000 and 2001–2007 at a coastal site in Hong Kong.

An increase of 30 ppb in $O_3$ concentration was reported in South America and Africa in the Southern Hemisphere (Zeng et al., 2008). An assessment of detailed continuous ground-level $O_3$ quantification was carried out at four sites in continental South Africa which showed that the $O_3$ concentrations frequently exceeded the South African air quality standard limit (Laban et al., 2018). Tropospheric $O_3$ has been attaining high concentrations in the city of São Paulo the greatest metropolitan area of Brazil (Orlando et al., 2010). In the city of Rio de Janeiro, Brazil between 2012 and 2013 based on measurements of air quality monitoring stations in the Bangu, Campo Grande and Irajá neighborhoods displayed that throughout the year concentrations of $O_3$ were higher during the weekends than during the weekdays. The Bangu station showed the lowest $O_3$ concentration differences between the weekdays and weekends and recorded the highest concentrations of $O_3$ and the highest number of $O_3$ exceedances (Martins et al., 2015). Urban air pollution levels were characterized including the $O_3$ measurements in Abidjan, the economic capital of Côte d'Ivoire in Africa during the dry season ranging 15 December 2015 to 16 February 2016, selecting 21 different sites. The highest $O_3$ concentration was recorded at the two coastal sites of Gonzagueville and Félix-Houphouët-Boigny International Airport located in southeast of the city, with mean concentrations $19.1 \pm 1.7$ and $18.8 \pm 3.0$ ppb, respectively (Bahino et al., 2018).

Projections indicate that $O_3$ concentration will reach considerably to a high level by 2100 (Mills et al., 2013). Mean surface $O_3$ concentration will increase by approximately 23 % till 2050 as anticipated by Morgan et al. (2006). By the end of 21st century $O_3$ will be a global problem as predicted by modeling analysis and that average $O_3$ concentration in the troposphere may rise by 20%–25% globally by the mid of this century (Van Dingenen et al., 2009). Intergovernmental Panel on Climate Change (IPCC) Assessment Report Five showed that if the present trend of emission continues then $O_3$ may probably rise by 20%–25% between 2015 and 2050 and increase further by 40%–60% by the end of this century (IPCC, 2014). A multimodal study carried out in Europe predicted an increase in the mean $O_3$ concentration of 0.9–3.6 ppb for 2040s in comparison to the concentrations that prevailed during the first decade of 21st century (Langner et al., 2012). Projections show that the incidence of years with higher values of cumulative $O_3$ could increase in Europe, chiefly in the northwestern regions (Fuhrer, 2009). Moreover, a 40% increase in $O_3$ concentration is projected in the South Asian region by 2050 (Tai et al., 2014). The effects of anticipated global changes in human-linked emissions and climate for the year 2050 relative to the year 2000 for trends in the risk of hazardous $O_3$ pollution episodes utilizing a global climate chemistry model, CAM-Chem were investigated. Moreover, specific pathways from the IPCC Special Report on Emissions Scenarios A1B, A1FI, and B1 were also considered. The peak $O_3$ concentration surpasses the limit in summer of 2050, based on 8 and 1-h standards, display that the threat of $O_3$ pollution episodes will escalate probably in the developing regions, whereas for North America, Europe, and East Asia under 8-h definition, show a hazardous risk under the A1FI scenario (Lei et al., 2012).

## 8.4 Biomonitoring of ozone utilizing different bioindicators species at various geographical locations

Biological monitoring or biomonitoring of $O_3$ is a process of $O_3$ monitoring by detecting quantitative changes in living organisms existing in a specific environment physically (Agathokleous et al., 2020). Often, very low concentrations of air pollutants are difficult to quantify with physical and chemical means, but plants can accumulate these pollutants to an amount that can be analyzed easily (Gupta and Kulshreshtha, 2016). Biomonitoring of air pollutants using vegetation has been a common practice for many decades (Temmerman et al., 2001).

Analytical observations of the native as well as cultivated species present in a given area can be utilized for biomonitoring studies. Biomonitoring conveys important evidence about the effects of pollution since these effects are manifested as visible injuries in sensitive plants, while tolerant plants accumulate pollutants as well, and hence can be used as bioindicators

(Gupta and Kulshreshtha, 2016). Usually, the biomonitoring methods are passive and active ones. In passive monitoring, plant growth is seen as a natural process of development and passive bioindicators are plants, already present at the study site (Falla et al., 2000). Moreover, the native species do not require any specific care and maintenance (Falla et al., 2000). It has been found that ambient $O_3$ can be estimated by passive samplers and clover biomonitoring mini stations (Manes et al., 2003). The detector plants *in situ* may display characteristic foliar injury symptoms to $O_3$ exposure. Nevertheless, the efficiency of native species as the indicator plant considering their antioxidant defense system responses is a better alternative (Nakazato et al., 2018). While in the active method, the bioindicator plants are imported on the experimental site (Falla et al., 2000) and the presence of pollutant is determined by the experimental plant for its response which is assessed indirectly by recording biochemical and physiological variations in the plant (Gupta and Kulshrestha, 2016). While many nonindigenous plant species are used too, consisting of various clones or cultivars, which manifest symptomatic foliar damage being reliable when exposed to ambient concentrations of $O_3$.

Ozone sensitivity of plants is variable among the species, and variances also exist amongst the cultivars or genotypes of the same species (Biswas et al., 2008; Singh et al., 2018). Injuries in form of visible symptoms are a typical indicator to evaluate the sensitivity of a plant towards $O_3$ (Biswas et al., 2008). Ozone uptake results in visible foliar symptoms in sensitive plants (Alves et al., 2016). Several plant species have been characterized for their $O_3$ injury symptoms and are utilized as bioindicators consequently indicating the $O_3$ concentrations according to the resulting injury. For long, the *Nicotiana tabacum* (tobacco) has been used worldwide as an $O_3$ indicator plant (Cheng and Sun, 2013). A sensitive tobacco cultivar Bel-W3 has been employed in several biomonitoring studies which indicate the presence of phytotoxic $O_3$ concentrations since the damage on the leaf surface can be correlated with the $O_3$ pollution dosage (Falla et al., 2000; Klumpp et al., 2006b). The characteristic necrosis detected on tobacco leaf is represented by small-sized black spots (Falla et al., 2000). Tobacco plant (cv. Bel-W3) was used to estimate the spatial distribution of $O_3$ in France (Gombert et al., 2006) (Table 8.1). Comparative studies on the two cultivars of tobacco Bel-W3 (sensitive cultivar) and Bel-B (resistant cultivar) to measure the air quality have also been carried out (Table 8.1). In Italy, the importance of hypersensitive Bel-W3 tobacco plants was highlighted, as well as the possibility of tobacco germlings instead of adult plants for

**Table 8.1** Biomonitoring programs carried out at different locations utilizing the bioindicator species *Nicotiana tabacum* L. and *Trifolium repens* L.

| S. No | Plant/cultivar/clone | Location | Ozone condition/ concentration | Plant response | References |
|---|---|---|---|---|---|
| *Nicotiana* | | | | | |
| 1. | *Nicotiana tabacum* L. cultivar Bel-W3 ($O_3$- sensitive) and Bel-B ($O_3$- resistant) | Different locations in Poland | Ambient air | Extensive $O_3$ injury was manifested on Bel-W3 tobacco plants with highest degree of injury, reaching 60% of the foliar area in many locations in southern and central Poland, No injury on Bel-B at any location | Bytnerowicz et al. (1993) |
| 2. | *Nicotiana tabacum* L. cultivar Bel-W3 ($O_3$-sensitive) and cv. Bel-B ($O_3$-tolerant) | Montague-Amherst and Mount Equinox sites in the north-east USA | Open-top chamber (OTC) and open plot having ambient $O_3$ | Bel-W3 was significantly more sensitive than Bel-B at both the sites. | Krupa et al. (1993) |
| 3. | *Nicotiana tabacum* L. cultivar Bel-W3 and Bel-B | Italy (Pisa, Mezzana, San Piero, Livorno), Capraia and Gorgona islands west of the Tuscan coast (central Italy) | Ambient air | Two week old Bel-W3 seedlings showed $O_3$ injury symptom on cotyledons varying between 2.2-3.7 on a sale of ozone injury (1-5) Bel-B cotyledons showed minimal lesions | Lorenzini et al. (1995) |
| 4. | *Nicotiana tabacum* L. cultivar Bel-W3, Bel-C and Bel-B | Catalonia (northeast Spain) | OTC [charcoal filtered air (CFA), non-filtered (NF) and non-filtered+40 ppb $O_3$ (NF+)] | No visible injury in CFA Foliar injury followed the pattern: Bel-W3 > Bel-C > Bel-B in NF and NF+ treatments | Ribas et al. (1998) |

*(Continued)*

**Table 8.1** Cont'd

| S. No | Plant/cultivar/clone | Location | Ozone condition/ concentration | Plant response | References |
|---|---|---|---|---|---|
| 5. | *Nicotiana tabacum* L. cultivar Bel-W3 | Catalonia (northeast Spain) | Ambient air | Foliar injury symptoms observed at each of the nine experimental site | Ribas and Penuelas, (2003) |
| 6. | *Nicotiana tabacum* L. cultivar Bel-W3 and Bel-B | Grenoble area (southeast France) | Ambient air | Bel-W3 foliar injury varied from 0 to 30% depending on the $O_3$ exposure period<br>Bel B plants did not show any visible $O_3$ injury | Gombert et al. (2006) |
| 7. | *Nicotiana tabacum* L. cv. Bel-W3 | Seven sites in Valencia (Spain) | Ambient air | The urban stations showed lower injury than the less traffic–exposed stations.<br>Leaf injury was significantly related to 24 and 12 h mean $O_3$ values | Calatayud et al. (2007) |
| 8. | *Nicotiana tabacum* cultivars Bel-W3 and Bel-B | Two sites (Ponzan city center and a suburb of Ponzan) | Ambient air | Visible foliar damage occurred as bifacial necrosis and was found only in Bel-W3<br>No foliar symptom on Bel-B | Bandurska et al., (2009) |
| 9. | *Nicotiana tabacum* cultivar Bel-W3 | São Paulo (Brazil) | Four $O_3$ exposure periods ranging between 21.1 and 60.5 | Leaf injury as necrosis appeared in almost all of the oldest leaves | Esposito et al. (2009) |
| 10. | *Nicotiana tabacum* L. cv. Bel-W3 and Bel-B | Maurienne valley in Southeastern France | 24 h average ranging between 25 and 44 ppb at the three reference site | Necroses observed on all Bel-W3 plants and symptoms intensity varied with the location of the plants<br>No necrosis observed on Bel-B tobacco plant | Laffray et al. (2007) |

| | | | | | |
|---|---|---|---|---|---|
| 11. | *Nicotiana tabaccum* L. cultivar Bel-W3 | City centre, suburbs, semi-rural/rural sites in and around Lahore city, Pakistan | Ambient air | Lowest leaf injury indices of 5-7% occurred at city centre sites while highest of 18-27% at the rural sites | Kafiatullah et al. (2012) |
| 12. | *Nicotiana tabacum* L. cultivar Bel-W3 and Bel-B | Poland | Ambient air | Injury occurred only on cultivar Bel-W3 | Borowiak et al. (2013) |
| *Trifolium* | | | | | |
| 1. | *Trifolium repens* L. clones NC-S ($O_3$-sensitive) and NC-R ($O_3$-resistant) | USA | OTCs (0.5 to 1.5 times the ambient $O_3$ concentrations) | Ozone routinely injured the leaves and reduced the growth of NC-S more than that of NC-R | Heagle et al. (1994) |
| 2. | *Trifolium repens* L. clones NC-S and NC-R | Southern Italy | Ambient air | Ozone injury was observed on NC-S but not on the NC-R in both years of study (1997 and 1998) | Postiglione et al. (2000) |
| 3. | *Trifolium repens* L. cultivar Regal clones NC-R and NC-S | Two sites in Po plain of Italy | Ambient air | Biomass reduction of up to 60% in NC-S clone relative to NC-R | Fumagalli et al. (2003) |
| 4. | *Trifolium repens* L. cultivar Regal clones NC-R and NC-S | Rome municipal area, Italy | Biological mini stations and passive samplers | NC-S had higher visible injury score than the NC-S and the clover clones responses were positively correlated to the $O_3$ concentrations | Manes et al. (2003) |

*(Continued)*

**Table 8.1** Cont'd

| S. No | Plant/cultivar/clone | Location | Ozone condition/ concentration | Plant response | References |
|---|---|---|---|---|---|
| 5. | *Trifolium repens* L. clones NC-S and NC-R | Coordination Centre of the ICP Vegetation at the Centre for Ecology and Hydrology, Bangor (United Kingdom) | Ozone treatment (60 ppb for 56 days, 5 h $day^{-1}$) | NC-S exhibited foliar injury and the mean injured leaves surpassed 50% of the samples<br>Ozone-induced lesions were rare on the resistant clone at the end of the $O_3$ exposure | Nali et al. (2005) |
| 6. | *Trifolium repens* L. cultivar Regal | Province of Novara (North Italy) | Ambient air | Leaf necrotic spots detected | Piraino et al. (2006) |
| 7. | *Trifolium repens* L. cultivar Regal clones NC-S and NC-R | Three sites in Hungry (Gödöllô, Bugacpuszta and Mátraháza) | Open field; cumulative AOT 40 for the season up to 12,000 ppb h | Typical $O_3$ injury symptoms found on NC-S on every site but the extent of $O_3$ injury symptom was not proportionate to the $O_3$ concentration | Villányi (2008) |
| 8. | *Trifolium repens* cultivar Regal clones NC-R and NC-S | Four sites in the Metropolitan Region of Campinas, Brazil | Ambient air | No significant linear regression between the NC-S/NC-R biomass ratio and $O_3$ concentration was found in two years of study (2012–2013) | Sandrin et al. (2018) |

biomonitoring systems were suggested (Nali et al., 2006). Furthermore, the factors affecting the $O_3$ sensitivity of tobacco Bel-W3 cultivar have been investigated by Cheng and Sun (2013). Ozone monitoring was carried out with *Nicotiana tabacum* cultivar Bel-W3 at 23 sites in Northern Italy from April 27, 1987 to October 27, 1987 exposed to ambient air, and it was found that $O_3$ induced typical leaf necrosis was observed at every site in tobacco plants (Mignanego et al., 1992). A biomonitoring approach carried out in 12 urban agglomerations in 8 European countries employed the ozone-sensitive bioindicator plant *Nicotiana tabacum* cultivar Bel-W3 to assess the occurrence of phytotoxic $O_3$ effects at urban, suburban, rural, and traffic-exposed sites from 2000 to 2002. Foliar injuries caused due to $O_3$ clearly showed an increasing gradient from Northern and North-Western Europe to Central and Southern European sites. The obtained data displayed a significant relationship between the degree of foliar injury and different parameters of $O_3$ pollution such as mean AOT20 and AOT40 values (Klumpp et al., 2006b). *Corchorus olitorius* cultivar local, *Trifolium alexandrinum* L. cultivar Masry, *Eruca sativa* cultivar local, and *Medicago sativa* L. cultivar local, ranked in order of decreasing sensitivity and manifested characteristic $O_3$ injury symptoms earlier and at lower concentrations of $O_3$ than Bel-W3 bioindicator in an extensive biomonitoring screening at Egypt (Madkour and Laurence, 2002). A compilation of investigations which have been carried out using *Nicotiana* and *Trifolium* as bioindicators have been summarized in Table 8.1.

*Trifolium repens* L. (white clover) has proved to be an efficient bioindicator species. *Trifolium repens* L. cultivar "Regal" was tested in the USA (North Carolina State University, Raleigh), comprising of two specific NC-S ($O_3$-sensitive) and NC-R ($O_3$-resistant) clones that differ in their biomass responses to ambient $O_3$. It is being broadly used in the USA and Europe to determine the $O_3$ concentrations (Heagle et al., 1994; Harmens et al., 2009. http://icpvegetation.ceh.ac.uk). The clones exhibit measurable differences in growth response to $O_3$, however, rapid response, may occur only in the early stages of their growth, needing a frequent reintroduction of new plants at the monitoring site (Manning, 1998). The NC-S/NC-R *Trifolium repens* biotypes were analyzed and the gas exchange properties of photosynthesis indicated that NC-S has a higher value of stomatal conductance (Nali et al., 2009). Apart from *Trifolium repens*, *Trifolium pratense* and *Trifolium alexandrinum* clover species have also been tested for their $O_3$ responses (Saviranta et al., 2010; Chaudhary and Agrawal, 2013). Ozone sensitivity of *Trifolium* species has been ranked as *Trifolium subterraneum*

> *Trifolium repens* > *Trifolium pratense* in studies having $O_3$ concentrations varying between approximately 30–70 ppb (Karlsson et al., 1995; Futsaether et al., 2009). Investigations have been carried out with *Phaseolus vulgaris* cultivars also (Burkey et al., 2005; ICP Vegetation, 2010; Chaudhary et al., 2013; Agathokleous et al., 2017; Leung et al., 2020). The sensitivity of beans of the genus *Phaseolus* to $O_3$ is well known and certainly ozone-sensitive (S156) and ozone-resistant (R123) *Phaseolus vulgaris* (Bush bean, French dwarf bean) genotypes have been used in bioindicator studies as those carried out in the framework of The International Cooperative Programme on Effects of Air Pollution on Natural Vegetation and Crops, ICP-Vegetation (Harmens et al., 2011). In Hong Kong, bioindicator species *Phaseolus vulgaris* (bush bean) with various degrees of $O_3$ sensitivity displayed that leaf shows a graduated scale of damage pattern that can be advantageous for indicating $O_3$ levels (Leung et al., 2020).

Other bioindicators are also used like *Centaurea jacea* (brown knapweed) as a model plant to evaluate the relationship between $O_3$ amount and resulting effects on the injury symptoms and the performance of plants which are used in the framework of European programs (Nali et al., 2006). Studies have found that *Coleus* has the potential to be used as an $O_3$ bioindicator plant (Padri and Umponstira, 2019). Target species for broader biomonitoring service in temperate China have been identified; and the best bioindicator tree species suggested for the Beijing area is *Ailanthus altissima* since it is a common species throughout the area and $O_3$ injury symptoms can be characterized very well (Feng et al., 2014). Comparative sensitivity of five Egyptian plants namely, *Malva parviflora*, *Medicago sativa*, *Melilotus indicus*, *Senecio vulgaris*, and *Sonchus oleraceus*, to elevated levels of $O_3$ (50 and 100 ppb for 5 h per day) were tested. As per the $O_3$ resistance (R%), quantified as relative growth rate, the test species were arranged in the descending order as *M. parviflora*>*M. indicus*>*S. vulgaris*>*S. oleraceus*>*M. sativa* (El-Khatib, 2003). In an experiment with *N. tabacum*, *P. guajava*, and *Ipomoea nil* in Brazil, authors concluded that the most suitable and least suitable bioindicator plant in subtropics for $O_3$ monitoring is *N. tabacum* and *P. guajava*, respectively (Alves et al., 2011). In a biomonitoring program (1994–2009) covering the northeast and north central sub regions including 450 field sites in 24 states of US, native species including the herbaceous plants like *Rubus allegheniensis*, *Asclepias spp*, *Apocynum androsaemifolium*, and trees *Prunus serotina* and *Fraxinus Americana* were sampled. Over the 16-year sampling period, injury indices altered annually with regard to the seasonal $O_3$ concentrations measured as SUM06 (cumulative ozone exposure) and

N100 (peak ozone concentrations) and moisture conditions prevailing at the site (Smith, 2012).

Moreover, several tree species have been categorized and successfully employed as bioindicator species. A number of clones of the Aspen differing in $O_3$ sensitivity were continuously exposed for 105 days with different $O_3$ treatments (53.7 $\mu g\ m^{-3}$, 100 $\mu g\ m^{-3}$, and 200 $\mu g\ m^{-3}$) and the fitness of the clones as potential replacements for tobacco as indicator plants in the forested areas was experimented (Keller, 1988). In sensitive clones, an after-effect was visible since the weight of new shoots and the lifespan of leaves was reduced with increasing $O_3$ stress in the preceding year (Keller, 1988). A biomonitoring program utilizing the forests trees was carried out from 1996 to 2000 in the United States. Ozone injury was found more frequently on biomonitoring sites in the eastern Unites States than in the interior or west-coast areas and the injury varied in accordance with the $O_3$ concentrations (Smith et al., 2003). It has been found that *Psidium guajava* can be used as an $O_3$ bioindicator in the tropics (Furlan et al., 2007); while *Solenostemon scutellaroidis* can also act as an $O_3$ indicator plant (Padri and Umponstira, 2019). Moreover, *Viburnum lantana* can be used for *in situ* bioindicator trials (Gottardini et al., 2010). To access the occurrence of phytotoxic $O_3$ effect at two monitoring sites one in the city area with dense traffic, and another at the mountain site, 2-year-old seedlings of *Quercus robur* L. ($O_3$-sensitive) and *Quercus rubra* Michx. ($O_3$-tolerant) were exposed to ambient air from June to August (Parvanova et al., 2009). It was proposed that the $O_3$ induced changes in the physiological activity of tree seedlings may serve as yearly indicators of leaf injury (Parvanova et al., 2009).

## 8.5 Visible injury symptoms as a tool for biomonitoring

Plants exposed to $O_3$ display visual injuries. Ozone exposure over a long period leads to leaf yellowing or early senescence, while the dark brown necrotic spots are usually related to acute $O_3$ stress (Kangasjärvi et al., 2005). Typical visible injury symptoms in plants include stippling, flecking, surface bleaching, necrosis, reddening, or pigmentation (bronzing) or chlorosis (Feng et al., 2014). The sensitivity of plants to $O_3$ toxicity can point toward its potential of being a significant bioindicator (Padri and Umponstira, 2019). Foliar injury symptoms on certain species called bioindicator plants can be used to identify and monitor $O_3$ stress in the natural environment (Smith et al., 2003). Many plants manifest typical foliar injury symptoms when exposed to ambient $O_3$, making them useful as bioindicators for a particular location or region (Heggestad, 1991; Ribas et al., 1998). Visible injuries on leaves can occur on both adaxial and abaxial surface and necrosis commonly appears between the veins. It has been found generally that the degree of $O_3$ injury is related to the $O_3$ concentration in the environment (Budka et al., 2014). Therefore biomonitoring provides a technique to record direct foliar injury regardless of the direct measure of $O_3$ uptake (Coulston et al., 2003).

*Nicotiana tabacum* L. is highly sensitive to $O_3$ exposure, chlorotic spots on its leaves appear even when $O_3$ exposure is for a relatively short time period, while progressively large necrotic patches develop leading to the withering of whole leaves (Esposito et al., 2009). Tobacco and its two cultivars Bel-W3 and Bel-B have been identified as very good bioindicators of $O_3$. Under appropriate conditions tobacco cultivar Bel-B displays the adaxial leaf surface flecking whereas Bel-W3 chiefly manifests the bifacial necrosis symptoms. Visualization of foliar injury of sensitive tobacco Bel-W3 can be utilized as a qualitative but not essentially a quantitative indicator of relative ambient $O_3$ concentration (Krupa et al., 1993). Visible foliar injuries as a result of $O_3$ were at least two to three times more than that observed in resistant cultivars (Krupa et al., 1998). Hence, the tobacco biomonitoring system depends on visual estimations of injury on leaves which can deliver rough estimates of relative seasonal $O_3$ concentrations (Mignanego et al., 1992).

*Trifolium repens* have also been used in various biomonitoring surveys. Data collected at a site in central Italy during 1997–2007 using the $O_3$-sensitive and $O_3$ resistant biotypes (NC-S/NC-R) white clover (*Trifolium repens*) were analyzed to show white clover as a cost-effective indicator of $O_3$ phytotoxicity (Nali et al., 2009). Relevant relationships between visible $O_3$ injury and accumulated AOT40 dose were reported (Nali et al., 2009). Futsaether et al. (2015) found that the $O_3$-induced leaf injuries differed significantly between the three clover species (white, red, and subterranean clover). Red clover plants displayed a significantly lower fraction of injured leaves than subterranean clover, as well as smaller $O_3$ scores per plant and $O_3$ scores per injured leaf as compared to white and subterranean clovers. *Phaseolus vulgaris* L. (Bush bean) mottling pattern on leaves has a wide geographical distribution, and the plant is able to grow in tropical, subtropical, and temperate climates. Genotypes of bush bean *Phaseolus vulgaris* with variable $O_3$ sensitivity were utilized to investigate the impacts of ambient $O_3$ using $O_3$-induced foliar injury index. The foliar damage due to $O_3$ showed a graduated scale of damage pattern and the authors proposed that this can be beneficial for indicating the $O_3$ levels (Leung et al., 2020). Chaudhary and Agrawal (2013) observed $O_3$ specific foliar injury symptoms on *Phaseolus vulgaris* (mung bean) as small pale yellow and brown flecks and bifacial necrosis upon 10 ppb elevated $O_3$ above the ambient $O_3$ level. Chlorotic spots, as well as chlorotic lesions, were witnessed in *Phaseolus vulgaris* exposed to a single pulse of $O_3$ ($150 \pm 10$ $mm^3$ $m^{-3} \times 3$ h) (Faoro and Iriti, 2005). Moreover, $O_3$ levels in China were high enough to cause foliar injury symptoms in the form of chlorotic mottling, interveinal chlorosis, stippling, flecking, and necrosis in 28 different species, and the target species for wide-ranging biomonitoring surveys in temperate China were identified (Feng et al., 2014).

Characteristic $O_3$ injury symptoms in form of diffuse discoloration and leaf stippling in sensitive as well as resistant cultivar has been observed in bioindicator plant *Centurea jacea* (Francini et al., 2008). Leaf injury percentage, interveinal chlorosis, and necrotic symptoms were examined in *Ipomoea nil* to employ the qualitative biomonitoring of $O_3$ (Ferreira et al., 2012). The wild tomato species *Lycopersicon pimpinellifolium* (currant tomato) was exposed to different $O_3$ concentrations, both in controlled environment fumigation facilities and in open-top chambers, to assess its sensitivity and to verify its potential as a bioindicator plant (Iriti et al., 2006). Ozone sensitivity of the plants was detected at an early stage of growth, responding with distinct chlorotic spots within 24 h after exposure to a single pulse of 50 ppb $O_3$ for 3 h, and differentiating peculiar symptoms, like bronzing, reddish necrotic stipples, and advanced necrosis, depending upon the $O_3$ concentration (Iriti et al., 2006). Ozone specific visible injury symptom was noticed on *Hibiscus syriacus* plants used as ornamental hedges (Paoletti et al., 2009). Moreover, its seeds were collected and the resulted seedlings were exposed to $O_3$ in filtered vs. not-filtered (+30 ppb) OTCs. It was found that the level of exposure inducing visible injury in the seedlings grown in OTC was lower than that in the ambient grown hedge plants (Paoletti et al., 2009). The visible injury manifestation in the OTC grown plants confirmed that sensitivity to $O_3$ was heritable and recommended that this deciduous shrub can be effectively utilized as an $O_3$ bioindicator plant (Paoletti et al., 2009).

Among the forest tree species, pine species have been potentially used as $O_3$ bioindicators (Coulston et al., 2003; Kefauver et al., 2014). *Tibouchina pulchra* are native Atlantic forest species that are used as a bioindicator of visible injuries of $O_3$. After 25 days of $O_3$ exposure (AOT40 = 3871 ppb h), interveinal red stippling appeared in plants in the NF + $O_3$ chamber (Furlan et al., 2008). Similarly, interveinal red stippling appeared in the *Psidium guajava* "Paluma" plants in non-filtered chamber with added 40 ppb $O_3$ (NF + $O_3$) after 5 days of exposure, while after 2 months of $O_3$ exposure, injured leaves per plant corresponded to 86% in NF + $O_3$ and 25% in the NF treatment and the degree of injured leaf area was explained primarily by the accumulated $O_3$ exposure (Furlan et al., 2007). The typical visible $O_3$ injury symptoms in *Fagus sylvatica* is bronzing and brown stippling; yellow interveinal stippling in *Acer campestre, Quercus faginea,* while brown interveinal stippling in *Fraxinus angustifolia, Populus nigra* and accelerated senescence with interveinal leaf chlorosis in *Salix alba* (http://www.ozoneinjury.org), for conifers the injury symptoms includes tip burning, flecking and diffuse chlorotic mottling (Innes et al., 2001;Vollenweider et al., 2013). While red spots on *Prunus serotina, Fraxinus americana,* brown spots on *Liriodendron tulipifera,* and purple ones on *Asclepia syriaca* are observed due to $O_3$ exposure (Skelly et al., 1987). A list of tree species depicting $O_3$ induced injury has been provided in Table 8.2.

## 8.6 Biomarkers for ozone stress

The bioindicator plants often show the damaging effects of $O_3$ even before visible symptoms develop (Alves et al., 2016) and the effects on physiological and biochemical parameters, provide a prompt and reliable detection of $O_3$ stress, even before the appearance of visible injury symptoms. Hence, the biochemical and physiological parameters constitute suitable biomarkers for identifying early $O_3$ induced changes and provide a potential tool for selecting genotypes for sensitivity to $O_3$ pollution and also allow the cause–effect relationships to be established at an early stage of pollution (Biolley et al., 1998). New methods of biomonitoring like to detect microscopic effects, mutagenicity, and biochemical disturbances are promising (Temmerman et al., 2001). Several studies have made attempts to correlate changes in the levels of antioxidant enzymes or compounds with the adaptation to $O_3$ induced oxidative stress and also suggested that different cultivars of the same species exhibit differences in the $O_3$ sensitivity (Scebba et al., 2003). The biomarkers of $O_3$ stress can be cellular markers or molecular markers.

**Table 8.2** Ozone-induced injury symptoms on different tree species in experiments carried out at various geographical locations.

| S. No. | Location | Months/year/ season | Plant | Ozone treatment | Ozone concentration/dose | Ozone-induced injury symptom | References |
|---|---|---|---|---|---|---|---|
| 1. | Switzerland | 1993–1996 | *Betula pendula, Morns nigra, Prunus serotina, Viburnum lantana* | Continuously Stirred Tank Reactor (CSTR) chamber system | Ozone exposures of 30, 60, 90, or 120 ppb for 7 h $d^{-1}$ for 5 d $week^{-1}$ | Based upon the severity of leaf reddening, upper surface stippling, and premature senescence the sensitivity trend was: *Viburnum lantana* > *Morns nigra* > *Betula pendula* > *Prunus serotina* | Skelly et al. (1998) |
| 2. | Tervuren-12 Km east of Brussels (Belgium) | April–September | *Fagus sylvatica* L. *Populus nigra* L. | OTCs [Charcoal-filtered (CF), non-filtered (NF) and non-filtered air+30 ppb (NF+)] | 24 h $day^{-1}$ mean and AOT40 (NF) = 18 ppb and 4055 ppb h respectively<br>24 h $day^{-1}$ mean and AOT40 (NF+) = 19 ppb and 8880 ppb h, respectively | In *Fagus sylvatica* up to 11 % of all leaves showed yellow discoloration and 5.5 % showed necrotic spots injury<br>In *Populus nigra* some of the leaves showed 60–70 % slight bronzing | Bortier et al. (2000) |
| 3. | Bavarian forest (Germany) | 1995–1997 | *Fagus sylvatica* L | Phytotrons | Seasonal $O_3$ mean (May to October) in 1995 was 44 ppb. The seasonal means were 30 ppb in 1996, and 40 ppb in 1997 respectively | Discoloration of leaves becoming yellowish-brown and tended to develop small dark-brown necrotic spots | Baumgarten et. al. (2000) |

*(Continued)*

**Table 8.2** Cont'd

| S. No. | Location | Months/year/ season | Plant | Ozone treatment | Ozone concentration/dose | Ozone-induced injury symptom | References |
|---|---|---|---|---|---|---|---|
| 4. | Canton Ticino in southern Switzerland | May–August, 2000 and 2001 | *Crataegus monogyna* Jacq. *Fraxinus excelsior* L. *Populus nigra* L. *Prunus avium* L. *Salix alba* L. *Tilia cordata* Miller *Tilia platyphyllos* Scop. *Viburnum lantana* L. *Viburnum opulus* L. | OTCs (filtered air chambers), non-filtered air chambers, and open plots) | 12 h mean ambient seasonal $O_3$ was 56.9 ppb in 2000 and 55.3 in 2001 12 h mean $O_3$ in non-filtered air chambers was 51.2 ppb in 2000 and 50.3 in 2001 | Ozone injury onset in response to ambient $O_3$ exposure in 2001: *Populus nigra* L. > *Viburnum lantana* L.> *Salix alba* > *Crataegus monogyna Jacq.* > *Viburnum opulus* > *Tilia platyphyllos Scop.*> *Prunus avium* > *Fraxinus excelsior* > *Tilia cordata* | Novak et al. (2003) |
| 5. | North–eastern USA and south-eastern Spain | 2000 and 2001 Summer season | Thirteen native species of north–eastern USA and 27 native species of southeastern Spain | Continuously Stirred Tank Reactor (CSTR) chambers in a greenhouse | Ozone exposures of 30, 60, 90, and 120 ppb were given for 7 h $day^{-1}$, 5 days $week^{-1}$ | < 30 ppb– in general no $O_3$ injury; at higher ozone exposures, visible injury progressed as chlorotic spotting, discoloration, reddening, pigmented stippling, necrosis, and premature senescence of leaves | Orendovici et al. (2003) |

| | | | | | | | |
|---|---|---|---|---|---|---|---|
| 6. | Central Pennsylvania, USA | 1998 and 1999 Summer season | *Acer rubrum* L. *Fraxinus Americana* L. *Prunus serotina* Ehrh., | OTCs and open plots [(charcoal–filtered air: 50% ambient $O_3$), (non-filtered air: 98% ambient $O_3$), (ambient air: 100% $O_3$),] | Average seasonal 24 h ambient $O_3$ concentrations: 1998 = 42.1 ppb 1999 = 40.6 ppb | *Prunus serotina* developed more $O_3$-induced injury (<34% affected leaf area) than *Fraxinus Americana* (<5% affected leaf area), while *Acer rubrum* rarely manifested foliar injury | Schaub et al. (2003) |
| 7. | Central and Eastern Europe | – | *Alnus incana* L. *Sorbus aucuparia* | Pass through chambers or continuous stirred tank reactors (CSTRs) | 60 or 80 ppb $O_3$ for 7h $day^{-1}$ /7days $week^{-1}$. | Percent injury severity *Alnus incana* = 62.5% *Sorbus aucuparia* = 81.25% | Manning and Godzik, (2004) |
| 8. | Po valley, (northern Italy) | 2004 and 2005 April-September | *Fagus sylvatica* L. *Quercus robur* L. | OTC (charcoal-filtered and non-filtered) and open plots | 50 % ambient $O_3$ in charcoal-filtered OTC (CF), 95 % ambient $O_3$ in non-filtered OTC (NF) and 100% ambient $O_3$ in open plots (OA) | Visible foliar injury was detected only in *Fagus sylvatica* inside NF and OA. In *Quercus robur* no characteristic symptoms were observed, but few trees in NF and OA treatments manifested slight yellowing of leaf at the end of the growing season | Gerosa et al. (2008) |

*(Continued)*

**Table 8.2** Cont'd

| S. No. | Location | Months/year/season | Plant | Ozone treatment | Ozone concentration/dose | Ozone-induced injury symptom | References |
|---|---|---|---|---|---|---|---|
| 9. | Tuscany (central Italy) | April to September | *Fraxinus ornus* L. | Indoor chambers (150 ppb, 8 h $d^{-1}$ for 7 weeks) | AOT40–21.3 ppm h and the maximum hourly $O_3$ concentration reached 93 ppb | leaf reddening, stippling, and necrosis | Paoletti et al. (2009) |
| 10. | São Paulo State, south-eastern Brazil | – | *Astronium graveolens* Jacq | Ambient air and green house containing charcoal filtered air | AOT40 for the period January 2013 to February 2014 = 18,986 ppb h | Adaxial leaf surface with dark intercostal stippling | Cassimiro et al. (2016) |
| 11. | Margone and Lasino (Northern Italy) | Five-month | *Viburnum lantana* L. | Open field | Margone-111 ± 17.0 ppb Lasino-89 ± 12.4 ppb | interveinal red-brown stipples | Gottardini et al. (2014) |
| 12. | Trentino, north Italy | 2010–2015 | *Viburnum lantana L.* | Observation plots designated as ViburNeT based on prior field survey | May-July mean hourly $O_3$ concentration (2010–2015) was 37.9–45.8 ppb | Reddish-brown to dark interveinal stippling and the frequency of symptomatic plants varied from 20.7% to 50.6%. | Gottardini et al. (2017) |

### 8.6.1 Cellular markers

Ozone may cause cellular functional alterations even before the onset of visible symptoms (Gottardini et al., 2014). *Nicotiana tabacum* "Bel-W3" used widely as an $O_3$ bioindicator plant, shows typical necrosis preceded by the microscopic markers of oxidative stress (Pedroso and Alves, 2015). *Nicotiana tabaccum* Bel-W3 and Bell B cultivars were exposed to $O_3$ in the tropical region and the activity of antioxidant enzymes like superoxide dismutase (SOD), guiacol peroxidase, and ascorbate peroxidase (APX), were examined (Bandurska et al., 2009). A significant correlation between APX activity within the two hours of $O_3$ exposure and the degree of leaf damage was detected in *Nicotiana tabaccum* (Bandurska et al., 2009). In tobacco cultivar, Bel-W3 comparative assessment of the redox state indicators like ascorbic acid, glutathione, SOD, GR, and APX were quantified. The redox state of tobacco reduced few days after their placement in the subtropical environment and caused them to become susceptible to oxidative stress caused due to chronic dose of $O_3$ (Dias et al., 2011). Reactive oxygen species (ROS) production is there in the $O_3$ sensitive tobacco cultivar Bel-W3 but not in the Bel B cultivar tolerant to $O_3$, which indicates that $O_3$ causes oxidative burst in the sensitive tobacco cultivar Bel-W3 which results in the generation of hypersensitive cell death-like lesions (Schraudner et al., 1998). Three $O_3$ bioindicator species *Nicotiana tabacum* "Bel-W3," Ipomoea nil "Scarlet O'Hara," and *Psidium guajava* "Paluma" were tested to detect hydrogen peroxide ($H_2O_2$) accumulation and cell death by exposing them to the chronic levels of $O_3$ in a subtropical area. A histocytochemical technique which was the combination of 3,3'-diaminobenzidine (DAB) with Evans blue staining was used in the experiment. All the three tested species accumulated $H_2O_2$, but a positive correlation with $O_3$ concentration was detected only in *Nicotiana tabacum* thus confirming it to be the best bioindicator species among the three with respect to the $H_2O_2$ biomarker assay. Moreover, a positive correlation between cell death and $O_3$ exposure was also observed in *Nicotiana tabacum* (Alves et al., 2011). In an experiment with three clover species white clover (*Trifolium repens*), red clover (*Trifolium pratense*), and subterranean clover (*Trifolium subterraneum*), it was suggested that the less-sensitive red clover cultivars might have had a higher constitutive defense response under $O_3$ stress than the more $O_3$-sensitive subterranean clover and white clover cultivars (Futsaether et al., 2015). The physiological and biochemical basis of chronic $O_3$ exposure (60 ppb for 56 days, 5 h $day^{-1}$) on NC-R (resistant) and NC-S (sensitive) clones of *Trifolium repens* was analyzed. NC-S exhibited foliar injury and had reduced photosynthetic pigments, while lipids peroxidation and solute leakage enhanced, while the guaiacol peroxidase induction was mildly linked with NC-S clone following the $O_3$ treatment (Nali et al., 2005).

Fatima et al. (2019) found thiols and ascorbic acid as potential biomarkers for $O_3$ stress in wheat cultivars differing in their relative sensitivity to $O_3$. It has been proposed that ascorbic acid (AA) plays a pivotal role in $O_3$ tolerance in various plants and therefore can be used as $O_3$ stress marker in sensitive plants. Frei et al. (2012) found that a moderately higher level of injury symptoms induced by $O_3$ exposure in rice cultivar ND6172 was correlated with the lower concentration of foliar ascorbate. Wheat cultivars Sculptur and Columbo showed differential $O_3$ injury symptoms; while Columbo displayed symptoms within a few weeks, Sculptur displayed no symptom weeks after exposure to $O_3$. It was found that one of the underlying mechanisms for this response was the higher content of the constitutive ascorbate pool in cultivar Sculptur (Picchi et al., 2017). The potential of *Ipomoea nil Scarlet* Hara for $O_3$ biomonitoring in the subtropics analyzing the ascorbic acid content, activity of SOD and peroxidases demonstrated that *Ipomoea nil* may be employed for qualitative biomonitoring of $O_3$ which also depends on the oxidative power of the environment (Ferreira et al., 2012). Furthermore, the $O_3$-resistant cultivar WXJ14 had 35.6% higher AA content than the $O_3$-sensitive cultivar SY63 when exposed to 1.5 times the concentration of ambient $O_3$ (Wang et al., 2013). Ascorbic acid is an important parameter in screening soybean cultivars for their $O_3$ stress responses (Jiang et al., 2018). *Glycine max* L. cultivars Sambaíba ($O_3$-sensitive) and Tracajá ($O_3$-tolerant) were exposed to 40 ppb $O_3$ concentration, and it was detected that Sambaíba had comparatively much higher AA content (85%) with respect to Tracajá (27%) (Bulbovas et al., 2014). *Glycine max* L. genotype Fiskeby III ($O_3$-tolerant) had higher AA level than the Mandarin Ottawa ($O_3$-sensitive) genotype when treated with 70 ppb $O_3$ (Roth, 2017). *Trifolium alexandrinum* L., cultivars Bundel, Wardan, JHB-146, Saidi, Fahli, and Mescavi were treated with 10 ppb above ambient $O_3$ level and it was reported that the most sensitive cultivar Wardan had 26% increased AA levels while the most resistant cultivar Fahli had 71.2% increased AA level (Chaudhary and Agrawal, 2013). A twofold increase in AA content was found in *Medicago truncatula*-resistant accession Marrakech as compared to the sensitive accession JE154 upon 300 ppb of $O_3$ exposure (Puckette et al., 2008). Thiol acts as a signaling molecule triggering the signaling routes from apoplast or cell membrane to the cytosol under $O_3$ stress (Gomez et al., 2004). Pandey et al. (2015) have recommended reduced glutathione (GSH) as one of the factors to screen for $O_3$ sensitivity/tolerance in rice cultivars. *Medicago truncatula* plants were given 300 ppb $O_3$ for 6 h and it was witnessed that resistant cultivar JE154 showed significant increases in GSH and GSSG levels (Puckette et al., 2008).

The phytotoxicity of $O_3$ has been shown to damage plant photosynthesis and induce early death of leaves and retired growth (Dias et al., 2011). The species, jute cultivar (local), clover cultivar (Masry), garden rocket cultivar (local), and alfalfa cultivar (local), ranked in order of decreasing sensitivity, displayed typical $O_3$ injury symptoms faster and at lower $O_3$ concentrations than that of tobacco cultivar Bel-W3. Three variables were tested in search of a reliable tool for the diagnosis and prediction of $O_3$ response prior to the appearance of visible foliar senescence which is the pigment degradation, stomatal conductance, and net photosynthesis (Madkour et al., 2002). A good correlation was found between stomatal conductance restriction and $CO_2$ flux into the mesophyll cells and $P_{net}$ while pigment degradation was found to be unreliable in predicting $O_3$ sensitivity of species (Madkour et al., 2002).

For biomonitoring study total chlorophyll, carotenoid, total sugar, and ascorbic acid were analyzed to study the impact of air pollutants on tree species and it was suggested that these parameters can be used as an indicator of air pollution for early diagnosis of stress or as a marker for physiological impairment to trees prior to the onset of visible injury symptoms (Thambavani and Prathipa, 2012). Moreover, chlorophyll content and chlorophyll A fluorescence transit was correlated with the development of leaf injury in *Viburnum lantana* L. plants (Gottardini et al., 2014). Several studies carried out in Southeastern Brazil regions target to establish the best native tree species and the respective appropriate biomarkers for future assessment of pollution effects on tropical forests (Ferreira et al., 2012). PAL is an important enzyme of the phenyl-propanoid pathway and is considered a biochemical marker involved in the activation of plant defense. *Centurea jacea* has been suggested as a potential bioindicator for $O_3$ stress and it was detected that the anthocyanins were quickly utilized by resistant clone as antioxidant compounds (Francini et al., 2008). PAL activity showed enhancement of 91% and 41% respectively in sensitive and resistant clones compared to controls after $O_3$ treatment of 200 ppb for 5 hour (Francini et al., 2008). The enzyme 3-deoxy-D-arabino-heptulosonate-7-phosphate synthase reached three to fivefold inductions after 4 hours of $O_3$ fumigation in bioindicator tobacco cultivar Bel-W3 (Janzik et al., 2005).

Histo-cytochemical methods can contribute to efficient analysis of susceptible species for the $O_3$ biomonitoring (Faoro and Iriti, 2009). The microscopic events observed like anticlinal cell wall protrusions, condensation of cytoplasm in parenchyma tissue, changes in chlorophyll autofluorescence, and chromatin condensation indicate the oxidative stress caused by $O_3$ exposure and can be utilized as markers for $O_3$ toxicity (Pedroso and Alves, 2015). Furthermore, microscopy is suggested as an effective tool to confirm and evaluate $O_3$ injury to Mediterranean vegetation (Faoro and Iriti, 2009). A 3, 3'-diaminobenzidine (DAB)-Evans blue staining was proposed to validate $O_3$ symptoms at the microscopic level and for a pre-visual identification of $O_3$ injury (Faoro and Iriti, 2009). Ozone effects on *Trifolium repens* L. cv. Regal clones NC-S (sensitive to $O_3$) and NC-R (resistant to $O_3$) were compared at the ultrastructural level. The resistant clones depicted much less damage like the chloroplast maintained an almost intact organization of thylakoid (Basile et al., 2010). The method has been tested positively in some of the important crop species like wheat, bean, tomato, and onion and, with some restriction, in forest species, which allows one to gain some useful insights into the mechanisms underlying the $O_3$ tolerance or sensitivity (Faoro and Iriti, 2009). *Lycopersicon pimpinellifolium* (currant tomato), the wild species of tomato was exposed to variable concentrations of $O_3$, both in controlled environment fumigation facilities and in open-top chambers, to validate its potential as a bioindicator plant (Iriti et al., 2006). Histo-cytochemical investigations with DAB, to localize $H_2O_2$, and Evans blue, to detect dead cells, suggested that sensitivity of tomato cultivar currant to $O_3$ could be due to a deficiency in the anti-oxidant pools. Moreover, Evans blue staining with its intensity showed various degrees of cell membrane degradation and/or various stages of cell death (Iriti and Faoro, 2008). This combination of stainings proved to be useful, either to predict visible symptoms early before their appearance and also to validate foliar $O_3$ injury (Iriti et al., 2006). Therefore differential response in species has been observed regarding the degree of sensitivity of plants against $O_3$ which is in accordance with the defense responses, differential constitutive levels or induction of enzymatic and or nonenzymatic antioxidants, pigments or metabolites which may be utilized for the biomonitoring studies of $O_3$.

### 8.6.2 Molecular markers

Techniques like genomics, transcriptomics, and proteomics have been utilized to investigate the molecular markers of $O_3$ stress in plants. Air pollution is one of the main reasons for DNA damage in living beings. It has been found that ROS are a major cause of oxidative damage to DNA (Labuschagne, 2007). Exposure of tobacco cultivars Bel B and Bel-W3 in the field with peak $O_3$ concentration reaching more than 80 ppb and the resulting comet assay analysis conducted on leaf tissues showed the effects of $H_2O_2$ on DNA damage in both the cultivars, which confer that certain component of the machinery that protects DNA integrity from oxidative stress might have been impaired in cultivar Bel-W3 resulting in significantly higher DNA damage in cultivar Bel-W3 (Restivo et al., 2002). Two clones one resistant and one sensitive of *Trifolium repens* cultivar Regal was used as a bioindicator plant to assess the genotoxicity of air pollutants in the Mediterranean environment of the Italian province of Novara. Genotoxicity was assessed with the help of AFLP molecular markers in response to ambient $O_3$ (Piraino et al., 2006). Analysis of transcriptome changes in $O_3$ sensitive and resistant cultivars provides important clues regarding the molecular changes governing the plant responses to $O_3$ stress (Puckette et al., 2008).

Gene expression studies have also provided some clues about the $O_3$ stress markers. The transcript accumulation in sensitive as well as resistant bioindicator plants like *Centurea jacea* has been observed (Francini et al., 2008). To evaluate $O_3$ induced foliar injury symptoms and the gene expression responses, 14-day old *Oryza sativa* L. cultivar Nipponbare leaves were exposed to an $O_3$ concentration of 200 ppb for 1, 12, 24, 48, and 72 h. It was detected that the expression of 146 genes was found to be induced with increased foliar injury up to 72 h, displaying a strong positive correlation between the subset of expressed genes and $O_3$-induced visible injury. It was therefore concluded that the genes associated with leaf injury symptoms may serve as a means of potential biomarkers to study the mechanisms of $O_3$ induced injury symptoms (Cho et al., 2012). Rice genotypes differing in $O_3$ sensitivity SL15 (sensitive to $O_3$) and SL41 (tolerant to $O_3$) were exposed to an $O_3$ concentration of 120 ppb, and the microarray- based gene expression profiling showed that genes showing altered expression were involved in enzymatic detoxification of ROS, biosynthesis of antioxidants and programmed cell death, and a total of 156 probes expressed differentially between the sensitive and tolerant cultivar (Frei et al., 2010). *Medicago truncatula* genotypes were evaluated for their differential $O_3$ stress responses, showing that sensitive genotype Jemalong showed a downregulation of defense response genes 12 hours post $O_3$ exposure (Puckette et al., 2008). Expression of *OsORAP1* gene was checked across different rice genotypes to discern the inter-varietal response of different rice genotypes. Authors found that 100 ppb $O_3$ exposure resulted in significantly high induction of *OsORAP1* expression across the selected 18 rice genotypes (Ashrafuzzaman et al., 2020). Therefore, the gene OsORAP1 can be selected as a useful molecular marker to gain an insight into $O_3$ induced stress responses in plants. *Centaurea jacea* a herbaceous bioindicator plant when exposed to an $O_3$ concentration of 200 ppb for 5 h showed no alteration in chalcone synthase transcripts in sensitive and resistant lines as well as after $O_3$ treatment (Francini et al., 2008). Moreover, post 4 h $O_3$ fumigation in *Nicotiana tabacum* L. cultivar Bel-W3, the transcript level of 3-Dehydroquinate dehydratase-shikimate, 5-dehydrogenase and Chorismate synthase showed induction pattern ranging from 1.8 to 2.6 times; while Shikimate kinase remained unchanged by $O_3$ (Janzik et al., 2005).

Molecular mechanisms linked to the differential $O_3$ tolerance of *Glycine max* (L.) Merr. genotypes Mandarin (Ottawa) and Fiskeby III was inspected, which showed that *Protein D1* expression were down-regulated in sensitive cultivar Mandarin (Ottawa) but remained unchanged in tolerant cultivar Fiskeby III (Chutteang et al., 2016). The air pollution stress variations using RuBisCO enzyme in plants at natural environmental conditions have also been detected (Govindaraju et al., 2018). Agrawal et al. (2002) in rice observed a significant decline in major proteins involved in photosynthesis including RuBisCO in leaves due to $O_3$ and the stimulation of some stress-related defense proteins, and Komatsu (2008) has suggested that these candidate proteins may be related to the stress–tolerance trait and therefore may be noted as potential biomarker proteins. A drastic reduction in RuBisCO larger and smaller subunits was noted due to elevated $O_3$ exposure in rice by Feng et al. (2008). Reductions in the accumulation of RuBisCO larger and smaller subunits have been noted in wheat (Sarkar et al., 2010), rice (Sarkar et al., 2015), and maize (Singh et al., 2014).

Gottardini et al. (2016) have suggested that $O_3$ treatment results in the induction of heat shock proteins (HSPs) which may aid in increasing the photosynthetic apparatus resistance to photoinhibition. Induction of HSPs due to $O_3$ exposure was observed in rice seedlings of Nipponbare cultivar (Cho et al., 2008). Interestingly, Hsp70 levels, which are an important stress marker, were only slightly increased in the $O_3$ sensitive clone (NC-S) with respect to the $O_3$-resistant clone (NC-R) (Basile et al., 2010). Some specific and rapidly accumulating PR proteins, ATP induced Clp protease induced 24 h post $O_3$ exposure in rice (Cho et al., 2008) perhaps may serve as potential marker proteins to indicate the $O_3$ linked injury in rice, as well as other plants in general (Kang et al., 2010).

## 8.7 Climatic impacts and meteorological conditions affect the biomonitoring approach

Often the visible leaf injury correlates with the $O_3$ concentration and displays a directly proportional relationship. However, usually the practice of indicator plants to monitor $O_3$ pollution becomes problematic when trying to correlate the ambient $O_3$ concentration with the injury symptoms severity (Cheng and Sun, 2013). A study of leaf injury indices as indicators of cumulative tropospheric $O_3$ resulted in visible leaf injuries in tobacco which were much higher in the year 2010 with respect to 2011 due to favorable meteorological conditions like higher temperature and solar radiation for $O_3$ formation (Cheng and Sun, 2013). Ribas et al. (1998) suggested that tobacco cultivars (Bel W3, Bel B, and Bel C) were good indicators of $O_3$ pollution but other environmental and meteorological factors need to be considered in using them as bioindicator plant. The $O_3$ phytotoxicity in rural areas of Catalonia through Bel-W3 tobacco was assessed by determining the $O_3$ induced leaf area injury percentage in plants exposed from spring to autumn in the years 1995–1999, simultaneously on nine field sites where high concentrations of $O_3$ prevailed (Ribas and Penuelas, 2003). The $O_3$ concentrations correlated fairly with foliar $O_3$ injury; however, their biomonitoring capacity also depended on diverse environmental factors predominantly those associated with stomatal behavior such as vapor pressure deficit (Ribas and Penuelas, 2003).

A variable degree of intraspecific $O_3$ sensitivity in *Centurea Jacea* L. was observed which highlighted the importance of climate-dependent influence of population-specific plant development on $O_3$ sensitivity (Bassin et al., 2004). Therefore, one of the greatest challenges of biomonitoring methodology is to establish a linear relationship between pollutant concentrations and markers measured in plants (Nakazato et al., 2018). The bioindicator responses can become mild or intensified by the variations in climatic conditions (Nakazato et al., 2018). An analysis of metrological parameters, $O_3$ concentrations, and visible injury showed that high relative humidity levels favor the $O_3$ phytotoxicity at the Catalan site (Gimeno et al., 1995). A biomonitoring study program in eight European countries at twelve urban agglomerations, employed the $O_3$-sensitive bioindicator plant *Nicotiana tabacum* cultivar Bel-W3 to evaluate the phytotoxic effects of $O_3$ at rural, suburban, urban, and traffic-exposed sites. Between the sites, differences within local networks were comparatively small, nevertheless inter-annual and seasonal differences were considerably higher, since some presented almost linear relationships, whereas other sites showed no association between $O_3$-induced effects and $O_3$ pollution, due to the variations in meteorological environments and related $O_3$ concentrations (Klumpp et al., 2006). This was because the actual $O_3$ flux into the leaf which determines the effects on plants was modified by various environmental factors, rather than ambient $O_3$ concentration (Klumpp et al., 2006). In southern Italy, the effects of ambient $O_3$ exposure on the morphology of leaves, water use, and yield on two clones of *Trifolium repens* L. cultivar "Regal" differing in $O_3$ tolerance was experimented. In both the years of study (1997–1998), $O_3$ injury was observed on the $O_3$-sensitive clone (NC-S) but not on the $O_3$-resistant one (NC-R), and simultaneously the leaf, as well as stolon dry matter production, was considerably lower in NC-S than in NC-R. However, the authors commented that it cannot be ruled out that other influences, like high temperature, interacted with the $O_3$ effects on the production of biomass (Postiglione et al., 2000). While sometimes $O_3$ induced responses in plants may or may not correlate with the prevailing climatic conditions. In Brazil, $O_3$ biomonitoring of NC-S and NC-R biotypes of *Trifolium repens* cultivar Regal was tested for the years 2012 and 2013 at four sites with neighboring crop production in the Metropolitan Region of Campinas (MRC). Biomass accumulation of both the clones was similar at four sites studied in the MRC, however substantial differences were observed between the monthly harvests during the trial period. During the first year, biomass of both the clones positively correlated with relative humidity and precipitation and negatively with vapor pressure deficit, particulate matter, $NO_2$, and $O_3$. While in the second year, biomass production of tested clones did not correlate significantly with fluctuating environmental conditions (Sandrin et al., 2018).

An analysis of biotic and abiotic variables reveals complex relationships with the great importance of meteorological factors in plant responses and under unstable climatic conditions the connection between $O_3$ exposure and foliar injury is weak and hence the qualitative nature of biomonitoring needs to be further confirmed (Silva et al., 2012). The meteorological and other factors like temperature, humidity, wind or altitude, need to be considered because they affect the intensity of $O_3$ injury in plants (Ribas et al., 1998; Penuelas et al., 1999). Therefore, the interactions between environmental conditions and $O_3$ exposure on plant response should be further studied for the establishment of appropriate critical levels and successful trials of biomonitoring (Gimeno et al., 1995).

## 8.8 Conclusion

Biomonitoring is an important technique to assess the negative effects of $O_3$ on vegetation. Several species are utilized in this process to monitor the effects of air pollutants which are designated as bioindicator plants. The indicator plants show characteristic responses and are used to detect the occurrence of phytotoxic levels of $O_3$ pollution which is regarded as a cost-effective method. Many of the pilot studies have been carried out globally using bioindicator species. Biomonitoring offers a methodology to document direct foliar injury irrespective of direct measurement of $O_3$ uptake. Moreover, cellular markers induced by physiological adjustments including oxidative burst, a hypersensitive response like reactions, defense responses and accelerated cell senescence may precede the visible injury symptoms. It must be considered that the standardization method of biomonitoring should be very sensitive for the development of air quality standards to prevent harmful effects on the plant health and ecosystem. Furthermore, the climatic impacts and interactions between environmental conditions and biomonitor plants should be considered critical because they affect the intensity of $O_3$ injury in plants and therefore may influence the biomonitoring process. It is therefore concluded that biomonitoring studies provide significant information that is complementary of ambient $O_3$ monitoring systems, and assist to raise the $O_3$ policy issues associated with the protection of crops and vegetation.

## References

Agathokleous, E., Saitanis, C.J., Burkey, K.O., Ntatsi, G., Vougeleka, V., Mashaheet, A.M., Pallides, A., 2017. Application and further characterization of the snap bean S156/R123 ozone biomonitoring system in relation to ambient air temperature. Sci. Total Environ. 580, 1046–1055.

Agathokleous, E., Saitanis, C.J., Feng, Z., De Marco, A., Araminiene, V., Domingos, M., Sicard, P., Paoletti, E., 2020. Ozone biomonitoring: a versatile tool for science, education and regulation. Curr. Opin. Environ. Sci. Health 18, 7–13.

Agrawal, G.K., Rakwal, R., Yonekura, M., Kubo, A., Saji, H., 2002. Proteome analysis of differentially displayed proteins as a tool for investigating ozone stress in rice (Oryza sativa L.) seedlings. Proteomics 2 (8), 947–959.

Agrawal, M., Singh, B., Rajput, M., Marshall, F., Bell, J.N.B., 2003. Effect of air pollution on peri-urban agriculture: a case study. Environ. Pollut. 126 (3), 323–329.

Ahammed, Y.N., Reddy, R.R., Gopal, K.R., Narasimhulu, K., Basha, D.B., Reddy, L.S.S., Rao, T.V.R., 2006. Seasonal variation of the surface ozone and its precursor gases during 2001–2003, measured at Anantapur (14.62 N), a semi-arid site in India. Atmos. Res. 80 (2-3), 151–164.

Alves, E.S., Moura, B.B., Pedroso, A.N., Tresmondi, F., Domingos, M., 2011. The efficiency of tobacco Bel-W3 and native species for ozone biomonitoring in subtropical climate, as revealed by histo-cytochemical techniques. Environ. Pollut. 159 (12), 3309–3315.

Alves, E.S., Moura, B.B., Pedroso, A.N.V., Tresmondi, F., Machado, S.R., 2016. Cellular markers indicative of ozone stress on bioindicator plants growing in a tropical environment. Ecol. Indic. 67, 417–424.

Andersen, S.O., Halberstadt, M.L., Borgford-Parnell, N., 2013. Stratospheric ozone, global warming, and the principle of unintended consequences—an ongoing science and policy success story. J. Air Waste Manage. Assoc. 63 (6), 607–647.

Ashmore, M.R., 2005. Assessing the future global impacts of ozone on vegetation. Plant Cell Environ. 28 (8), 949–964.

Ashrafuzzaman, M., Ueda, Y., Frei, M., 2020. Natural sequence variation at the OsORAP1 locus is a marker for ozone tolerance in Asian rice. Environ. Exp. Bot. 178, 104153.

Bahino, J., Yoboué, V., Galy-Lacaux, C., Adon, M., Akpo, A., Keita, S., Liousse, C., Gardrat, E., Chiron, C., Ossohou, M., Gnamien, S., 2018. A pilot study of gaseous pollutants' measurement ($NO_2$, $SO_2$, $NH_3$, $HNO_3$ and $O_3$) in Abidjan, Côte d'Ivoire: contribution to an overview of gaseous pollution in African cities. Atmos. Chem. Phys. 18 (7), 5173–5198.

Bandurska, H., Borowiak, K., Miara, M., 2009. Effect of two different ambient ozone concentrations on antioxidative enzymes in leaves of two tobacco cultivars with contrasting ozone sensitivity. Acta Biol. Cracoviensia Series Botanica 51, 37–44.

Basile, A., Esposito, S., Sorbo, S., Napolitano, E., Conte, B., Salvatori, E., Manes, F., 2010. Ultrastructural alterations induced by tropospheric ozone: comparison between resistant and sensitive clones of Trifolium repens L. CV. Regal. Int. J. Environ. Health 4 (2-3), 260–277.

Bassin, S., Kölliker, R., Cretton, C., Bertossa, M., Widmer, F., Bungener, P., Fuhrer, J., 2004. Intra-specific variability of ozone sensitivity in Centaurea jacea L., a potential bioindicator for elevated ozone concentrations. Environ. Pollut. 131 (1), 1–12.

Baumgarten, M., Werner, H., Häberle, K.H., Emberson, L.D., Fabian, P., Matyssek, R., 2000. Seasonal ozone response of mature beech trees (Fagus sylvatica) at high altitude in the Bavarian forest (Germany) in comparison with young beech grown in the field and in phytotrons. Environ. Pollut. 109 (3), 431–442.

Biolley, J.P., Lauga, B., Cagnon, C., Duran, R., Salvado, J.C., Goulas, P., 1998. Phenolic pattern of bean (Phaseolus vulgaris L.) as an indicator of chronic ozone stress. Water Air Soil Pollut. 106 (3), 355–368.

Biswas, D.K., Xu, H., Li, Y.G., Sun, J.Z., Wang, X.Z., Han, X.G., Jiang, G.M., 2008. Genotypic differences in leaf biochemical, physiological and growth responses to ozone in 20 winter wheat cultivars released over the past 60 years. Glob. Change Biol. 14 (1), 46–59.

Borowiak, K., 2013. Morphological changes in two tobacco and petunia cultivars under exposure to tropospheric ozone. Acta Biol. Cracoviensia. Series Botanica 55 (1), 58–66.

Bortier, K., De Temmerman, L., Ceulemans, R., 2000. Effects of ozone exposure in open-top chambers on poplar (Populus nigra) and beech (Fagus sylvatica): a comparison. Environ. Pollut. 109 (3), 509–516.

Brauer, M., Freedman, G., Frostad, J., Van Donkelaar, A., Martin, R.V., Dentener, F., Dingenen, R.V., Estep, K., Amini, H., Apte, J.S., Balakrishnan, K., 2016. Ambient air pollution exposure estimation for the global burden of disease 2013. Environ. Sci. Technol. 50 (1), 79–88.

Budka, A., Kayzer, D., Borowiak, K., Zbierska, J., Wolna-Maruwka, A., Schroeter-Zakrzewska, A., Chlebowska, A., 2014. Visible tobacco leaf injury indices as indicators of cumulative tropospheric ozone effect. Arch. Environ. Prot. 40 (4), 53–65.

Bulbovas, P., Souza, S.R., Esposito, J.B.N., Moraes, R.M., Alves, E.S., Domingos, M., Azevedo, R.A., 2014. Assessment of the ozone tolerance of two soybean cultivars (Glycine max cv. Sambaíba and Tracajá) cultivated in Amazonian areas. Environ. Sci. Pollut. Res. 21 (17), 10514–10524.

Burkey, K.O., Miller, J.E., Fiscus, E.L., 2005. Assessment of ambient ozone effects on vegetation using snap bean as a bioindicator species. J. Environ. Qual. 34 (3), 1081–1086.

Bytnerowicz, A., Manning, W.J., Grosjean, D., Chmielewski, W., Dmuchowski, W., Grodzinska, K., Godzik, B., 1993. Detecting ozone and demonstrating its phytotoxicity in forested areas of Poland: a pilot study. Environ. Pollut. 80 (3), 301–305.

Calatayud, V. Ozone injury in European Forest Species. http://www.ozoneinjury.org.

Calatayud, V., Sanz, M.J., Calvo, E., Cerveró, J., Ansel, W., Klumpp, A., 2007. Ozone biomonitoring with Bel-W3 tobacco plants in the city of Valencia (Spain). Water Air Soil Pollut. 183 (1-4), 283–291.

Cassimiro, J.C., Moura, B.B., Alonso, R., Meirelles, S.T., Moraes, R.M., 2016. Ozone stomatal flux and O3 concentration-based metrics for Astronium graveolens Jacq., a Brazilian native forest tree species. Environ. Pollut. 213, 1007–1015.

Chaudhary, N., Agrawal, S.B., 2013. Intraspecific responses of six Indian clover cultivars under ambient and elevated levels of ozone. Environ. Sci. Pollut. Res. 20 (8), 5318–5329.

Chaudhary, N., Singh, S., Agrawal, S.B., Agrawal, M., 2013. Assessment of six Indian cultivars of mung bean against ozone by using foliar injury index and changes in carbon assimilation, gas exchange, chlorophyll fluorescence and photosynthetic pigments. Environ. Monit. Assess. 185 (9), 7793–7807.

Chen, Z., Li, R., Chen, D., Zhuang, Y., Gao, B., Yang, L., Li, M., 2020. Understanding the causal influence of major meteorological factors on ground ozone concentrations across China. J. Cleaner Prod. 242, 118498.

Cheng, J.F., Sun, E.J., 2013. Factors affecting ozone sensitivity of tobacco Bel-W3 seedlings. Bot. Stud. 54 (1), 1–7.

Chevalier, A., Gheusi, F., Delmas, R., Ordóñez, C., Sarrat, C., Zbinden, R., Thouret, V., Athier, G., Cousin, J.M., 2007. Influence of altitude on ozone levels and variability in the lower troposphere: a ground-based study for western Europe over the period 2001–2004. Atmos. Chem. Phys. 7 (16), 4311–4326.

Cho, K., Kubo, A., Shibato, J., Agrawal, G.K., Saji, H., Rakwal, R., 2012. Global identification of potential gene biomarkers associated with ozone-induced foliar injury in rice seedling leaves by correlating their symptom severity with transcriptome profiling. Int. J. Life Sci. 6 (1), 1–13.

Cho, K., Shibato, J., Agrawal, G.K., Jung, Y.H., Kubo, A., Jwa, N.S., Tamogami, S., Satoh, K., Kikuchi, S., Higashi, T., Kimura, S., 2008. Integrated transcriptomics, proteomics, and metabolomics analyses to survey ozone responses in the leaves of rice seedling. J. Proteome Res. 7 (7), 2980–2998.

Cho, K., Tiwari, S., Agrawal, S.B., Torres, N.L., Agrawal, M., Sarkar, A., Shibato, J., Agrawal, G.K., Kubo, A., Rakwal, R., 2011. Tropospheric ozone and plants: absorption, responses, and consequences. Rev. Environ. Contam. Toxicol. 212, 61–111.

Chutteang, C., Booker, F.L., Na-Ngern, P., Burton, A., Aoki, M., Burkey, K.O., 2016. Biochemical and physiological processes associated with the differential ozone response in ozone-tolerant and sensitive soybean genotypes. Plant Biol. 18, 28–36.

Coates, J., Mar, K.A., Ojha, N., Butler, T.M., 2016. The influence of temperature on ozone production under varying NOx conditions—a modelling study. Atmos. Chem. Phys. 16 (18), 11601–11615.

Coulston, J.W., Smith, G.C., Smith, W.D., 2003. Regional assessment of ozone sensitive tree species using bioindicator plants. Environ. Monit. Assess. 83 (2), 113–127.

Dias, A.P., Dafré, M., Rinaldi, M.C., Domingos, M., 2011. How the redox state of tobacco 'Bel-W3' is modified in response to ozone and other environmental factors in a sub-tropical area?. Environ. Pollut. 159 (2), 458–465.

Elampari, K., Debaje, S.B., Jeyakumar, S.J., Chithambarathanu, T., 2013. Measurements of ozone and its precursor nitrogen dioxide and crop yield losses due to cumulative ozone exposures over 40 ppb (AOT40) in rural coastal southern India. J. Atmos. Chem. 70 (4), 357–371.

El-Khatib, A.A., 2003. The response of some common Egyptian plants to ozone and their use as biomonitors. Environ. Pollut. 124 (3), 419–428.

Emberson, L.D., Büker, P., Ashmore, M.R., Mills, G., Jackson, L.S., Agrawal, M., Atikuzzaman, M.D., Cinderby, S., Engardt, M., Jamir, C., Kobayashi, K., 2009. A comparison of North American and Asian exposure–response data for ozone effects on crop yields. Atmos. Environ. 43 (12), 1945–1953.

Esposito, M.P., Ferreira, M.L., Sant'Anna, S.M., Domingos, M., Souza, S.R., 2009. Relationship between leaf antioxidants and ozone injury in Nicotiana tabacum 'Bel-W3' under environmental conditions in São Paulo, SE–Brazil. Atmos. Environ. 43 (3), 619–623.

European Environmental Agency (EEA), 2016. Air Quality in Europe. https://www.eea.europa.eu/publications/air-quality-in-europe-2016.

Falla, J., Laval-Gilly, P., Henryon, M., Morlot, D., Ferard, J.F., 2000. Biological air quality monitoring: a review. Environ. Monit. Assess. 64 (3), 627–644.

Faoro, F., Iriti, M., 2005. Cell death behind invisible symptoms: early diagnosis of ozone injury. Biol. Plant. 49 (4), 585–592.

Faoro, F., Iriti, M., 2009. Plant cell death and cellular alterations induced by ozone: key studies in Mediterranean conditions. Environ. Pollut. 157 (5), 1470–1477.

Fatima, A., Singh, A.A., Mukherjee, A., Agrawal, M., Agrawal, S.B., 2019. Ascorbic acid and thiols as potential biomarkers of ozone tolerance in tropical wheat cultivars. Ecotoxicol. Environ. Saf. 171, 701–708.

Felzer, B.S., Cronin, T., Reilly, J.M., Melillo, J.M., Wang, X., 2007. Impacts of ozone on trees and crops. C.R. Geosci. 339 (11-12), 784–798.

Feng, Y., Komatsu, S., Furukawa, T., Koshiba, T., Kohno, Y., 2008. Proteome analysis of proteins responsive to ambient and elevated ozone in rice seedlings. Agric. Ecosyst. Environ. 125 (1–4), 255–265.

Feng, Z., Sun, J., Wan, W., Hu, E., Calatayud, V., 2014. Evidence of widespread ozone-induced visible injury on plants in Beijing, China. Environ. Pollut. 193, 296–301.

Ferreira, M.L., Esposito, J.B.N., de Souza, S.R., Domingos, M., 2012. Critical analysis of the potential of Ipomoea nil 'Scarlet O'Hara'for ozone biomonitoring in the sub-tropics. J. Environ. Monit. 14 (7), 1959–1967.

Francini, A., Nali, C., Pellegrini, E., Lorenzini, G., 2008. Characterization and isolation of some genes of the shikimate pathway in sensitive and resistant Centaurea jacea plants after ozone exposure. Environ. Pollut. 151 (2), 272–279.

Frei, M., Tanaka, J.P., Chen, C.P., Wissuwa, M., 2010. Mechanisms of ozone tolerance in rice: characterization of two QTLs affecting leaf bronzing by gene expression profiling and biochemical analyses. J. Exp. Bot. 61 (5), 1405–1417.

Frei, M., Wissuwa, M., Pariasca-Tanaka, J., Chen, C.P., Südekum, K.H., Kohno, Y., 2012. Leaf ascorbic acid level—is it really important for ozone tolerance in rice? Plant Physiol. Biochem. 59, 63–70.

Fuhrer, J., 2009. Ozone risk for crops and pastures in present and future climates. Naturwissenschaften 96 (2), 173–194.

Fumagalli, I., Mignanego, L., Mills, G., 2003. Ozone biomonitoring with clover clones: yield loss and carryover effect under high ambient ozone levels in northern Italy. Agric. Ecosyst. Environ. 95 (1), 119–128.

Furlan, C.M., Moraes, R.M., Bulbovas, P., Domingos, M., Salatino, A., Sanz, M.J., 2007. Psidium guajava 'Paluma' (the guava plant) as a new bio-indicator of ozone in the tropics. Environ. Pollut. 147 (3), 691–695.
Furlan, C.M., Moraes, R.M., Bulbovas, P., Sanz, M.J., Domingos, M., Salatino, A., 2008. Tibouchina pulchra (Cham.) Cogn., a native Atlantic Forest species, as a bio-indicator of ozone: visible injury. Environ. Pollut. 152 (2), 361–365.
Futsaether, C.M., Vollsnes, A.V., Kruse, O.M., Indahl, U.G., Kvaal, K., Eriksen, A.B., 2015. Daylength influences the response of three clover species (Trifolium spp.) to short-term ozone stress. Boreal Environ. Res. 20, 90–104.
Futsaether, C.M., Vollsnes, A.V., Kruse, O.M.O., Otterholt, E., Kvaal, K., Eriksen, A.B., 2009. Effects of the Nordic photoperiod on ozone sensitivity and repair in different clover species studied using infrared imaging. AMBIO J. Hum. Environ. 38 (8), 437–443.
Gerosa, G., Marzuoli, R., Desotgiu, R., Bussotti, F., Ballarin-Denti, A., 2008. Visible leaf injury in young trees of Fagus sylvatica L. and Quercus robur L. in relation to ozone uptake and ozone exposure. An open-top chambers experiment in South Alpine environmental conditions. Environ. Pollut. 152 (2), 274–284.
Ghosh, A., Pandey, B., Agrawal, M., Agrawal, S.B., 2020. Interactive effects and competitive shift between Triticum aestivum L.(wheat) and Chenopodium album L.(fat-hen) under ambient and elevated ozone. Environ. Pollut. 265, 114764.
Ghosh, A., Singh, A.A., Agrawal, M., Agrawal, S.B., 2018. Ozone toxicity and remediation in crop plants. Sustainable Agriculture Reviews. 27. Springer, Cham, pp. 129–169.
Ghude, S.D., Kulkarni, S.H., Kulkarni, P.S., Kanawade, V.P., Fadnavis, S., Pokhrel, S., Jena, C., Beig, G., Bortoli, D., 2011. Anomalous low tropospheric column ozone over Eastern India during the severe drought event of monsoon 2002: a case study. Environ. Sci. Pollut. Res. 18 (8), 1442.
Gimeno, B.S., Penuelas, J., Porcuna, J., Reinert, R.A., 1995. Biomonitoring ozone phytotoxicity in eastern Spain. Water Air Soil Pollut. 85 (3), 1521–1526.
Gombert, S., Asta, J., Seaward, M.R.D., 2006. Lichens and tobacco plants as complementary biomonitors of air pollution in the Grenoble area (Isère, southeast France). Ecol. Indic. 6 (2), 429–443.
Gomez, L.D., Noctor, G., Knight, M.R., Foyer, C.H., 2004. Regulation of calcium signalling and gene expression by glutathione. J. Exp. Bot. 55 (404), 1851–1859.
Gorai, A.K., Tuluri, F., Tchounwou, P.B., Ambinakudige, S., 2015. Influence of local meteorology and $NO_2$ conditions on ground-level ozone concentrations in the eastern part of Texas, USA. Air Qual. Atmos. Health 8 (1), 81–96.
Gottardini, E., Cristofolini, F., Ferretti, M., 2017. Foliar symptoms on Viburnum lantana reflect annual changes in summer ozone concentration in Trentino (northern Italy). Ecol. Indic. 78, 26–30.
Gottardini, E., Cristofori, A., Cristofolini, F., Bussotti, F., Ferretti, M., 2010. Responsiveness of Viburnum lantana L. to tropospheric ozone: field evidence under contrasting site conditions in Trentino, northern Italy. J. Environ. Monit. 12 (12), 2237–2243.
Gottardini, E., Cristofori, A., Cristofolini, F., Nali, C., Pellegrini, E., Bussotti, F., Ferretti, M., 2014. Chlorophyll-related indicators are linked to visible ozone symptoms: evidence from a field study on native Viburnum lantana L. plants in northern Italy. Ecol. Indic. 39, 65–74.
Gottardini, E., Cristofori, A., Pellegrini, E., La Porta, N., Nali, C., Baldi, P., Sablok, G., 2016. Suppression substractive hybridization and NGS reveal differential transcriptome expression profiles in wayfaring tree (Viburnum lantana L.) treated with ozone. Front. Plant Sci. 7, 713.
Govindaraju, M., Ganeshkumar, R.S., Suganthi, K., Suganthi, P., 2018. Functional analysis of rubisco protein in selected plants grown under air pollution stress environment. Indian J. Plant Physiol. 23 (2), 201–208.

Gupta, G.P., Kulshrestha, U., 2016. Biomonitoring and remediation by plantsPlant Responses to Air Pollution. Springer, Singapore, pp. 119–132.
Harmens, H., Mills, G., Hayes, F., Jones, L., Norris, D., Cooper, D., 2009. Air pollution and vegetation: ICP Vegetation annual report 2008/2009.
Harmens, H., Mills, G., Hayes, F., Norris, D., 2011. Air Pollution and Vegetation: ICP Vegetation Annual Report 2010/2011. NERC/Centre for Ecology & Hydrology.
Hayes, F., Mills, G., Harmens, H., Norris, D., 2007. Evidence of Widespread Ozone Damage to Vegetation in Europe (1990-2006). ICP Vegetation Programme Coordination Centre. CEH, Bangor, UK.
Heagle, A.S., Miller, J.E., Sherrill, D.E., 1994A White Clover System to Estimate Effects of Tropospheric Ozone on Plants23. American Society of Agronomy, Crop Science Society of America, and Soil Science Society of America, pp. 613–621.
Heggestad, H.E., 1991. Origin of Bel-W3, Bel-C and Bel-B tobacco varieties and their use as indicators of ozone. Environ. Pollut. 74 (4), 264–291.
ICPVegetation, 2010. Yield Response and Ozone Injury on Phaseolus Vulgaris: Experimental Protocol. CEH, Bangor, UK ICP Vegetation Programme Coordination Centre.
Innes, J.L., Skelly, J.M., Schaub, M., 2001. Ozone and Broadleaved Species. A Guide to the ideNtification of Ozone-Induced Foliar Injury. Paul Haupt Publishing, Bern, Switzrland, p. 136.
IPCC, 2014. Climate Change 2014: Synthesis Report. In: Pachauri, R.K., Meyer, L.A. (Eds.), Contribution of Working Groups I, II and III to the Fifth Assessment Report of the Intergovernmental Panel on Climate Change. IPCC, Geneva, Switzerland, p. 151.
Iriti, M., Belli, L., Nali, C., Lorenzini, G., Gerosa, G., Faoro, F., 2006. Ozone sensitivity of currant tomato (Lycopersicon pimpinellifolium), a potential bioindicator species. Environ. Pollut. 141 (2), 275–282.
Iriti, M., Faoro, F., 2008. Oxidative stress, the paradigm of ozone toxicity in plants and animals. Water Air Soil Pollut. 187, 285–301.
Jacob, D.J., Winner, D.A., 2009. Effect of climate change on air quality. Atmos. Environ. 43 (1), 51–63.
Janzik, I., Preiskowski, S., Kneifel, H., 2005. Ozone has dramatic effects on the regulation of the prechorismate pathway in tobacco (Nicotiana tabacum L. cv. Bel W3). Planta 223 (1), 20–27.
Jenkin, M.E., 2008. Trends in ozone concentration distributions in the UK since 1990: local, regional and global influences. Atmos. Environ. 42 (21), 5434–5445.
Jiang, L., Feng, Z., Dai, L., Shang, B., Paoletti, E., 2018. Large variability in ambient ozone sensitivity across 19 ethylenediurea-treated Chinese cultivars of soybean is driven by total ascorbate. J. Environ. Sci. 64, 10–22.
Kafiatullah, Wahid, A., Sheikh, S.A., Shamsi, S.R.A., 2012. Ozone biomonitoring in Pakistan using tobacco cultivar Bel-W3. Pak. J. Bot. 44 (2), 717–723.
Kang, S., Chen, S., Dai, S., 2010. Proteomics characteristics of rice leaves in response to environmental factors. Front. Biol. 5 (3), 246–254.
Kangasjärvi, J., Jaspers, P., Kollist, H., 2005. Signalling and cell death in ozone-exposed plants. Plant Cell Environ. 28 (8), 1021–1036.
Karlsson, G.P., Sellden, G., Skärby, L., Pleijel, H., 1995. Clover as an indicator plant for phytotoxic ozone concentrations: visible injury in relation to species, leaf age and exposure dynamics. New Phytol. 129 (2), 355–365.
Kefauver, S.C., Penuelas, J., Ribas, A., Díaz-de-Quijano, M., Ustin, S., 2014. Using Pinus uncinata to monitor tropospheric ozone in the Pyrenees. Ecol. Indic. 36, 262–271.
Keller, T.H., 1988. Growth and premature leaf fall in American aspen as bioindications for ozone. Environ. Pollut. 52 (3), 183–192.
Kleanthous, S., Vrekoussis, M., Mihalopoulos, N., Kalabokas, P., Lelieveld, J., 2014. On the temporal and spatial variation of ozone in Cyprus. Sci. Total Environ. 476, 677–687.

Klumpp, A., Ansel, W., Klumpp, G., Vergne, P., Sifakis, N., Sanz, M.J., Rasmussen, S., Ro-Poulsen, H., Ribas, A., Penuelas, J., Kambezidis, H., 2006a. Ozone pollution and ozone biomonitoring in European cities Part II. Ozone-induced plant injury and its relationship with descriptors of ozone pollution. Atmos. Environ. 40 (38), 7437–7448.

Klumpp, A., Ansel, W., Klumpp, G., Vergne, P., Sifakis, N., Sanz, M.J., Rasmussen, S., Ro-Poulsen, H., Ribas, A., Penuelas, J., Kambezidis, H., 2006b. Ozone pollution and ozone biomonitoring in European cities Part II. Ozone-induced plant injury and its relationship with descriptors of ozone pollution. Atmos. Environ. 40 (38), 7437–7448.

Komatsu, S., 2008. Research on the rice proteome: the contribution of proteomics technology in the creation of abiotic stress-tolerant plants. Rice 1 (2), 154–165.

Krupa, S.V., Manning, W.J., Nosal, M., 1993. Use of tobacco cultivars as biological indicators of ambient ozone pollution: an analysis of exposure-response relationships. Environ. Pollut. 81 (2), 137–146.

Krupa, S.V., Tonneijck, A.E.G. & Manning, W. (1998). Ozone. In: Recognition of Air Pollution Injury to Vegetation: A Pictorial Atlas. Flager R.B. (Ed.). Air & Waste Management Association, Pittsburgh.

Kumar, A., Singh, D., Singh, B.P., Singh, M., Anandam, K., Kumar, K., Jain, V.K., 2015. Spatial and temporal variability of surface ozone and nitrogen oxides in urban and rural ambient air of Delhi-NCR, India. Air Qual. Atmos. Health 8 (4), 391–399.

Kumar, N., Middey, A., Rao, P.S., 2017. Prediction and examination of seasonal variation of ozone with meteorological parameter through artificial neural network at NEERI, Nagpur, India. Urban Clim. 20, 148–167.

Laban, T.L., Zyl, P.G.V., Beukes, J.P., Vakkari, V., Jaars, K., Borduas-Dedekind, N., Josipovic, M., Thompson, A.M., Kulmala, M., Laakso, L., 2018. Seasonal influences on surface ozone variability in continental South Africa and implications for air quality. Atmos. Chem. Phys. 18 (20), 15491–15514.

Labuschagne, C.F., 2007. The Effect of Autohaemotherapy With Different Ozone Concentrations on the Oxidant/Antioxidant Status and DNA Integrity of Baboons. North-West University Doctoral dissertation.

Laffray, X., Rose, C., Garrec, J.P., 2007. Estimation of ozone concentration in a Valley of the Alps mountains based on Bel-W3 tobacco leaf injury. Water Air Soil Pollut. 186 (1), 29–42.

Lal, D.M., Ghude, S.D., Patil, S.D., Kulkarni, S.H., Jena, C., Tiwari, S., Srivastava, M.K., 2012. Tropospheric ozone and aerosol long-term trends over the Indo-Gangetic Plain (IGP), India. Atmos. Res. 116, 82–92.

Langner, J., Engardt, M., Baklanov, A., Christensen, J.H., Gauss, M., Geels, C., Hedegaard, G.B., Nuterman, R., Simpson, D., Soares, J., Sofiev, M., 2012. A multi-model study of impacts of climate change on surface ozone in Europe. Atmos. Chem. Phys. 12 (21), 10423–10440.

Lei, H., Wuebbles, D.J., Liang, X.Z., 2012. Projected risk of high ozone episodes in 2050. Atmos. Environ. 59, 567–577.

Leung, F., Pang, J., Tai, A.P., Lam, T., Tao, D.K., Sharps, K., 2020. Evidence of ozone-induced visible foliar injury in Hong Kong using Phaseolus vulgaris as a bioindicator. Atmosphere 11 (3), 266.

Lorenzini, G., Nali, C., Biagioni, M., 1995. Long range transport of photochemical ozone over the Tyrrhenian Sea demonstrated by a new miniaturized bioassay with ozone-sensitive tobacco seedlings. Sci. Total Environ. 166 (1-3), 193–199.

Lorenzini, G., Saitanis, C., 2003. Ozone: a novel plant "pathogen". In Abiotic stresses in plants, Springer, Dordrecht, pp. 205–229.

Madkour, S.A., Laurence, J.A., 2002. Egyptian plant species as new ozone indicators. Environ. Pollut. 120 (2), 339–353.

Mahmood, F., Khokhar, M.F., Mahmood, Z., 2020. Examining the relationship of tropospheric ozone and climate change on crop productivity using the multivariate panel data techniques. J. Environ. Manage. 272, 111024.
Manes, F., De Santis, F., Giannini, M.A., Vazzana, C., Capogna, F., Allegrini, I., 2003. Integrated ambient ozone evaluation by passive samplers and clover biomonitoring mini-stations. Sci. Total Environ. 308 (1-3), 133–141.
Manninen, S., Huttunen, S., Tømmervik, H., Hole, L.R., Solberg, S., 2009. Northern plants and ozone. AMBIO J. Hum. Environ. 38 (8), 406–413.
Manning, W.J., 1998. The use of plants as bioindicators of ozoneAir Pollution and Climate Change Effects on Forest Ecosystems, 19.
Manning, W.J., Godzik, B., 2004. Bioindicator plants for ambient ozone in Central and Eastern Europe. Environ. Pollut. 130 (1), 33–39.
Martins, E.M., Nunes, A.C., Corrêa, S., 2015. Understanding ozone concentrations during weekdays and weekends in the urban area of the city of Rio de Janeiro. J. Braz. Chem. Soc. 26 (10), 1967–1975.
Mauzerall, D.L., Wang, X., 2001. Protecting agricultural crops from the effects of tropospheric ozone exposure: reconciling science and standard setting in the United States, Europe, and Asia. Annu. Rev. Energy Env. 26 (1), 237–268.
Mignanego, L., Biondi, F., Schenone, G., 1992. Ozone biomonitoring in northern Italy. Environ. Monit. Assess. 21 (2), 141–159.
Mills, G., Wagg, S., Harmens, H., 2013. Ozone Pollution: Impacts on Ecosystem Services and Biodiversity. NERC/Centre for Ecology & Hydrology.
Mishra, A.K., Rai, R., Agrawal, S.B., 2013. Differential response of dwarf and tall tropical wheat cultivars to elevated ozone with and without carbon dioxide enrichment: growth, yield and grain quality. Field Crops Res. 145, 21–32.
Monks, P.S., Archibald, A.T., Colette, A., Cooper, O., Coyle, M., Derwent, R., Fowler, D., Granier, C., Law, K.S., Mills, G.E., Stevenson, D.S., 2015. Tropospheric ozone and its precursors from the urban to the global scale from air quality to short-lived climate forcer. Atmos. Chem. Phys. 15, 8889–8973.
Morgan, P.B., Ainsworth, E.A., Long, S.P., 2003. How does elevated ozone impact soybean? A meta-analysis of photosynthesis, growth and yield. Plant Cell Environ. 26 (8), 1317–1328.
Morgan, P.B., Mies, T.A., Bollero, G.A., Nelson, R.L., Long, S.P., 2006. Season-long elevation of ozone concentration to projected 2050 levels under fully open-air conditions substantially decreases the growth and production of soybean. New Phytol. 170 (2), 333–343.
Naja, M., Akimoto, H., 2004. Contribution of regional pollution and long-range transport to the Asia-Pacific region: Analysis of long-term ozonesonde data over Japan. J. Geophys. Res.: Atmos. 109 (D21).
Nakazato, R.K., Esposito, M.P., Cardoso-Gustavson, P., Bulbovas, P., Pedroso, A.N.V., de Assis, P.I.L.S., Domingos, M., 2018. Efficiency of biomonitoring methods applying tropical bioindicator plants for assessing the phytoxicity of the air pollutants in SE, Brazil. Environ. Sci. Pollut. Res. 25 (20), 19323–19337.
Nali, C., Francini, A., Lorenzini, G., 2009. White clover clones as a cost-effective indicator of phytotoxic ozone: 10 years of experience from central Italy. Environ. Pollut. 157 (5), 1421–1426.
Nali, C., Pucciariello, C., Mills, G., Lorenzini, G., 2005. On the different sensitivity of white clover clones to ozone: physiological and biochemical parameters in a multivariate approach. Water Air Soil Pollut. 164 (1-4), 137–153.
Nali, C., Francini, A., Lorenzini, G., 2006. Biological monitoring of ozone: the twenty-year Italian experience. J. Environ. Monit. 8 (1), 25–32.
Nishanth, T., Kumar, M.S., Valsaraj, K.T., 2012. Variations in surface ozone and NOx at Kannur: a tropical, coastal site in India. J. Atmos. Chem. 69 (2), 101–126.

Novak, K., Skelly, J.M., Schaub, M., Kräuchi, N., Hug, C., Landolt, W., Bleuler, P., 2003. Ozone air pollution and foliar injury development on native plants of Switzerland. Environ. Pollut. 125 (1), 41–52.
Ohara, T.A.H.K., Akimoto, H., Kurokawa, J.I., Horii, N., Yamaji, K., Yan, X., Hayasaka, T., 2007. An Asian emission inventory of anthropogenic emission sources for the period 1980–2020. Atmos. Chem. Phys. 7 (16), 4419–4444.
Ojha, N., Naja, M., Singh, K.P., Sarangi, T., Kumar, R., Lal, S., Lawrence, M.G., Butler, T.M., Chandola, H.C., 2012. Variabilities in ozone at a semi-urban site in the Indo-Gangetic Plain region: Association with the meteorology and regional processes. J. Geophys. Res. Atmos. (D20), 117.
Orendovici, T., Skelly, J.M., Ferdinand, J.A., Savage, J.E., Sanz, M.J., Smith, G.C., 2003. Response of native plants of northeastern United States and southern Spain to ozone exposures; determining exposure/response relationships. Environ. Pollut. 125 (1), 31–40.
Orlando, J.P., Alvim, D.S., Yamazaki, A., Corrêa, S.M., Gatti, L.V., 2010. Ozone precursors for the São Paulo Metropolitan area. Sci. Total Environ. 408, 1612–1620.
Padri, M., Umponstira, C., 2019. Coleus cultivars (Solenostemon scutellarioides (L.) Codd.) as potential bioindicators of chronic ozone exposure. In IOP Conference Series: Earth and Environmental Science, 308. IOP Publishing, 012004.
Pal, P., 2014. Biomonitoring with special reference to visible damages in different plant species due to air pollution-A review. Int. Lett. Nat. Sci. 11 (1), 32–37.
Pandey, A.K., Majumder, B., Keski-Saari, S., Kontunen-Soppela, S., Mishra, A., Sahu, N., Pandey, V., Oksanen, E., 2015. Searching for common responsive parameters for ozone tolerance in 18 rice cultivars in India: Results from ethylenediurea studies. Sci. Total Environ. 532, 230–238.
Paoletti, E., De Marco, A., Beddows, D.C., Harrison, R.M., Manning, W.J., 2014. Ozone levels in European and USA cities are increasing more than at rural sites, while peak values are decreasing. Environ. Pollut. 192, 295–299.
Paoletti, E., Ferrara, A.M., Calatayud, V., Cerveró, J., Giannetti, F., Sanz, M.J., Manning, W.J., 2009. Deciduous shrubs for ozone bioindication: Hibiscus syriacus as an example. Environ. Pollut. 157 (3), 865–870.
Parvanova, P., Tzvetkova, N., Bratanova-Doncheva, S., Fikova, R., Lyubenova, M., 2009. Active Biomonitoring of Ecotoxicological Effects from Tropospheric Ozone Impact on Representative Tree Species. Biotechnol. Biotechnol. Equip. 23 (supp1), 333–336.
Pedroso, A.N.V., Alves, E.S., 2015. Temporal dynamics of the cellular events in tobacco leaves exposed in São Paulo, Brazil, indicate oxidative stress by ozone. Environ. Sci. Pollut. Res. 22 (9), 6535–6545.
Pellegrini, E., Francini, A., Lorenzini, G., Nali, C., 2011. PSII photochemistry and carboxylation efficiency in Liriodendron tulipifera under ozone exposure. Environ. Exp. Bot. 70 (2-3), 217–226.
Peñuelas, J., Ribas, A., Gimeno, B.S., Filella, I., 1999. Dependence of ozone biomonitoring on meteorological conditions of different sites in Catalonia (NE Spain). Environ. Monit. Assess. 56 (2), 221–224.
Picchi, V., Monga, R., Marzuoli, R., Gerosa, G., Faoro, F., 2017. The ozone-like syndrome in durum wheat (Triticum durum Desf.): mechanisms underlying the different symptomatic responses of two sensitive cultivars. Plant Physiol. Biochem. 112, 261–269.
Piraino, F., Aina, R., Palin, L., Prato, N., Sgorbati, S., Santagostino, A., Citterio, S., 2006. Air quality biomonitoring: assessment of air pollution genotoxicity in the Province of Novara (North Italy) by using Trifolium repens L. and molecular markers. Sci. Total Environ. 372 (1), 350–359.
Postiglione, L., Fagnano, M., Merola, G., 2000. Response to ambient ozone of two white clover (Trifolium repens L. cv. "Regal") clones, one resistant and one sensitive, grown in a Mediterranean environment. Environ. Pollut. 109 (3), 525–531.

Puckette, M.C., Tang, Y., Mahalingam, R., 2008. Transcriptomic changes induced by acute ozone in resistant and sensitive Medicago truncatulaaccessions. BMC Plant Biol. 8 (1), 46.

Rai, R., Singh, A.A., Agrawal, S.B., Agrawal, M., 2016. Tropospheric $O_3$: a cause of concern for terrestrial plantsPlant Responses to Air Pollution. Springer, Singapore, pp. 165–195.

Reddy, B.S.K., Kumar, K.R., Balakrishnaiah, G., Gopal, K.R., Reddy, R.R., Sivakumar, V., Lingaswamy, A.P., Arafath, S.M., Umadevi, K., Kumari, S.P., Ahammed, Y.N., 2012. Analysis of diurnal and seasonal behavior of surface ozone and its precursors (NOx) at a semi-arid rural site in southern India. Aerosol Air Qual. Res. 12 (6), 1081–1094.

Restivo, F.M., Laccone, M.C., Buschini, A., Rossi, C., Poli, P., 2002. Indoor and outdoor genotoxic load detected by the Comet assay in leaves of Nicotiana tabacum cultivars Bel B and Bel W3. Mutagenesis 17 (2), 127–134.

Ribas, A., Peñuelas, J., 2003. Biomonitoring of tropospheric ozone phytotoxicity in rural Catalonia. Atmos. Environ. 37 (1), 63–71.

Ribas, A., Filella, I., Gimeno, B.S., Peñuelas, J., 1998. Evaluation of tobacco cultivars as bioindicators and biomonitors of ozone phytotoxical levels in Catalonia. Water Air Soil Pollut. 107 (1), 347–365.

Roth, A.R., 2017. Differential Ozone Response in Gas Exchange and Antioxidant Levels in Ozone-Tolerant and Sensitive Soybean Genotypes.

Society, R., 2008. Ground-Level Ozone in the 21st Century: Future Trends, Impacts and Policy Implications. The Royal Society, London Science Policy report 15/08 October.

Saitanis, C.J., Panagopoulos, G., Dasopoulou, V., Agathokleous, E., Papatheohari, Y., 2015. Integrated assessment of ambient ozone phytotoxicity in Greece's Tripolis Plateau. J. Agric. Meteorol. 71 (2), 55–64.

Sandrin, C.Z., Gonçalves da Silva Engela, M.R., Sacramento, G.N., Domingos, M., 2018. Ozone biomonitoring with white clover clones: the first experience in Brazil. CLEAN–Soil Air Water 46 (12), 1800377.

Sarkar, A., Rakwal, R., Bhushan Agrawal, S., Shibato, J., Ogawa, Y., Yoshida, Y., Kumar Agrawal, G., Agrawal, M., 2010. Investigating the impact of elevated levels of ozone on tropical wheat using integrated phenotypical, physiological, biochemical, and proteomics approaches. J. Proteome Res. 9 (9), 4565–4584.

Sarkar, A., Singh, A.A., Agrawal, S.B., Ahmad, A., Rai, S.P., 2015. Cultivar specific variations in antioxidative defense system, genome and proteome of two tropical rice cultivars against ambient and elevated ozone. Ecotoxicol. Environ. Saf. 115, 101–111.

Saviranta, N.M., Julkunen-Tiitto, R., Oksanen, E., Karjalainen, R.O., 2010. Leaf phenolic compounds in red clover (Trifolium pratense L.) induced by exposure to moderately elevated ozone. Environ. Pollut. 158 (2), 440–446.

Scebba, F., Soldatini, G., Ranieri, A., 2003. Ozone differentially affects physiological and biochemical responses of two clover species; Trifolium repens and Trifolium pratense. Environ. Pollut. 123 (2), 209–216.

Schraudner, M., Moeder, W., Wiese, C., Camp, W.V., Inzé, D., Langebartels, C., Sandermann Jr, H., 1998. Ozone-induced oxidative burst in the ozone biomonitor plant, tobacco Bel W3. Plant J. 16 (2), 235–245.

Schaub, M., Skelly, J.M., Steiner, K.C., Davis, D.D., Pennypacker, S.P., Zhang, J., Ferdinand, J.A., Savage, J.E., Stevenson, R.E., 2003. Physiological and foliar injury responses of Prunus serotina, Fraxinus americana, and Acer rubrum seedlings to varying soil moisture and ozone. Environ. Pollut. 124 (2), 307–320.

Sicard, P., De Marco, A., Troussier, F., Renou, C., Vas, N., Paoletti, E., 2013. Decrease in surface ozone concentrations at Mediterranean remote sites and increase in the cities. Atmos. Environ. 79, 705–715.

Sicard, P., Serra, R., Rossello, P., 2016. Spatiotemporal trends in ground-level ozone concentrations and metrics in France over the time period 1999–2012. Environ. Res. 149, 122–144.

Silva, D.T., Meirelles, S.T., Moraes, R.M., 2012. Relationship between ozone, meteorological conditions, gas exchange and leaf injury in Nicotiana tabacum Bel-W3 in a subtropical region. Atmos. Environ. 60, 211–216.
Singh, A.A., Agrawal, S.B., 2017. Tropospheric ozone pollution in India: Effects on crop yield and product quality. Environ. Sci. Pollut. Res. 24 (5), 4367–4382.
Singh, A.A., Agrawal, S.B., Shahi, J.P., Agrawal, M., 2014. Assessment of growth and yield losses in two Zea mays L. cultivars (quality protein maize and nonquality protein maize) under projected levels of ozone. Environ. Sci. Pollut. Res. 21 (4), 2628–2641.
Singh, A.A., Fatima, A., Mishra, A.K., Chaudhary, N., Mukherjee, A., Agrawal, M., Agrawal, S.B., 2018. Assessment of ozone toxicity among 14 Indian wheat cultivars under field conditions: growth and productivity. Environ. Monit. Assess. 190 (4), 190.
Singh, A.A., Singh, S., Agrawal, M., Agrawal, S.B., 2015. Assessment of ethylene diurea-induced protection in plants against ozone phytotoxicity. Rev. Environ. Contam. Toxicol. 233, 129–184.
Singh, S., Bhatia, A., Tomer, R., Kumar, V., Singh, B., Singh, S.D., 2013. Synergistic action of tropospheric ozone and carbon dioxide on yield and nutritional quality of Indian mustard (Brassica juncea (L.) Czern.). Environ. Monit. Assess. 185 (8), 6517–6529.
Sinha, B., Singh Sangwan, K., Maurya, Y., Kumar, V., Sarkar, C., Chandra, B.P., Sinha, V., 2015. Assessment of crop yield losses in Punjab and Haryana using 2 years of continuous in situ ozone measurements. Atmos. Chem. Phys. 15 (16), 9555–9576.
Skelly, J.M., Davis, D.D., Merrill, W., Cameron, E.A., Brown, H.D., Drummond, D.B., Leon, S., 1987. Diagnosing Injury to Eastern Forest Trees. The Pennsylvania State University, Washington, DC, pp. 1–122 for the USDA Forest Service. p. 122.
Skelly, J.M., Innes, J.L., Snyder, K.R., Savage, J.E., Hug, C., Landolt, W., Bleuler, P., 1998. Investigations of ozone induced injury in forests of southern Switzerland: field surveys and open-top chamber experiments. Chemosphere 36 (4-5), 995–1000.
Smith, G., 2012. Ambient ozone injury to forest plants in Northeast and North Central USA: 16 years of biomonitoring. Environ. Monit. Assess. 184 (7), 4049–4065.
Smith, G., Coulston, J., Jepsen, E., Prichard, T., 2003. A national ozone biomonitoring program–results from field surveys of ozone sensitive plants in northeastern forests (1994–2000). Environ. Monit. Assess. 87 (3), 271–291.
Staehelin, J., Poberaj, C.S., 2008. Long-term tropospheric ozone trends: A critical review. Climate Variability and Extremes During the Past 100 Years, 271–282.
Tai, A.P., Martin, M.V., Heald, C.L., 2014. Threat to future global food security from climate change and ozone air pollution. Nat. Clim. Change 4 (9), 817–821.
Temmerman, L.D., Bell, N.B., Garrec, J.P., Klumpp, A., Krause, G.H., Tonneijck, A.E., 2001. Biomonitoring of air pollutants with plants—considerations for the future in EuroBionet – Urban Air Pollution. Bioindic. Environ. Awareness, 337–373.
Thambavani, S.D., Prathipa, V., 2012. Biomonitoring of air pollution around urban and industrial sites. J. Res. Biol. 1, 007–014.
Tiwari, S., Agrawal, M., 2018. Ozone Biomonitoring, Biomass and Yield Response Tropospheric Ozone and Its Impacts on Crop Plants. Springer, Cham, pp. 115–166.
Tiwari, S., Rai, R., Agrawal, M., 2008. Annual and seasonal variations in tropospheric ozone concentrations around Varanasi. Int. J. Remote Sens. 29 (15), 4499–4514.
U.S. Environmental Protection Agency (USEPA) 1980-2008 average annual emissions, all criteria pollutants in MS Excel. National Emissions Trend Data, Office of Air Quality planning and Standards 2009.
U.S. Environmental Protection Agency (USEPA), 2006. Air quality criteria for resistance and a possible metric. Atmos. Environ. 38, 2323–2337.
Van Dingenen, R., Dentener, F.J., Raes, F., Krol, M.C., Emberson, L., Cofala, J., 2009. The global impact of ozone on agricultural crop yields under current and future air quality legislation. Atmos. Environ. 43 (3), 604–618.

Villányi, V., 2008. Ozone biomonitoring at mountainous and lowland areas in Hungary. Acta Biol. Szeged. 52 (1), 209–212.
Vingarzan, R., 2004. A review of surface ozone background levels and trends. Atmos. Environ. 38 (21), 3431–3442.
Vollenweider, P., Fenn, M.E., Menard, T., Günthardt-Goerg, M. and Bytnerowicz, A., 2013. Structural injury underlying mottling in ponderosa pine needles exposed to ambient ozone concentrations in the San Bernardino Mountains near Los Angeles, California. Trees, 27(4), pp. 895-911.
Wang, J., Zeng, Q., Zhu, J., Liu, G., Tang, H., 2013. Dissimilarity of ascorbate–glutathione (AsA–GSH) cycle mechanism in two rice (Oryza sativa L.) cultivars under experimental free-air ozone exposure. Agric. Ecosys. Environ. 165, 39–49.
Wang, T., Wei, X.L., Ding, A.J., Poon, C.N., Lam, K.S., Li, Y.S., Chan, L.Y., Anson, M., 2009. Increasing surface ozone concentrations in the background atmosphere of Southern China, 1994–2007. Atmos. Chem. Phys. 9 (16), 6217–6227.
Wang, Y., Zhang, Y., Hao, J., Luo, M., 2011. Seasonal and spatial variability of surface ozone over China: contributions from background and domestic pollution. Atmos. Chem. Phys. 11 (7), 3511–3525.
Winiwarter, W., Klimont, Z., 2011. The role of N-gases ($N_2O$, $NO_x$, $NH_3$) in cost-effective strategies to reduce greenhouse gas emissions and air pollution in Europe. Curr. Opin. Environ. Sustain. 3 (5), 438–445.
Yadav, D.S., Mishra, A.K., Rai, R., Chaudhary, N., Mukherjee, A., Agrawal, S.B., Agrawal, M., 2020. Responses of an old and a modern Indian wheat cultivar to future $O_3$ level: physiological, yield and grain quality parameters. Environ. Pollut. 259, 113939.
Yan, Y., Pozzer, A., Ojha, N., Lin, J., Lelieveld, J., 2018. Analysis of European ozone trends in the period 1995–2014. Atmos. Chem. Phys. 18 (8), 5589–5605.
Zeng, G., Pyle, J.A., Young, P.J., 2008. Impact of climate change on tropospheric ozone and its global budgets. Atmos. Chem. Phys. 8 (2), 369–387.

# CHAPTER 9

# Statistical modeling and evaluation of research tools employed in biomonitoring of plant species for air pollution abatement: A case study of Nigeria

**Winifred Uduak Anake[a], Oluwole Akinwumi Odetunmibi[b], Timothy Ashibel Anake[b]**
[a]Department of Chemistry, College of Science & Technology, Covenant University, Ota, Nigeria
[b]Department of Mathematics, College of Science & Technology, Covenant University, Ota, Nigeria

## 9.1 Introduction

The air pollution mitigation capacity of trees in adverse environmental conditions has received increasing attention in the recent decade (Kwak et al., 2020; Khanoranga, 2019; Selmi et al., 2016). This is due to the severe nature of urban air pollution, which requires continuous air quality monitoring and timely dissemination of information on air quality. However, this is not feasible in some cases, following the high cost of instrument acquisition, insecurity, lack of manpower, and sustainable funding. Research tools that assemble physiological, biochemical, and socioeconomic parameters of plants could provide a cost-effective and eco-friendly alternative (Chaudhary and Rathore, 2019).

Plants act as biofilters and green tools for air pollution monitoring and control, taking advantage of the large surface area of their leaves for air pollutants impression, absorption, and accumulation (Chaudhary and Rathore, 2019; Alhesnawi et al., 2018; Wei et al., 2017). A significant amount of particulate matter pollutants adsorbs on leaf surfaces while gaseous pollutants are absorbed through the leaf stomata (Karmakar et al., 2021; Anake et al., 2018). Oxidant pollutants such as particulates bound with redox-cycling substances, ozone, and nitrogen oxides are prominent sources of oxidative stress to plants, animals, and humans in the environment (Cross et al., 2002).

*New Paradigms in Environmental Biomonitoring Using Plants*
DOI: https://doi.org/10.1016/B978-0-12-824351-0.00011-0

Biomonitoring is a method applied to measure living organisms' response to changes in their environment or suspected endangered environment over time and space. This technique has gained increasing attention over the years and complements the physicochemical approaches for air quality evaluation (Khairallah et al., 2018; Cuny, 2019). Biological monitoring of variations in the environmental conditions due to air pollutants can be performed using passive or active bioindicators. Passive bioindicators are plants already growing in the proposed study sites. On the contrary, active bioindicators are plants imported from an unpolluted environment into the proposed study sites by transplantation and exposed over a stipulated time frame (Cuny, 2019; Badamasi, 2017).

The impact of air pollutants such as sulfur dioxide, nitrogen dioxide, ozone, particulate matter, heavy metals, fluoro-compounds, volatile organic compounds etc., on the environment has been determined from the monitoring of lichens, mosses, tree leaves, tree bark, tree rings, shrubs and herbaceous plants. Unlike plants, lichens do not have roots; hence, they are excellent indicators of airborne pollutants since they absorb nutrients directly from the atmosphere. In urban and industrial areas, where there is a paucity of lichens, leaves of trees are specially chosen as bioindicators due to their direct contact with air pollutants (Karmakar et al., 2021; Badamasi, 2017; Maatoug et al., 2012). However, knowledge gaps exist, such as dearth of information on plant species specific removal rates for different pollutants and a clear understanding of the impact the local vegetation structure has on air quality.

Biomonitoring studies often use statistical models to identify key relationships between variables. Statistical model refers to applying statistical methods for the analysis of a set of experimental or previously monitored air quality data (Konishi and Kitagawa, 2008; Asmussen and Kroese, 2006). It presents a collection of observed data in mathematical equations to predict and make a decision. Statistical models predict future occurrence of cases under investigation using the current sets of data, which helps the nonanalysts and stakeholders make essential and informed decisions (Botev and Kroese, 2012; Kroese and Chan, 2014). Several statistical models like linear regression model, logistic model, log-linear model etc. could be employed or utilized. Most standard techniques are categorized as supervised learning, including regression and classification models and unsupervised learning, such as clustering algorithms and association rules (Asmussen and Kroese, 2006; Kroese and Chan, 2014). Regression models are often used to determine the independent variables that influence the dependent variables. This information can be leveraged to make informed decisions.

Many research tools have been employed in identifying suitable plants for air pollution abatement. The most common tools presented in Section 9.2 include air pollution tolerance index (APTI), anticipated performance index (API), and dust capturing potentials (DCP).

## 9.2 Research tools used in screening suitable plants for air pollution abatement

APTI, API, and DCP are some research tools used in identifying suitable plants for air pollution control. Identifying suitable plant species for green belt development is most effective when a combination of both APTI and API indices is applied (Karmakar et al., 2021; Ogunkunle et al., 2015; Leghari et al., 2019).

### 9.2.1 APTI

The APTI indicates plants' potential to act as sensitive or tolerant species against air pollution (Girish et al., 2017; Shrestha et al., 2021; Anake et al., 2019; Molnár et al., 2020). Plants' ability to respond to pollutants in the air depends on a combination of different physiological and biochemical parameters, such as pH of leaf extract, ascorbic acid contents, relative water content, and total chlorophyll content (Roy et al., 2020; Singh and Rao, 1983). Ascorbic acid is a strong reducing agent that activates the defense mechanism of the plant. It increases at a higher pH value and reduces at a lower pH value. Plants with a higher amount of ascorbic acid are regarded as tolerant to air pollution. The reverse applies to sensitive plants (Ogunkunle et al., 2015; Singh et al., 1991). Relative water content (RWC) reveals the potential of a plant's cell membrane in maintaining its permeability. Increased RWC implies more tolerant plants since it promotes physiological balance in extreme environmental conditions (Anake et al., 2019; Roy et al., 2020). pH promotes the photosynthetic efficiency of plants due to the conversion of hexose sugar into ascorbic acid increases at high pH. Increased pH in plants is attributed to their tolerance capacity (Khanoranga, 2019). Also, the total chlorophyll content is significant in plant metabolism, a reduction in the total chlorophyll content in the polluted environment reflects an increase in plant sensitivity and vice versa (Ogunkunle et al., 2015; Singh et al., 1991).

These parameters are used in APTI computation, as proposed by Singh and Rao (Singh and Rao, 1983). Plants with APTI values less than or equals to 11, 12–16 and, greater than or equals to 17 are grouped as sensitive, intermediately tolerant or sensitive, and tolerant to air pollution, respectively. Based on APTI grading, plants in the sensitive range are effective bioindicators, while those classified as tolerant species are suitable for green belt development (Ogunkunle et al., 2015; Anake et al., 2019; Zhang et al., 2020; Nadgorska Socha et al., 2017). Current scientific evidence indicates that APTI index has been applied by several researchers globally including China (Zhang et al., 2020; Zhang et al., 2016), France (Selmi et al., 2016), Ghana (Uka et al., 2019), Hungary (Molnár et al., 2020), India (Karmakar et al., 2021; Shrestha et al., 2021; Sahu et al., 2020; Prajapati and Tripathi, 2008; Pathak et al., 2011; Kaur and Nagpal, 2017), Indonesia (Sulistijorini Mas'ud et al., 2008), Iran (Javanmard et al., 2019; Ghafari et al., 2021), Iraq (Alhesnawi et al., 2018), Nigeria (Ogunkunle et al., 2015; Anake et al., 2019; Ogbonna et al., 2019), and Pakistan (Khanoranga, 2019) to screen sensitive and tolerant plants for air pollution control.

### 9.2.2 API

API involves grading plants following relevant socioeconomic and biological parameters such as leaf surface and texture, canopy structure, tree height and hardness, and the APTI.

The criteria employed for calculating the API of various plant species can be found in the method given by Prajapati and Tripathi (2008), Ogunkunle et al. (2015), Pandey et al. (2015), and Leghari et al. (2019). Tree species are allotted 16 points and scaled to percentages to ascertain the performance ability based on the score obtained. Following the API scores (%) plants are further ranked as follows: best = 91–100, excellent = 81–90, very good = 71–80, good = 61–70, moderate = 51–60, poor = 41–50, very poor = 31–40, and not recommended <30 (Karmakar et al., 2021; Jain et al., 2019; Kashyap et al., 2018).

### 9.2.3 DCP

Trees have higher dust or particulate absorbing and trapping capacity than shorter vegetations. This potential depends on plant height, leaf area, size, texture, shape, orientation, and petioles of the leaf, absence or presence of hair, leaf micromorphology etc. Other contributing factors include meteorological factors such as wind speed and wind direction (Roy et al., 2020; Popek et al., 2018; Wang et al., 2017; Sahu and Sahu, 2015). Particular emphasis has been placed on the small size, complex shape and hairy/waxy surface, foliage longevity, canopy density, and significant leaf characteristics for accumulated dust or particulate matter. Although dust depositions on plants induce an adverse effect on physiological and micro-morphological characteristics of trees such as low chlorophyll pigment, mutations, chromosomal damage, yellowing of leaves (chlorosis), browning of leaves (necrosis), reduced pollen viability and pollen size (Kaur et al., 2017; Esfahani et al., 2013). Some researchers attribute increased dust tolerating potentials in some plants species to the synthesis of carotenoids and proteins, resulting in the nonenzymatic resistance of plants to numerous abiotic stresses (Prajapati and Tripathi, 2008; Singh and Pal, 2017).

## 9.3 Case study of Nigeria

Our study constitutes an attempt to identify suitable bioindicators via the application of different research tools along with common statistical models for air pollution monitoring and control in Ogun state, Nigeria. The novelty of the present study is reflected in the application of dust capturing potential research tool and statistical models in this geographic area. In recent years, the semi-urban area of Ogun State, Nigeria, has experienced increasing pollution due to improper gaseous and particulate emission control from industrial processes, urbanization, vehicular traffic, and loss of natural vegetation (Anake et al., 2017; Anake et al., 2018; Etim, 2012). The absence of adequate enforcement of pollution control by authorized agencies has necessitated the deployment of sustainable and greener measures to curb the increasing environmental ambient air pollution. Hence, the potential of ten selected plant species from three sites (Industrial, Academic and agricultural (control)) has been investigated to (1) determine their air pollution tolerance levels using APTI approach and DCP, (2) screen their performance ability for green belt design, and (3) model the relationship between APTI and each of the plant biochemical parameters.

### 9.3.1 Materials and methods

#### *9.3.1.1 Study area and sample collection*

The sampling sites were marked as three locations: agricultural site, Atan (control); Industrial site, Ota; and Academic site, Canaan land, in Ogun State, Nigeria (Fig. 9.1). Commonly occurring tree species were selected from each site, and GPS72H, GARMIN, Taiwan was used to identify the sampling locations. Simple random sampling method was adopted in the collection of leaves from each species. Matured plant leaves of *Elaesis guineensis, Terminalia catappa, Mangifera indica, Carica papaya, Manihot esculenta Crantz, Tectona grandis, Anacardium occidentale, Cassia fistular, Musanga*

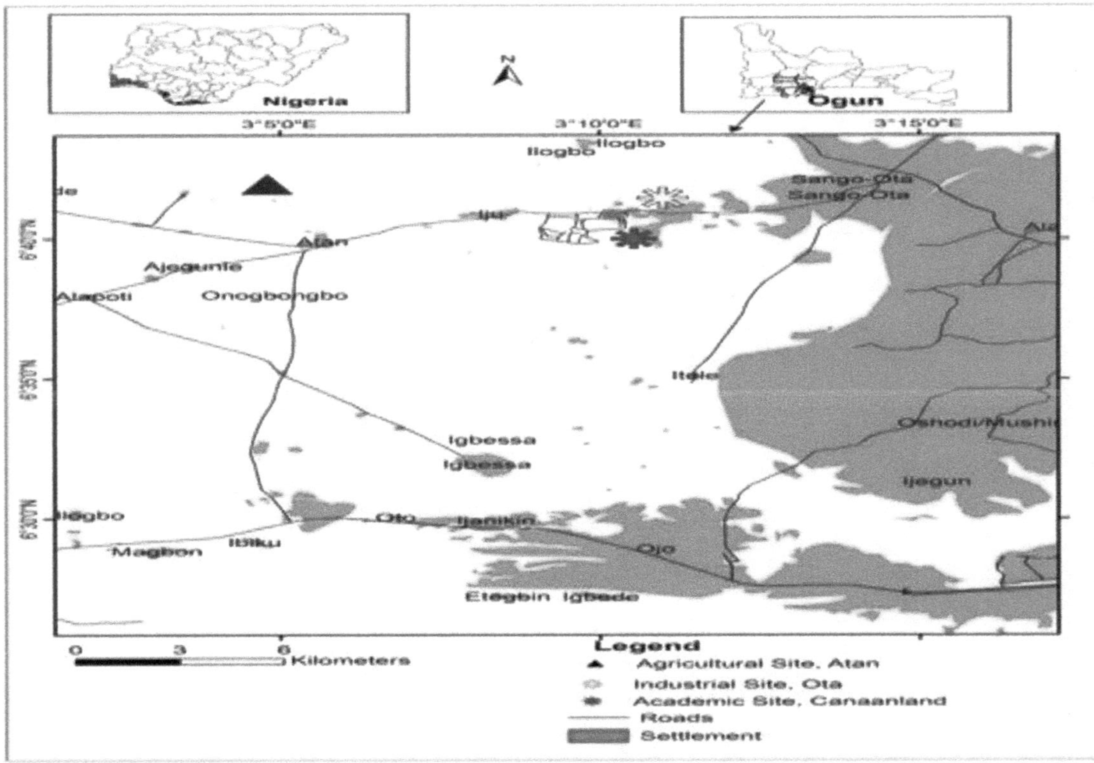


**Fig. 9.1** ***Map showing the sampling sites of tree species in Ogun state, Nigeria.***

*cecropioides, and Adenanthera pavonine* were sampled in triplicates. Leave samples were collected along with the stalk in the morning hours between January and March 2018, preserved in an ice chest. Processing and storage were done immediately after plant identification.

#### 9.3.1.2 Estimation of DCP

The leaf sample area was calculated, after which the individual leaf weight was obtained gravimetrically and substituted into Eq. (9.1) for DCP estimation (Sahu and Sahu, 2015; Bharti et al., 2018).

$$W = \frac{W2 - W1}{A} \tag{9.1}$$

Where: W = dust content ($g/cm^2$), W1 = initial weight of beaker (g), W2 = final weight of beaker with dust (g), and A = total area of the leaf ($cm^2$).

#### 9.3.1.3 Determination of biochemical parameters

The matured and healthy plant samples were examined for four biochemical parameters. For the pH of leaf extract (P), 5 g of grounded fresh leaf material was mixed in 50 mL of distilled water. The filtrate leaf extract pH was determined using a calibrated digital glass electrode pH meter (Pandey et al., 2015). Relative water content denoted as (R) was obtained by first recording the fresh weight of the leaves sample. The same samples were dipped in water for 24 hours, blotted to dryness and reweighed to determine the leaves' turgid weight. The turgid leaves were then oven-dried for 12 hours at 60°C, and re-weighed (Pathak et al., 2011). The ascorbic acid content of leaves denoted as (A) was analyzed using the spectrophotometric method. 1 g of the ground fresh leaves was extracted for 15 minutes using 4 mL of oxalic acid-EDTA, 1 mL of orthophosphoric acid, 2 mL of 5% tetraoxosulfate (VI) acid, 4 mL of ammonium molybdate, and 3 mL of water. The absorbance measurement of the filtered extract was taken at 760 nm with an ultraviolet spectrophotometer. Levels of ascorbic acid were further extrapolated from a standard ascorbic acid curve (Prajapati and Tripathi, 2008; Bharti et al., 2018). Also, total chlorophyll content was determined according to the method by Arnon (Arnon, 1949). The extracts were centrifuged for proper separation, after which the absorbance was measured at 643, 645, and 663 nm using a UV spectrophotometer and calculated according to Arnon's equation (Bharti et al., 2018; Arnon, 1949). Quality control was done with analytical grade chemicals and all reagents were prepared with deionized water.

#### 9.3.1.4 Estimation of APTI

The obtained biochemical parameters values of P, R, A, and T from Section 9.3.1.3 was used in the computation of APTI based on the method proposed by Singh and Rao (Singh and Rao, 1983) as shown in Eq. 9.2.

$$\mathrm{APTI} = \mathrm{A(T+P)} + \mathrm{R}/10 \tag{9.2}$$

Where:
A = ascorbic acid content (mg/g),
T = total chlorophyll content (mg/g),
P = pH of the leaf extract,
R= relative water content of leaf (%).

#### 9.3.1.5 Evaluation of API

The APTI value obtained for plant samples and various plant parameters, including but not limited to canopy structure, plant type, size, texture, hardiness, economic importance, etc., was employed in the estimation of API following the grade distribution pattern (Pathak et al., 2011; Patel and Kumar, 2018). Computation of API is as shown in Eq. 9.3.

$$\mathrm{API} = \frac{\text{No. of } ''+'' \text{ obtained}}{\text{Total No. of } ''+''} \times 100 \tag{9.3}$$

#### 9.3.1.6 Statistical modeling methods

Statistical analyses in this work were implemented with Microsoft Excel 2013 and SPSS statistical software version 23.0. Analyses were performed for individual samples and expressed as mean ± SD. Each of the biochemical parameters was compared with APTI using simple linear regression analysis, multiple regression, and two-way analysis of variance. Multiple linear regression was used to predict dependent variables' estimates from the original values of independent variables. R-square values for the data were obtained to investigate the variability level of the data under investigation.

### 9.3.2 Results and discussion

#### 9.3.2.1 DCP

The leaf DCP of 10 plant species under study in the 3 sites are shown in Table. 9.1. DCP of the individual plants varied from 0.6 to 4.3 mg/cm$^2$ in the industrial site (IND), 1.6 to 7.1 mg/cm$^2$ in the academic site (ASC), and 3.1 to 5.4 mg/cm$^2$ in agricultural sites (control site). *Musanga cecropioides* (7.1 mg/cm$^2$) from ASC and *Tectona grandis* (0.6 mg/cm$^2$) from IND had the highest and lowest dust capturing potential respectively across the

**Table 9.1** Dust capturing potentials, biochemical parameters, and air pollution tolerance indices (APTI) of investigated plants.

| Study site | Common name | Taxon | DCP ($mg/cm^2$) | A (mg/g) | P | T(mg/g) | R (%) | APTI | Rating |
|---|---|---|---|---|---|---|---|---|---|
| Industrial Site, Ota (IND) | Cassava | *Manihot esculenta* | 0.91 | 1.18 ± 0.003 | 5.92 ± 0.046 | 1.80 ± 0.012 | 78.9 ± 9.74 | 8.80 | Sensitive |
| | Teak | *Tectona grandis* | 0.6 | 1.54 ± 0.005 | 4.27 ± 0.255 | 0.304 ± 0.006 | 87.6 ± 8.25 | 9.46 | Sensitive |
| | Pawpaw | *Carica Papaya* | 4.3 | 3.02 ± 0.008 | 6.97 ± 0.081 | 0.923 ± 0.014 | 78.9 ± 0.58 | 10.3 | Sensitive |
| | Cashew | *Anacardium occidentale* | 3.9 | 0.682 ± 0.005 | 5.94 ± 0.017 | 0.542 ± 0.003 | 20.8 ± 2.00 | 2.52 | Sensitive |
| Academic Site, Canaan land (ASC) | Golden shower tree | *Cassia fistular* | 3.8 | 0.739 ± 0.019 | 6.38 ± 0.006 | 0.455 ± 0.008 | 89.9 ± 8.3 | 9.49 | Sensitive |
| | Umbrella tree | *Musanga cecropioides* | 7.1 | 1.33 ± 0.138 | 6.52 ± 0.015 | 0.342 ± 0.006 | 96.6 ± 6.01 | 10.6 | Sensitive |
| | Red sandal wood | *Adenanthera pavonina* | 3.2 | 2.00 ± 0.008 | 6.49 ± 0.085 | 1.06 ± 0.007 | 84.7 ± 12 | 9.98 | Sensitive |
| | Oil palm | *Elaesis guineensis* | 1.6 | 1.45 ± 0.002 | 5.74 ± 0.309 | 0.304 ± 0.002 | 89.8 ± 6.31 | 9.85 | Sensitive |
| Agricultural Site, Atan (ASA) | Pawpaw | *Carica papaya* | 4.7 | 2.95 ± 0.056 | 6.51 ± 0.176 | 0.507 ± 0.024 | 19.0 ± 12.1 | 3.97 | Sensitive |
| | Oil palm | *Elaesis guineensis* | 3.1 | 1.60 ± 0.014 | 6.57 ± 0.119 | 0.549 ± 0.5 | 87.9 ± 8.61 | 9.92 | Sensitive |
| | Mango | *Mangifera indica* | 4.5 | 1.67 ± 0.005 | 6.16 ± 0.070 | 0.630 ± 0.01 | 99.8 ± 0.22 | 11.1 | Sensitive |
| | Almond | *Terminalia catappa* | 5.4 | 1.27 ± 0.002 | 6.69 ± 0.150 | 0.634 ± 0.025 | 87.7 ± 4.21 | 9.70 | Sensitive |

study locations. This work corroborates with Bharti et al. (2018), which recorded that DCP of different investigated plants varied from 0.1 to 5.7, and 0.07 to 5.4 mg/cm$^2$ at industrial and control sites, respectively, while Rai and Panda (2014) showed a variation of DCP from 1.02 mg/cm$^2$ to 0.54 mg/cm$^2$ in roadside tree vegetation.The trend of dust captured among the sample location and plant species were (ASC) *Musanga cecropioides* > (ASA) *Terminalia catappa* > (ASA) *Carica papaya* > (ASA) *Mangifera indica* > (IND) *Carica Papaya* > (IND) *Anacardium occidentale* > (ASC) *Cassia fistula* > (ASC) *Adenanthera pavonina*> (ASA) *Elaesis guineensis* > (ASC) *Elaesis guineensis* > (IND) *Manihot esculenta* > (IND) *Tectona grandis*. Nine species recorded higher dust accumulating potentials (3.1 to 7.1 mg/cm$^{2)}$, thus improving the ambient air quality (Roy et al., 2020; Bharti et al., 2018; Rai and Panda, 2014).

#### *9.3.2.2 APTI*

APTI indicates the susceptibility of each plant species toward air pollution (Singh et al., 1991). APTI result was obtained from the mean values of all four biochemical parameters. Table 9.1 shows the APTI of plants from industrial sites ranging from 2.52 (*Anacardium occidentale*) to 10.3 (*Carica papaya*). From the academic site, the APTI of plants ranged from 9.49 (*Cassia fistular*) to 10.6 (*Musanga cecropioides*). Also, the control site, designated as an agricultural site, the APTI of plants ranged from 3.97(*Carica papaya*) to 11.1 (*Mangifera indica*). The lowest APTI of 2.52 across the three study sites was recorded for Anacardium occidentale from the industrial site indicating that it is the most sensitive to air pollution. Across the investigated sites, APTI highest score was in the order agricultural site (APTI 11.1) > academic site (APTI 10.6) > industrial site (APTI 10.3). Among the study locations, Industrial site plants recorded a decreased APTI of 10.3 compared to the 11.1 APTI value in the control site (Agricultural). This may be associated with longer pollution exposure. This study corroborates with Rai and Panda (2014) work where an increased APTI was obtained in the control area plants compared to the roadside plants. APTI are categorized as sensitive (1–11), intermediate sensitive or tolerant (12–16) and tolerant (≥17) (Ogunkunle et al., 2015; Anake et al., 2019; Bharti et al., 2018). In this work, all plants were found within the APTI range of 1–11, rated as sensitive. They can be recommended as bioindicators for ambient air pollution monitoring and control (Roy et al., 2020; Prajapati and Tripathi, 2008).

From the APTI indices, it was observed that individual plants varied in their response to air pollution. This has been reported in several previous studies, and the variation has been connected to the different geographical areas (Hajizadeh et al., 2019; Raza et al., 1985; Karmakar and Padhy, 2019; Barwise and Kumar, 2020). It is also attributed to their ability to undergo physical and chemical adaptation, thus mitigating the stress induced by pollutants based on individual plants' antioxidant (Manjunath and Reddy, 2019).

### 9.3.2.3 Statistical modeling of bioindicators responses

Fig. 9.2A–D presents a linear regression analysis of the different biochemical parameters with APTI. A strong positive correlation was observed between APTI and relative water content ($R^2$ = 0.9521). On the contrary, pH of plant leaf extract ($R^2$ = 0.0042), total chlorophyll content ($R^2$ = 0.0042), and ascorbic acid content ($R^2$ = 0.0004) exhibited a positive but insignificant correlation with APTI. This result suggests that all the biochemical characteristics have an important role when considering air pollution mitigation potentials of plant species.

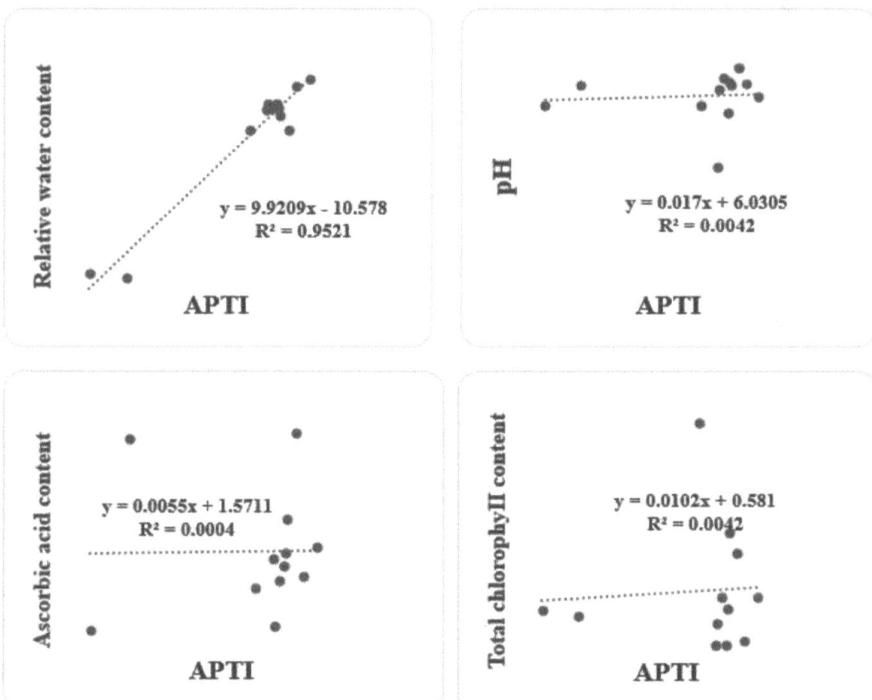


**Fig. 9.2** Linear regression analysis of APTI with individual variables: (A) relative water contents; (B) pH of leaf extract; (C) ascorbic acid contents; and (D) total chlorophyll contents.

**Table 9.2** Overall model for biochemical parameters and APTI.

| Model | Sum of squares | df | Mean square | F | *P* value |
|---|---|---|---|---|---|
| Regression | 78.999 | 4 | 19.750 | 7545.925 | 0.000[a] |
| Residual | 0.018 | 7 | 0.003 | | |
| Total | 79.018 | 11 | | | |

Dependent Variable: APTI.
[a]Predictors: (Constant), Relative water content (%), pH of leaf extract, Ascorbic acid content (mg/g), Total chlorophyll content (mg/g).

Table 9.2 presents the overall model for the multiple linear regression equation, and it shows that the model is significant since the p-value is less than 0.05. Table 9.3 presents the model coefficient value of each of the independent variables against the dependent variable of APTI. The result shows that each of the independent variables is significant, with the *P* values less than 0.05. A unit change in the pH of leaf extract, total chlorophyll, relative water contents and ascorbic acid content values brings about an increase in APTI of plant species by 0.173, 0.146, 0.100, and 0.727, respectively. The findings of this study agree with the results presented by other researchers (David, 2005; Rencher and Christensen, 2012; Berk, 2007).

The two-way ANOVA results presented in Table 9.4 showed that the Taxon parameters are not significant with a *P* value of 0.479. In contrast,

**Table 9.3** Multiple regression (linear function model) of each biochemical parameters on APTI.

| Regression coefficient | Coefficients values | Standard error | t-stat | *P* value |
|---|---|---|---|---|
| Constant ($\beta_0$) | -1.208 | 0.146 | -8.269 | 0.000 |
| pH of leaf extract ($\beta_1$) | 0.173 | 0.024 | 7.297 | 0.000 |
| Total chlorophyll content ($\beta_2$) | 0.146 | 0.037 | 3.938 | 0.006 |
| Relative water content ($\beta_3$) | 0.100 | 0.001 | 173.123 | 0.000 |
| Ascorbic acid content ($\beta_4$) | 0.727 | 0.022 | 32.363 | 0.000 |

**Table 9.4** Two-way analysis of variance for Taxon and biochemical parameters.

| Source | Type II sum of squares | df | Mean square | F | Sig. |
|---|---|---|---|---|---|
| Model | 73,295.672[a] | 15 | 4886.378 | 26.177 | 0.000 |
| Taxon | 2021.619 | 11 | 183.784 | 0.985 | 0.479 |
| Bioparameter | 49,461.091 | 3 | 16,487.030 | 88.322 | 0.000 |
| Error | 6160.068 | 33 | 186.669 | | |
| Total | 79,455.740 | 48 | | | |

[a]R Squared = 0.922 (Adjusted R Squared = 0.887).

biochemical parameters are significant, with a *P* value of 0.000 (Table 9.4). The model $R^2$ value was 0.922, which indicates a high level of variability among the data for the analysis and that the data fit the model well.

#### 9.3.2.4 API

The API results for the study are presented in Tables 9.5 and 9.6. The API results indicate *Tectona grandis* from the industrial site and *Mangifera indica* from the agricultural site as very good performers with a grade point of 75%. *Musanga cecropioides*, located in the academic site, was identified as a good performer (62.5%), while *Terminalia catappa* from the agricultural was ranked as a moderate performer with a grade point of 56.25%. Based on API grading, plants with the highest API scores are suitable for green belt development in the study area (Ogunkunle et al., 2015; Roy et al., 2020; Pathak et al., 2011). *Manihot esculenta, Cassia fistular and Anacardium occidentale L, Adenanthera pavonine, Elaesis guineensis* were categorized under the very poor and poor API grade respectively as such are not recommended for green belt programs. The APTI, biological characteristics and socio-economic grade points of screened plant species indicate that plant species present within the API range of best to moderate performers can be recommended for cultivation as green belts species (Pathak et al., 2011; Patel and Kumar, 2018; Banerjee et al., 2018; Khalid et al., 2019). The limitation of this study is that similar plant species were not found in all the study sites as such sample selection was done based on the dominance of common plant species in each study site.

## 9.4 Conclusion

Screening of suitable plant species for green belt development to reduce ambient air pollution in specific geographical areas can be enhanced by a combination of different research tools along with common statistical models. APTI results identified all the 10 investigated plants as sensitive species. This sensitive species can serve as bioindicators of air pollution in their respective study sites. Higher dust accumulating potentials (3.1–7.1 mg/$cm^2$), was recorded by nine species thus improving the ambient air quality. Based on API grading of very good to moderate performers, *Tectona grandis* (industrial site), *Mangifera indica and Terminalia catappa* (agricultural site), and *Musanga cecropioides (*academic site) were recommended for green belt development. Linear regression analysis revealed that relative water content had the highest relationship ($R^2$ = 0.9521) with APTI. Also, all the studied

**Table 9.5** Evaluation of individual plants following APTI values, socioeconomic importance, and biological parameters.

| | | | | | | | | | Grade allotted | |
|---|---|---|---|---|---|---|---|---|---|---|
| **Taxon** | **APTI** | **PH** | **CS** | **TT** | **LS** | **LT** | **EI** | **Hardiness** | **Total plus (+)** | **Scoring %** |
| *Manihot esculenta* | + | - | + | + | - | + | ++ | - | 6 | 37.5 |
| *Tectona grandis* | + | ++ | ++ | + | + | ++ | ++ | + | 12 | 75 |
| *Carica papaya* | - | ++ | + | + | + | - | ++ | + | 8 | 50 |
| *Anacardium occidentale L* | - | + | + | + | + | - | ++ | + | 7 | 43.75 |
| *Cassia fistula* | + | + | + | - | + | - | + | + | 6 | 37.5 |
| *Musanga cecropioides* | + | + | ++ | + | ++ | - | ++ | + | 10 | 62.5 |
| *Adenanthera pavonina* | + | ++ | + | + | - | - | ++ | + | 8 | 50 |
| *Elaesis guineensis* | + | ++ | + | + | - | + | ++ | + | 8 | 50 |
| *Mangifera indica* | + | ++ | ++ | + | ++ | + | ++ | + | 12 | 75 |
| *Terminalia catappa* | + | ++ | ++ | - | + | - | ++ | + | 9 | 56.25 |

APTI: Air pollution tolerance index; PH: Plant habit; CS: Canopy structure; TT: Type of tree; LS: Laminar size; LT: Laminar texture; EI: Economic importance; and H: Hardiness.

**Table 9.6** API value of the screened plant species.

| Taxon | Grade allotted | | API value | Assessment |
|---|---|---|---|---|
| | Total plus (+) | Percentage % | | |
| *Manihot esculenta* | 6 | 37.5 | 1 | Very poor |
| *Tectona grandis* | 12 | 75 | 5 | Very good |
| *Carica papaya* | 8 | 50 | 2 | Poor |
| *Anacardium occidentale L* | 7 | 43.75 | 2 | Poor |
| *Cassia fistula* | 6 | 37.5 | 1 | Very poor |
| *Musanga cecropioides* | 10 | 62.5 | 4 | Good |
| *Adenanthera pavonina* | 8 | 50 | 2 | Poor |
| *Elaesis guineensis* | 8 | 50 | 2 | Poor |
| *Mangifera indica* | 12 | 75 | 5 | Very good |
| *Terminalia catappa* | 9 | 56.25 | 3 | Moderate |

biochemical parameters were significant with APTI, as such a unit change in each of the biochemical parameters resulted in an increase in APTI. Therefore, more of the identified bioindicators and green belt development plants should be cultivated in the respective study sites to abate the increasing ambient air pollution. Also, intensive screening of more plants to identify tolerant species, best and excellent performers is recommended for future work

## References

Alhesnawi, A.M., Alsalman, I., Najem, NA., 2018. Evaluation of air pollution tolerance index of some plants species in Kerbala city. Iraq J. Pharm. Sci. Res. 10 (6), 1386–1390.

Anake, W.U., Ana, G.R.E.E., Williams, A.B., Fred-Ahmadu1, O.H., Benson, NU., 2017. Chemical speciation and health risk assessment of fine particulate bound trace metals emitted from Ota Industrial Estate, Nigeria, IOP Conf. Series: Earth and Environmental Science.

Anake, W.U., Bayode, F.O., Omonhinmin, C.A., Williams, AB., 2018. Ambient air pollution control using air pollution tolerance index and anticipated performance index of trees. Int. J. Civil Eng. Technol. 9 (12), 417–425.

Anake, W.U., Eimanehi, J.E., Omonhinmin, CA., 2019. Evaluation of air pollution tolerance index and anticipated performance index of selected plant species. Indones. J. Chem. 19 (1), 239–244. http://doi.org/10.22146/ijc.35270.

Anake, W.U., Eimanehi, J.E., Riman, H.S., Omonhinmin, CA., 2018. Tree species for biomonitoring and green belt design: a case study of Ota Industrial Estate, Nigeria, IOP Conf. Series: Earth and Environmental Science.

Arnon, DI., 1949. Copper enzymes in isolated chloroplasts: polyphenol oxidise in Beta vulgaris. Plant Physiol. 2, 1–15.

Asmussen, S., Kroese, DP., 2006. Improved algorithms for rare event simulation with heavy tails. Adv. Appl. Probab. 38 (2), 545–558.

Badamasi, H., 2017. Biomonitoring of air pollution using plants. MAYFEB J. Environ. Sci. 2, 27–39.

Banerjee, S., Banerjee, A., Palit, D., Roy, P., 2018. Assessment of vegetation under air pollution stress in urban industrial area for greenbelt development. Int. J. Environ. Sci. Technol. 16 (10), 1–14.

Barwise, Y., Kumar, P., 2020. Designing vegetation barriers for urban air pollution abatement: a practical review for appropriate plant species selection. Clim. Atmos. Sci. 3, 12. http://doi.org/10.1038/s41612-020-0115-3.

Berk, RA., 2007. Regression analysis: a constructive critique. Crim. Justice Rev. 32 (3), 301–302. http://doi.org/10.1177/0734016807304871.

Bharti, K.S., Trivedi, A., Kumar, N., 2018. Air pollution tolerance index of plants growing near an industrial site. Urban Clim. 24, 820–829.

Botev, Z.I., Kroese, DP., 2012. Efficient Monte Carlo simulation via the generalised splitting method. Stat. Comput. 22 (1), 1–16.

Chaudhary, I.J., Rathore, D., 2019. Dust pollution: its removal and effect on foliage physiology of urban trees. Sustain. Cities Soc. 51, 101696. doi.org/10.1016/j.scs.2019.101696.

Cross, C.E., Valacchi, G., Schock, B., Wilson, M., Weber, S., Eiserich, J., 2002. Environmental oxidant pollutant effects on biologic systems a focus on micronutrient antioxidant–oxidant interactions. Am. J. Respir. Crit. Care. Med. 166, S44–S50. http://doi.org/10.1164/rccm.2206015.

Cuny, D., 2019 Plants and lichens sentinels of air quality, Encyclopédie de l'Environnement, http://www.encyclopedie-environnement.org/?p=6782. [Accessed January 31, 2021].

David, AF., 2005. Statistical Models: Theory and Practice, 2nd ed. Cambridge University Press, New York Chapter 2.

Esfahani, A.A., Amini, H.N., Kar, S.S., Hoodaji, M., Shirvani, M., Porsakhi, K., 2013. Assessment of air pollution tolerance index of higher plants suitable for green belt development in East of Esfahan City, Iran. J. Ornam. Hortic. Plants 3 (2), 87–94.

Etim, E.U., 2012. Estimation of pollution load from an industrial estate, South-western Nigeria. Afr. J. Environ. Sci. Technolgoy. 61, 25–29.

Ghafari, S., Kaviani, B., Sedaghathoor, S., Allahyari, MS., 2021. Assessment of air pollution tolerance index (APTI) for some ornamental woody species in green space of humid temperate region (Rasht, Iran). Environ. Dev. Sustain. 23, 1579–1600. https://doi.org/10.1007/s10668-020-00640-1.

Girish, L., Krishnankutty, K., Vaidya, S., 2017. Air pollution tolerance index of selected plants growing near road side of Navi Mumbai, Maharashtra. IJCR 9 (9), 57807–57811.

Hajizadeh, Y., Mokhtari, M., Faraji, M., Abdolahnejad, A., Mohammadi, A., 2019. Biomonitoring of airborne metals using tree leaves: protocol for biomonitor selection and spatial trend. MethodsX 6, 1694–1700.

Jain, S., Bhattacharya, T., Chakraborty, S., 2019. Comparison of plant tolerance towards air pollution of rural, urban and mine sites of Jharkhand: a biochemical approach to identify air pollutant sink. In: Kalamdhad, A, Singh, J, Dhamodharan, K (Eds.), Advances in Waste Management. Springer, Singapore, pp. 123–142.

Javanmard, Z., Kouchaksaraei, M.A., Bahrami, H., Hosseini, S.M., Sanavi, M., SAM, S.D., 2019. Dust collection potential and air pollution tolerance indices in some young plant species in arid regions of Iran. Biogeosci. For. 12, 558–564. http://doi.org/10.3832/ifor3063-012.

Karmakar, D., Deb, K., Padhy, PK., 2021. Ecophysiological responses of tree species due to air pollution for biomonitoring of environmental health in urban area. Urban Clim. 35, 100741. https://doi.org/10.1016/j.uclim.2020.100741.

Karmakar, D., Padhy, PK., 2019. Air pollution tolerance, anticipated performance, and metal accumulation indices of plant species for greenbelt development in urban industrial area. Chemosphere 237, 124522.

Kashyap, R., Sharma, R., Uniyal, SK., 2018. Bioindicator responses and performance of plant species along a vehicular pollution gradient in western Himalaya. Environ. Monit. Assess. 190, 1e17.

Kaur, M., Nagpal, AK., 2017. Evaluation of air pollution tolerance index and anticipated performance index of plants and their application in development of green space along the urban areas. Environ. Sci. Pollut. Res. 24, 18881–18895.

Kaur, M., Sharma, A., Katnoria, J.K., AKaur, N., 2017. Air pollution tolerance index (APTI): an important determinant for the development of green space in and around industrial/ urban areas. In: Rathoure, AK (Ed.), Bioremediation: Current Research and Applications. IK International Publishing House Pvt. Ltd, New Delhi, India, pp. 422–449.

Khairallah, Y., Houri, T., Osta, B., Romanos, D., Haddad, G., 2018. Biomonitoring airborne pollution: a case study of "Urginea maritima" species in Bentael natural reserve – Lebanon. J. Taibah. Univ. Med. Sci. 12 (6), 723–729. http://doi.org/10.1080/16583655.2018.1507418.

Khalid, N., Masood, A., Noman, A., Aqeel, M., Qasim, M., 2019. Study of the responses of two biomonitor plant species (Datura alba s& Ricinus communis) to roadside air pollution. Chemosphere 235, 832–841.

Khanoranga, KS., 2019. Phytomonitoring of air pollution around brick kilns in Balochistan province Pakistan through air pollution index and metal accumulation index. J. Cleaner Prod. 229, 727–738.

Konishi, S., Kitagawa, G., 2008. Information Criteria and Statistical Modeling, 1st ed. Springer-Verlag New York, New York Chapter 1.

Kroese, D.P., Chan, J.C.C., 2014. Statistical Modeling and Computation. Springer Science + Business Media, New York Chapter 4.

Kwak, M.J., Lee, J.K., Park, S., Lim, Y.J., Kim, H., Kim, K.N., et al., 2020. Evaluation of the importance of some east Asian tree species for refinement of air quality by estimating air pollution tolerance index, anticipated performance index, and air pollutant uptake. Sustainability 12 (7), 3067. http://doi.org/10.3390/su12073067.

Leghari, S.K., Akbar, A., Qasim, S., Ullah, S., Asrar, M., Rohail, H., 2019. Estimating anticipated performance index and air pollution tolerance index of some trees and ornamental plant species for the construction of green belts. Pol. J. Environ. Stud 28 (3), 1759–1769.

Maatoug, M., Taïbi, K., Akermi, A., Achir, M., Mestrari, M., 2012. Bio-monitoring of air quality using leaves of tree and lichens in urban environments. In: Khare, M (Ed.), Air Pollution – Monitoring, Modelling and Health. InTech, Croatia, pp. 223–224.

Manjunath, B.T., Reddy, J., 2019. Comparative evaluation of air pollution tolerance of plants from polluted and non-polluted regions of Bengaluru. J. Appl. Biol. Biotech. 7 (03), 63–68. http://doi.org/10.7324/JABB.2019.70312.

Molnár, V.É., Simon, E., Tóthmérész, B., Ninsawatd, S., Szabó, S., 2020. Air pollution induced vegetation stress – The Airs Pollution Tolerance Index as a quick tool for city health evaluation. Ecol. Indic. 113, 106234. http://doi.org/10.1016/j.ecolind.2020.106234.

Nadgorska Socha, A., Kandziora-Ciupa, M., Trzȩsicki, M., Barczyk, G., 2017. Air pollution tolerance index and heavy metal bioaccumulation in selected plant species from urban biotopes. Chemosphere 183, 471–482.

Ogbonna, C.E., Nwafor, F.I., Ugbogu, EA., 2019. Physiochemical properties and anticipated performance of selected plant species in Lokpaukwu Quarry Site in Abia State, Nigeria. J. Environ. Pollut. Hum. Health 7 (1), 7–14. http://doi.org/10.12691/jephh-7-1-2.

Ogunkunle, C.O., Suleiman, L.B., Oyedeji, S., Awotoye, O.O., Fatoba, PO., 2015. Assessing the air pollution tolerance index and anticipated the performance index of some tree species for biomonitoring environmental health. Agroforest Syst. 89, 447–454.

Pandey, A.K., Pandey, M., Mishra, A., Tiwary, S.M., Tripathib, BD., 2015. Air pollution tolerance index and anticipated performance index of some plant species for development of urban forest. Urban For. Urban Green. 14, 866–871.

Patel, D., Kumar, JI., 2018. An evaluation of air pollution tolerance index and anticipated performance index of some selected plant species. Open J. Air Pollut. 7, 1–13.
Pathak, V., Tripathi, B.D., Mishra, V.K., 2011. Evaluation of Anticipated Performance Index of some tree species for green belt development to mitigate traffic generated noise. Urban For. Urban Green. 10, 61–66.
Popek, R., Przybysza, A., Gawrońskaa, H., Klamkowskic, K., Gawroński, SW., 2018. Impact of particulate matter accumulation on the photosynthetic apparatus of roadside woody plants growing in the urban conditions. Ecotoxicol. Environ. Saf. 163, 56–62.
Prajapati, S.K., Tripathi, BD., 2008. Anticipated performance index of some tree species considered for green belt development in and around an urban area: a case study of Varanasi city, India. J. Environ. Manage. 88, 1343–1349.
Rai, P.K., Panda, LL., 2014. Dust capturing potential and air pollution tolerance index (APTI) of some road side tree vegetation in Aizawl, Mizoram, India: an IndoBurma hot spot region. Air Qual. Atmos. Health 7 (1), 93e101. http://doi.org/10.1007/s11869-013-0217-8.
Raza, S.N., Vijaykumari, N., Murthy, MS., 1985. Air pollution tolerance index of certain plants of HyderabadSymp Biomonitoring State of Environment. Indian National Science Academy, New Delhi, pp. 243–245.
Rencher, A.C., Christensen, WF., 2012. Methods of Multivariate Analysis, 3rd ed. John Wiley & Sons, Hoboken Chapter 10.
Roy, A., Bhattacharya, T., Kumari, M., 2020. Air pollution tolerance, metal accumulation and dust capturing capacity of common tropical trees in commercial and industrial sites. Sci. Total Environ., 137622. http://doi.org/10.1016/j.scitotenv.2020.
Sahu, C., Basti, S., Sahu, SK., 2020. Air pollution tolerance index (APTI) and expected performance index (EPI) of trees in Sambalpur town of India. SN Appl. Sci. 2, 1327. http://doi.org/10.1007/s42452-020-3120-?13.
Sahu, C., Sahu, SK., 2015. Air pollution tolerance index (APTI), anticipated performance index (API), carbon sequestration and dust collection potential of Indian trees species – A Review. Int. J. Emerg. Res. Maint. Technol. 4, 37–40.
Selmi, W., Weber, C., Rivière, E., Blond, N., Mehdi, L., Nowak, D., et al., 2016. Air pollution removal by trees in public green spaces in Strasbourg city. France Urban For. Urban Green 17, 192–201.
Shrestha, S., Baral, B., Dhital, N.B., Yang, H., 2021. Assessing air pollution tolerance of plant species in vegetation traffic barriers in Kathmandu Valley, Nepal. Sustain. Environ. Res. 31 (3), 1–9.
Singh, P., Pal, A., 2017. Response of dust accumulation on roadside plant species due to open cast mining at Jhansi-Allahabad NH-76, Uttar Pradesh, India. Trop. Plant Res. 4 (3), 461–467.
Singh, S.K., Rao, D.N., Agrawal, M., Pandey, J., Naryan, D., 1991. Air pollution tolerance index of plants. J. Environ. Manag. 32, 45–55.
Singh, S.K., Rao, DN., 1983. Evaluation of plants for their tolerance to air pollution, Proceedings of the Symposium on air Pollution Control, 218–224.
Sulistijorini Mas'ud, Z.A., Nasrullah, N., Bey, A., Tjitrosemito, S., 2008. Tolerance levels of roadside trees to air pollutants based on relative growth rate and air pollution tolerance index. HAYATI J. Biosci. 15 (3), 123–129. http://doi.org/10.4308/hjb.15.3.123.
Uka, U.N., Belford, E.J.D., Hogarh, J.N., 2019. Roadside air pollution in a tropical city: physiological and biochemical response from trees. Bull. Natl. Res. Centre 43 (90). http://doi.org/10.1186/s42269-019-0117-7.
Wang, R., Yao, X., Liu, Y., Guo, Y., Wang, S., Wang, B., 2017. Research on the dust-capturing capacities of environmental friendly materials based on the typical garden trees and the morphological structure of leaves in Suzhou. Key Eng. Mater. 737, 412–416.

Wei, X., Lyu, S., Yu, Y., Wang, Z., Liu, H., Pan, D., 2017. Phylloremediation of air pollutants: exploiting the potential of plant leaves and leaf-associated microbes. Front. Plant Sci. 8, 1318.

Zhang, P.Q., Liu, Y.J., Chen, X., Yang, Z., Zhu, M-h, Li, Y.P., 2016. Pollution resistance assessment of existing landscape plants on Beijing streets based on air pollution tolerance index method. Ecotoxicol. Environ. Saf. 132, 212–223. http://doi.org/10.1016/j.ecoenv.2016.06.003.

Zhang, W., Zhang, Y., Gong, J., Yang, B., Zhang, Z., Wang, B., Zhu, C., Shi, J., Yue, K., 2020. Comparison of the suitability of plant species for greenbelt construction based on particulate matter capture capacity, air pollution tolerance index, and antioxidant system. Environ. Pollut. 263, 114615.

# CHAPTER 10

# Biomarkers of arsenic stress in plants

**Pradyumna Kumar Singh[a], Sonali Dubey[c], Manju Shri[d], Sanjay Dwivedi[a,b], Debasis Chakrabarty[a,b], Rudra Deo Tripathi[a,b]**
[a]Council of Scientific and Industrial Research-National Botanical Research Institute (CSIR-NBRI), Lucknow, Uttar Pradesh, India
[b]Academy of Scientific and Innovative Research (AcSIR), Ghaziabad, Uttar Pradesh, India
[c]Institute of Biotechnology, Amity University, Noida, Uttar Pradesh, India
[d]Gujarat Biotechnology Research Centre, Gandhi Nagar, Gujarat, India

## 10.1 Introduction

Arsenic (As) is an environmental toxic metalloid, a food chain toxin and nonthreshold carcinogen which pollutes soil and groundwater globally from natural geological and anthropogenic resources (Tripathi et al., 2007; Hughes et al., 2011). As is continuously contaminating the environment due to anthropogenic activity such as mining, use of As containing herbicides, and industrialization. Unfortunately, As is a nonbiodegradable, highly toxic contaminant, accumulated in both plants and animals, hence a food chain contaminant. Sessile organisms like plants cannot escape the places where metalloid contamination happens. However, plants have various interesting detoxification and tolerance mechanisms to survive in heavy metal contaminated environments. Reduced uptake of As, synthesis of metal-binding proteins such as phytochelatins (PCs) and metallothionein (MT), sequestration of As inside the vacuole and extrusion of As outside the plants, are some of the detoxification mechanisms that plants are equipped to combat the As toxicity. Along with these mechanisms, plants are also possessed with a very strong antioxidant system to reduce oxidative stress due to As toxicity. Production and regulation of synthesis and accumulation of plant defense protein and osmolytes such as proline, mannitol, and glycine betaine, also play a vital role in reducing the As toxicity. The conversion of inorganic As to organic form or volatile form by enzymatic metabolism reduced the toxicity of As in the living system. The modulation of enzymes and metabolites to cope with the As toxicity may provide initial information of the As-induced toxicity or tolerance in the plants.

*New Paradigms in Environmental Biomonitoring Using Plants*
DOI: https://doi.org/10.1016/B978-0-12-824351-0.00014-6

In the animal system, the roles of antioxidant enzymes and MT have been identified as a biomarker during various contaminants exposure (Bebianno et al., 2004; Geret et al., 2003). The As tolerance and toxicity mechanism may, however, indicate ultimate pathways or potential products that could be used as plant biomarker.

Biomarkers can be defined as "Change in the biological or cellular component of a process, structure or function that can be measured in a biological system" (NRC, USA 1989). Various reports are available for the role of biomarkers in the animal system (Trukhachev et al., 2021; Klepp et al., 2019). However, in plants, specific biomarkers are rare due to the complexity of specific metabolic pathways in response to specific stresses. A particular biomarker is the synthesis of fluorocitrate with exposure of plants to fluorides, present in the soil or the atmosphere (Meyer et al., 1992). Different environmental stresses such as the presence of heavy metals result in the same responses such as oxidative stress, which further results in reactive oxygen species (ROS) build-up and increased MDA production and enhanced antioxidant pathways recruiting various antioxidant enzymes and increased metabolic content. As and other heavy metals induce a range of alterations in the expression of the gene and modulation in the induction of various proteins that help the plant to cope up with heavy metal stress. Regardless of the shared genes and proteins induced by heavy metals for their detoxifications, there is a range of specific genes and proteins that plant recruits to combat stress. ACR3 of *Pteris vittata* (As-hyperaccumulator plant) is an extrusion pump that extrudes out the AsIII from the cellular compartment to the vacuole is very specific to As may be used as a biomarker (Ellis et al., 2006; Indriolo et al., 2010). Further, Bianucci et al. (2017) showed that glutathione S-transferase could be used as a biomarker in peanut plants for As contamination. Along with these genes and proteins, this book chapter deals with the biomarkers that can be used for As contamination in the environment.

## 10.2 Markers for As-induced oxidative stress

### 10.2.1 Antioxidant enzymatic systems

The presence of As in the plants resulted in oxidative stress. To reduce the As induced oxidative damage, the antioxidant defense system plays a vital role. The antioxidant defense system comprises various enzymes with a broad role in managing the build-up of ROS. Superoxide dismutase, catalase, ascorbate peroxidase and glutathione reductase are necessary enzymes that participate in the antioxidant defense system.

**Table 10.1** Various superoxide dismutases found in plants microbes.

| S.No. | Name | Cellular compartment | Inhibition | Occurrence |
|---|---|---|---|---|
| 1. | Mn-SOD | Mitochondria, peroxisome | KCN and $H_2O_2$ insensitive | Plants |
| 2. | Fe-SOD | Chloroplast | KCN insensitive, $H_2O_2$ sensitive | Plants |
| 3. | Cu/Zn-SOD | Cytosol, chloroplast, mitochondria, peroxisome | KCN and $H_2O_2$ sensitive | Plants |
| 4. | Ni-SOD | - | - | Bacteria |

#### *10.2.1.1 Superoxide dismutase*

Superoxide dismutase (SOD) is a metalloenzyme. As stress induces ROS and creates an oxidative environment in the cell. SOD enzyme reduces the level of superoxide to protect the plants from oxidative stress. Superoxide contains the unpaired radical electron. It contributes to the negativity and reactivity of the superoxide anion. The SOD has been classified based on the presence of metal on the active site of the enzymes, i.e., Mn-SOD (manganese-superoxide dismutase), Cu/Zn-SOD (copper/zinc superoxide dismutase), Fe-SOD (Iron-superoxide dismutase), and Ni-SOD (nickel-superoxide dismutase). The Mn-SOD, Fe-SOD, and Cu/Zn-SOD are commonly present in various organelles of plant cells (Table 10.1). Cu/Zn SOD is prevalently found in plant cells. Ni-SOD is mostly present in bacteria (Perry et al., 2010). SOD carry out dismutation of the superoxide molecules to hydrogen peroxide ($H_2O_2$) and oxygen molecule ($O_2$) and provide the first line of defense against the oxidative stress during stress condition.

As stress induces superoxide radicals by disrupting the electron transfer chain in plant cell organelles such as mitochondria and chloroplast. The Mn-SOD and Cu/Zn-SOD are the key isoenzymes that take part in the dismutation of superoxide radicals in rice plants during As stress (Shri et al., 2009). It has been shown that Mn-SOD and Fe-SOD enzyme's activities were higher in mitochondrial alternative oxidase (*aox1a*) mutant *Arabidopsis thaliana* under AsV stress (Demircan et al., 2020). But the higher concentrations of As stress negatively affect the SOD enzyme's activities. The expression analysis of genes related to SOD indicated the contrast expression pattern between Cu/Zn-SOD (at2g28190, at1g08830) and Fe-SOD in *Arabidopsis thaliana* under AsV exposure (Abercrombie et al., 2008). Moreover, in *Isatis cappadocica*, a hyperaccumulator of As, the SOD enzyme activity was also induced during exposure to a wide range of As concentrations (0–1200 μM, AsV) (Souri et al., 2018). Both hyperaccumulator (*Pteris vittata*) and sensitive fern (*Nephrolepis exaltata* L.) showed the higher activity of SOD enzyme in frond exposed to 0, 150, or 300 μM of As (Srivastava et al., 2005). Inductions in activities of SOD during a range of As concentration exposure in various plants make it a suitable biomarker (Fatima and Ahmad, 2005; Rahman et al., 2015; Siddiqui et al., 2015).

#### 10.2.1.2 Catalase

Catalase enzymes have been identified in both prokaryotes and eukaryotes (Zamocky et al., 2008; Frugoli et al., 1996). CAT isozymes are mainly found in peroxisomes and primarily deal with the conversion of $H_2O_2$ without any reductant into water and oxygen molecules and reduce the ROS-induced oxidative stress in plants (Frugoli et al., 1996; Kato et al., 2004). As stress negatively affects plant growth and development. The various isoforms of CAT (cat1, cat2, and cat3) enzymes play a vital role in the detoxification of ROS in *Arabidopsis thaliana* during As stress. The study showed that the transcript level of cat1, cat2, and cat3 were induced in developing embryo of maize plant during As stress (0.01–10 mM) (Mylona et al., 1998). CAT activity was also induced in *Triticum aestivum*, colonized by arbuscular mycorrhizal fungi *Rhizoglomus intraradices* and *Glomus etunicatum* during As stress. Mycorrhizal colonization enhanced the antioxidant capacity of wheat plants (Sharma et al., 2017). Further, CAT enzyme's isoforms (cat1, cat2, and cat3) showed differential expression behavior during As exposure. In *Brassica juncea*, CAT3 was downregulated during AsV stress (Srivastava et al., 2015). Moreover, the expression profile indicated expression behavior of the catalase gene might also depend on developmental stages, organs and type of abiotic stresses (Surgun-Acar, & Zemheri-Navruz, 2019, Yang et al., 2019; Dvorak et al., 2020; Du et al., 2008; Anjum et al., 2016). The 14 genes of catalase enzymes have been identified in *Brassica napus*. Catalase genes expression differs in the root, shoot, leaf and silique (Raza et al., 2021). In the As-hyperaccumulator plant *Pteris vittata*, the catalase activity increased up to 300% in frond than the control plant during As stress. However, As-sensitive plant, *Pteris ensiformis*, showed 133% increase in catalase activity in frond on the As exposure (Kertulis-Tartar et al., 2009). The rapid leaf slices test study showed that As exposure altered the catalase enzyme activity in the pinna of ferns related to Pteridaceae (*Pteris vittata* L. and *Pteris cretica* L.) and Polypodiaceae (*Polypodiodes microrhizoma*) family found in the Western Himalayan (Kumar et al., 2021). Interestingly, the comparative study indicated that As-tolerant rice cultivar

IR-36 (accumulated higher level of As content) contained higher catalase activity than As-sensitive cultivar IET-4786 (low As accumulating cultivar) (Rai et al., 2011). The above study illustrated that catalase continuously performed to detoxify the As induces oxidative stress in various plant species (related to hyperaccumulator and nonhyperaccumulator of As categories).

#### *10.2.1.3 Ascorbate peroxidase*

Alternatively, glutathione S-transferase and the enzymes of ascorbate-glutathione pathways such as APX (ascorbate peroxidase) and GR (glutathione reductase) also play an important role in the regulation of oxidative burst during As exposure. APX is the ascorbate-dependent heme-containing peroxidases. APX reduced the $H_2O_2$ to $H_2O$ by using ascorbate as elector donor substrate in various organelles such as cytosol, chloroplasts, mitochondria, and peroxisomes in plants (Hasanuzzaman et al., 2020). The activity of the APX has been detected in several plant species during As stress. In monocot plants such as rice, APX activity was significantly induced after the supplementation of AsIII (Khan & Gupta 2018). The rice variety Triguna, inoculated with algae, *Chlorella vulgaris* and *Nannochlropsis* sp., had higher activity of APX during AsIII exposure (Upadhyay et al., 2016). Exposure of AsIII, at the higher toxic concentration, increased the APX activity in the root of rice plants (Shri et al., 2009). Rai et al. (2011) measured the APX activity in the seeding of four rice varieties (Triguna, IR-36, PNR-519, and IET-4786) exposed to AsV and AsIII for 7 days. In general, APX activity was induced in Triguna, IR-36, PNR-519 variety of rice during AsV (10–50 μM) and AsIII (5–25 μM). However, IET-4786 showed a reduction in APX activity on AsV exposure (10–50 μM) and AsIII exposure (5–25 μM). The differential variation in APX activity was also reported in the fenugreek (*Trigonella foenum-graecum* L. cv. Azad) on exposure of AsIII (10–30 mgAsIII/kg soil) (Talukdar 2013). The studies indicated that both AsV and AsIII exposure exhibited changes in the activity of APX in *Brassica juncea* (Ahmad, & Gupta 2013; Ahmad et al., 2021; Kanwar et al., 2015). Similarly, in both As hyperaccumulator and nonhyperaccumulator plant *P. vittata* and *V. Zizanioides*, the APX activity was modulated in root & frond exposed at various concentrations of As (Cao et al., 2004; Tiwari & Sarangi 2017).

#### *10.2.1.4 Glutathione reductase*

Glutathione reductase (GR) is an NADPH-dependent enzyme responsible for reducing GSSG to GSH thus maintaining the required level of GSH pool for the proper functioning of cells. Role of GR in mitigating the toxic effect of As in *Isatis cappadocica* has been demonstrated by increased activity of GR (Karimi et al., 2009). Along with *Isatis cappadocica*, the importance of GR in the other plants during As stress has been documented (Begum et al., 2016; Degola et al., 2015). In presence of As the levels of GSH are reduced as GSH is utilized as a substrate for the biosynthesis of PCs. The higher demand of GSH is fulfilled by the increased activity of GR in plants during As stress (Karimi et al., 2009). Studies indicated that enhancement in As supply increased the activities of antioxidant enzymes in various plant species, such as *Glycine max* L., *Helianthus annuus*, *Triticum aestivum*, and *Oryza sativa* (Singh et al., 2017; Chandrakar et al., 2017; Saidi et al., 2017; Sharma et al., 2017). The As-hyperaccumulating fern species have a much higher tolerance against As due to higher activities of antioxidant enzymes and low concentrations of ROS (Singh et al., 2006; Tiwari & Sarangi 2017).

### 10.2.2 Product of lipid peroxidation

Besides antioxidants enzymes activities, the malondialdehyde (MDA), the product of lipid peroxidation is another key marker of As-induced oxidative stress. The level of MDA content is directly dependent on the As stress. As stress-induced ROS level, which resulted in a higher level of MDA content in plants (Singh et al., 2016; Álvarez-Robles et al., 2020). Further, the As-hyperaccumulator plant *Isatis cappadocica* showed an increase in MDA contentment during the exposure of AsV stress (50, 200, 800, & 1200 μmol $L^{-1}$AsV) (Souri et al., 2018). As exposure induced the ROS and MDA level in both As-hyperaccumulator *Pteris vittata* and non–As-hyperaccumulator *Pteris ensiformis*, however, induction was more in *Pteris ensiformis* than *Pteris vittata* (Singh et al., 2006). Similarly, the level of MDA and ROS were higher in the two *Brassica napus* L. cultivars (ZS 758, and Zheda 622) during the exposure of AsIII (Farooq et al., 2016). The above studies indicated that variation in antioxidants enzymes activities, metabolites (MDA), and enzymes related to As detoxification mechanism used as a biochemical marker that indicated particular or group of environmental stress in plants. The modulation in biochemical markers also depends on the concentration, type, and duration of exposure in plants.

## 10.3 Markers related to As metabolism

The As-induced consequences include oxidative stress, DNA damage, proteins dysfunctions, and metabolic and transcriptomic deviation in plants (Potdukhe et al., 2018). The potential mechanisms of As detoxification are to minimize As uptake by root, an increase of As accumulation in the vacuole and volatilization of As by areal plant's parts (Table 10.2).

**Table 10.2** Gene and protein expression under arsenic stress in plants.

| S. No. | Sources (genes/proteins) | Transgenic plants/ microorganism | Gene function/response | References |
|---|---|---|---|---|
| 1. | *Oryza sativa* OsPht1;8 (OsPT8) | *Oryza sativa* | AsV transporter (High-affinity phosphate transporter) | Wu et al. (2011) |
| 2. | *Oryza sativa* OsPT4 | *Oryza sativa* | AsV transporter (High-affinity phosphate transporter) | Ye et al. (2017) |
| 3. | *Brassica napus* BnaPHT1 | *Brassica napus* | AsV transporter (High-affinity phosphate transporter) | Li et al. (2019) |
| 4. | *Oryza sativa* OsNIP2;1 (Lsi1) | *Oryza sativa* | AsIII uptake and efflux transport (aquaporins) | Ma et al. (2008); Zhao et al. (2010) |
| 5. | *Arabidopsis thaliana* AtNIP3;1 | *Arabidopsis thaliana* | AsIII efflux transport (aquaporins) | Xu et al. (2015) |
| 6. | *Oryza sativa* OsPIP2;4, OsPIP2;6, and OsPIP2;7 | *Arabidopsis thaliana* | Enhance AsIII tolerance and higher biomass accumulation | Mosa et al. (2012) |
| 7. | *Oryza sativa* OsHAC1;1 OsHAC1;2 | *Oryza sativa* | AsV reductase (conversion of AsV to AsIII in the plant cells and As accumulation) | Shi et al. (2016) |
| 8. | *Arabidopsis thaliana* AtACR2 | - | Conversion of AsV to AsIII in the plant cells and As accumulation in *Arabidopsis thaliana* | Nahar et al. (2017) |
| 9. | *Ceratophyllum demersum* CdPCS1 | *Oryza sativa* | Phytochelatin synthase (PCs synthesis) | Shri et al. (2014) |
| 10. | *Oryza sativa* OsNRAMP1 | *Arabidopsis thaliana* | NRAMP transporter (As transport to root xylem) | Tiwari et al. (2014) |

*(Continued)*

**Table 10.2** Cont'd

| S. No. | Sources (genes/proteins) | Transgenic plants/ microorganism | Gene function/response | References |
|---|---|---|---|---|
| 11. | *Oryza sativa* OsABCC1 | *Oryza sativa* | ABC (ATP-binding cassette) transporter (vacuolar sequestration of As) ABC transporter G family member 26 | Song et al. (2014b) |
| 12. | *Oryza sativa* (Glutaredoxin) OsGrx_C7 OsGrx_C2.1 | *Arabidopsis thaliana* | Overexpression of glutaredoxin reduces As accumulation and increases As tolerance of *A. thaliana* | Verma et al. (2016) |
| 13. | *Pteris vittata* PvACR3 gene | - | AsIII transporter in *Pteris vittata*. This gene is present in mosses, pteridophytes and gymnosperms but lost in angiosperms; that is why angiosperm is not able to accumulate and tolerate a high level of arsenic | Indriolo et al. (2010) |
| 14. | *Pteris vittata* PvACR3 gene | *Arabidopsis thaliana* | PvACR3 overexpression leads to higher AsIII accumulation and tolerance. | Chen et al. (2013) |
| 15. | *Pteris vittata* PvACR3 gene | *Arabidopsis thaliana* | Hyperaccumulation of As can be enhanced in transgenic *A. thaliana* by knocking out HAC1 (arsenate reductase) and overexpressing PvACR3 heterologous expression of PvACR3 markedly increased As tolerance and root-to-shoot As translocation in *A. thaliana*, with PvACR3 being localized to the plasma membrane | Wang et al. (2018) |

| S. No. | Sources (genes/proteins) | Transgenic plants/ microorganism | Gene function/response | References |
|---|---|---|---|---|
| 16. | *Pteris vittata* PvGSTF1 (PvGAPC1, PvOCT4) | - | Glyceraldehyde 3-P dehydrogenase (PvGAPC1), organic cation transporter (PvOCT4), and GST (PvGSTF1, also has arsenate reductase activity, are upregulated by As and required for As tolerance AsV to reduce AsIII). Essential for arsenic tolerance in *Pteris vittata* gametophyte | Cai et al. (2019) |
| 17. | *Oryza sativa* OsCytP450 (Os01g43740) | - | Highly expressed in the root of *Oryza sativa* during As stress | Chakrabarty et al. (2009) |
| 18. | *Oryza sativa* glutaredoxin (Os01g26912) | - | Highly expressed in the shoot of *Oryza sativa* during As stress | Chakrabarty et al. (2009) |
| 19. | *Oryza sativa* OsMT-I | - | Class I metallothionein, (protection against As toxicity by sequestration in *Oryza sativa*) | Gautam et al. (2012) |
| 20. | *Arachis hypogaea* glutathione S-transferase | - | Protection against As toxicity by sequestration in *Arachis hypogaea* | Bianucci et al. (2017) |
| 21. | *Ceratophyllum demersum* CdPCS1 | *Nicotiana tabacum* | Enhanced accumulation of and As without significant decrease in plant growth | Shukla et al. (2012) |

### 10.3.1 Transporter of As uptake and translocation

In natural surroundings, As is accumulated in root and transported to shoot actively or passively in plants. Generally, various transporter proteins in plants are involved in the uptake of an inorganic As (Chen et al., 2017). The concentration gradient between sink and source is the key stimulator for As uptake. Several plant species use the different mechanisms of As uptake based on the As species such as AsV and AsIII. Several studies reported that AsV could enter into plant cells through numerous phosphate transport proteins as AsV is a chemical analog of phosphate (Li et al., 2016; Awasthi et al., 2017). Enhanced cotransport of AsV and inorganic phosphate (Pi) was observed in the growth medium containing AsV or during phosphorous deficit (Ye et al., 2017). Plants take up AsV through different PHTs (Pi transporter proteins). Both low-affinity and high-affinity phosphate transporter have been reported in plants where the involvement of PHT1 proteins (high-affinity phosphate transporter) in AsV uptake was observed (Catarecha et al., 2007; Li et al., 2016). In contrast, the proteins associated with low-affinity phosphate transport involved in AsV uptake are not identified yet. Ye et al. (2017) reported the involvement of OsPT4 (Pht1 family) during uptake and transport of AsV in rice, where they identified the up-regulated expressions of OsPT1, OsPT4, and OsPT8 under AsV exposure. Transgenic rice plants over-expressing OsPT4 exhibited sensitivity with increased AsV accumulation when grown in different AsV concentrations during hydroponic culture. However, the OsPT4 CRISPR rice plants displayed decreased As accumulation. Hence, the study suggested the role of the OsPT4 as a biomarker gene for AsV transport and generating low As rice (Ye et al., 2017). Conversely, it has been reported that various nodulin-26-like intrinsic proteins (NIPs) are involved in the uptake of AsIII in plants. The involvement of NIP transporters has been reported in bidirectional AsIII transport while PHT transporters are unidirectional. As a result, bidirectional movement of AsIII could be possible based on the concentration gradient of As between the plant cells and growth medium (Amna et al., 2020).

Disentangling the uptake and transport mechanism of As is critically essential to understand and regulate its toxic effects. NIP2;1 (OsLsi1), Si (silicon) influx transporter has been reported to be involved in AsIII uptake by rice roots (Ma et al., 2008). In *Oryza sativa*, OsLsi1 is confined to the distal side of the endodermis, and exodermis cells facilitated AsIII influx successively (Ma et al., 2008). Constrictions within a pore, aromatic/arginine (ar/R), and asparagines-proline-alanine boxes are involved in regulating substrate transport through aquaporin (Venisse et al., 2021). It has been observed that distinct amino acid residues at the ar/R selectivity filter result in various substrate selections of NIPs. NIP-I, NIP-II, and NIP-III are three functional groups of NIPs proteins that have been classified according to Ar/R tetrad amino acids and have a distinctive pore size. The NIP-I proteins were identified in permitting the transport of glycerol, lactic acid and water. NIP-II proteins are permeable to boric acid glycerol instead of water as a result of larger pore size. Moreover, the widest pore of NIP-III results in the permeation of larger molecules, for example, silicic acid, but could not be penetrable to the moderately smaller solute such as glycerol (Sabir et al., 2020a). Likewise, similar substrate profiles of NIPs belonging to NIP-I (VvTnNIP1;1), NIP-II (VvTnNIP6;1), and NIP-III (VvTnNIP2;1) were identified in *Vitis vinifera* L. (Sabir et al., 2020b). The significant part of the ar/R region in different plant MIPs subfamilies could be demonstrated by replacing essential amino acids (Khabudaev et al., 2014). Transport of silicon, boron and AsIII was examined by expressing native and site-directed mutagenized variants of OsNIP2;1 gene from rice and AtNIP5;1 gene from *Arabidopsis* in *Xenopus oocytes* (Deshmukh et al., 2015; Mitani et al., 2011). The transport activities for AsIII, B and Si have been observed unaffected after replacing the amino acid of OsLsi1 at the ar/R H2 helix place, but the fractional loss of AsIII and complete loss of Si and boron transport activities was detected at the H5 location. It has been studied that silicon transport activity was not facilitated by alterations of the ar/R selectivity filter (AtNIP5;1) and the NPA motifs (OsLsi1). Boron transport activity was observed unaltered in AtNIP5;1 H2 mutant, whereas partial loss has been identified in H5 mutant. However, AsIII transport activity remains unaffected in AtNIP5;1 single and double mutations at the H2/ H5 place. Less selectivity for AsIII has been discovered at the ar/R filter H5 position (OsLsi1 and AtNIP5;1) in comparison to silicon and boron where residue at H5 place play a crucial role in the permeability of these minerals and AsIII transport activity has not been regulated by an ar/R selectivity filter (Mitani-Ueno et al., 2011).

Channel proteins such as SIP, PIP, and TIP are not permeable for AsIII. Furthermore, Lsi1 (Si influx) and Lsi2 (Si efflux) transporters were reported to be involved in AsIII transport as As uptake was reported to be significantly decreased in *Oryza sativa* during mutations in Lsi1 (Ma et al., 2008). In addition to the previously discussed Lsi1 transporter, which catalyzes uptake of AsIII and Sb(III), a second $Si(OH)_4$ transporter, Lsi2, was reported in rice (Ma et al., 2008). ArsB (bacterial AsIII efflux) transporters and plant citrate-efflux transporters have been observed to share homology with Lsi2. In rice, OsLsi2 has localized proximally in the exodermis and endodermis cell's plasma membranes, leading to the AsIII efflux from root cells to the xylem (Ma et al., 2007 & 2008). Reduction in AsIII transport and accumulation in the shoots has been observed during mutations in the Lsi2 gene (Ma et al., 2008). Though, the efflux of MAs or DMAs is not regulated by rice Lsi2, probably due to the dissociation of MAs and DMAs at cytoplasmic pH (Garbinski et al., 2019).

### 10.3.2 Transporters of As sequestration

In addition to this, the involvement of OsABCC1 (C-type ABC transporter) in reducing the transport of As in rice grain and detoxification was reported by Song et al. (2014b). OsABCC1 is localized to the phloem region of vascular bundles. Moreover, PC regulated As tolerance in yeast was conferred by OsABCC1, which is localized to the tonoplast (Song et al., 2014a & 2014b). Likewise, OsABCC1 is involved in As tolerance in rice by reducing the As accumulation to the grain at the reproductive phase. As tolerance in rice was confirmed by knocking out the OsABCC1 gene, which leads to the reduced As concentrations in the grain. As tolerance by OsABCC1 could be achieved due to its sequestration in the vacuoles of phloem cells thus restricting the dispersal of As to the grain (Song et al., 2014b). Therefore, over-expression of ABC transporter and PCS could be used as a biomarker for producing As-free plants with reducing accumulation in edible parts and thus decreasing As exposure to humans.

## 10.4 Enzymes and thiolic metabolite for As detoxification

As soon as AsV enters the plant cell, it is reduced or converted to more toxic form AsIII, and this step is thought of as the first step in detoxification of As in plants. AsIII can be extruded out from the plant cell to the external medium. The AsIII forms the complexes with either GSH, MT, or the PCs and gets sequestered inside the vacuole as a detoxification mechanism.

### 10.4.1 Arsenate reductase

The reduction of AsV to AsIII occurs through the enzyme arsenate reductase utilizing GSH as an electron donor. This conversion is a one-step reaction in which AsV is reduced to AsIII (Finnegan and Chen, 2012). Arsenate reductase gene has been cloned and well-characterized in *Arabidopsis* AtHAC1/ATQ1 (Chao et al., 2014; Sánchez-Bermejo et al., 2014), rice

OsHAC1;1 and OsHAC1;2 (Shi et al., 2016), and *Holcuslanatus* HlAsr (Bleeker et al., 2006). Arsenate reductase enzyme has a significant role in As hyperaccumulator plants such as *Pteris vittata.* The enzymatic activity of arsenate reductase is found much higher (7 fold higher) in hyperaccumulator *Pteris vittata* plants compared with the nonaccumulator *Oryza sativa* or *Arabidopsis thaliana* plant (Danh et al., 2014). Higher expression and activity of arsenate reductase and efficient transportation of As from root to shoot is thought to play a very significant role in the accumulation of the higher amount of As in the fronds of the *Pteris vittata* plant (Song et al., 2010).

## 10.4.2 Thiolic ligands and enzymatic systems

Once the AsV is reduced to AsIII, the second step is the sequestration of As inside the vacuole as a detoxification mechanism. Plants utilize the sulfhydryl group containing small peptides and proteins as a quencher of AsIII. AsIII has a higher affinity for binding to the protein, having reduced the thiol group. The small peptides such as GSH and PCs work efficiently in quenching or forming the complexes with the AsIII. GSH is a vital small peptide with plenty of vital roles in the life of the plants. GSH is a low molecular weight peptide with general formula ($\gamma$ Glu-Cys-Gly) and is synthesized with the help of $\gamma$-glutamyl cysteine synthetase ($\gamma$-ECS) and GSH synthetase (GS) enzymes. GSH directly forms complexes with AsIII by utilizing its thiol group (Jocelyn, 1972; Scott et al., 1993; Gailer & Lindner, 1998). Once the complex of AsIII-GSH is formed, the next step is its transportation and sequestration inside the vacuole. The transporters present on the tonoplast, such as AtABCC1 and AtABCC2, have been reported to transport GSH-organic molecules inside the vacuole (Wanke & Kolukisaoglu, 2010).

The interaction between As stress and the plant may be pointed up at different levels by using various biochemical activities such as modulation in antioxidant enzymes, oxidative stress products, stress proteins, and other parameters related to As metabolism. GST (glutathione S-transferase) enzymes are multifunctional, ubiquitous enzymes encoded by a multigene family in plants. GST enzymes play an essential role in transferring glutathione moiety to the exogenous or endogenous toxic substrate. Conjugation of GSH moiety to the toxic substrate triggers the solubilization, transportation, and vacuolar sequestration of GSH-conjugated toxic substrate or (Yan et al., 2019; Pawlik-Skowrońska et al., 2004; Zhao et al., 2002). GST also participated in detoxification of ROS during As stress. Srivastava et al. (2019) showed that overexpression of OsGSTU30 provided tolerance against AsV, AsIII, and Cr (VI) in *Arabidopsis thaliana*. The exposure of AsV and AsIII differentially modulated GST expression in As contrasting rice varieties (Dubey et al., 2016). Jain et al. (2010) analyzed the expression behavior of GST genes in rice plants (As-tolerant rice cultivar Azucena and As-sensitive cultivar Bala) exposed to AsV stress for 7 days. A total of 25 genes (out of 32) related to the Tau class of the GST gene family were differentially modulated during As stress. The *E. coli* (mutant arsC, arsenate reductase) cells transformed with pPvGST35, pPvGST41, and pPvGST57 harbor the AsV tolerance by enhancing the GSH-dependent reduction of AsV into AsIII (Khan et al., 2018). GST also enables the binding of As with GSH for detoxification of As. However, it has been shown that exposure of As stress induced the expression of several GST isoforms involved in As-GSH complexation in plants, which indicated that GST-mediated complexation of As and thiol prevented the As-induced damages in cells displaying a mechanism of As detoxification (Tiwari et al., 2022; Yan et al. (2019).

Phytochelatins are small peptides with the molecular formula of (γ-Glu-Cys)n-Gly, where n varies from 2 to 11 (Grill et al., 1987). PCs are posttranslationally synthesized with the help of enzymes phytochelatin synthase. AsIII has a high affinity to thiol groups and forms stable complexes with PCs in plant cells (Schmöger et al., 2000). Various reports are available for PCs mediated AsIII complexation and sequestration as key mechanisms for providing tolerance and protection against As in both As hyperaccumulator and nonaccumulator plants under constitutive and adaptive conditions (Gupta et al., 2011). Overexpression of genes that regulate Cys, GSH, and PCs biosynthesis have resulted in more significant, and efficient tolerance in plants against As toxicity (Tripathi et al., 2007; Wojas et al., 2010). In some As-hyperaccumulator plants such as *H. Lanatus*, *Silene paradoxa*, *Pteris vittata*, and *Pteris cretica*, PCs have not been responsible for a significant role in providing As-tolerance (Raab et al., 2007; Arnetoli et al., 2008). However, in some other, As hyperaccumulator plants such as in *Isatis cappadocica*, hyperaccumulation and tolerance are achieved by PCs complexation (>50%) which is observed to be a constitutive tolerance mechanism (Karimi et al., 2009). Sequestration of the AsIII-PCs complexes in vacuoles is an essential step in As detoxification metabolism. OsABCC1 transports AsIII-PCs across the tonoplast, its overexpression in yeast and *Arabidopsis* increased As-tolerance, while knockout mutants were hypersensitive to the metalloid (Song et al., 2014a). Overexpression of AtPCS1 has resulted in the higher accumulation of As and providing more tolerance to As in comparison to control (nontransgenic) plants. In contrast to *Arabidopsis*, rice possess two active As tolerance PCS i.e., OsPCS1 and OsPCS2. Recently, Hayashi et al. (2017) had shown PC-dependent pathway mediated by OsPCS1 and OsABCC1 rather exclusively accounts for a significant fraction of the As detoxification capacity in rice. Yamazaki et al. (2018) had demonstrated the involvement of OsPCS2 in providing tolerance to As in rice. These results indicated that PCs are vital and central molecules providing tolerance against As in plants. PCs can be called a qualitative biochemical biomarker (Ernst, WHO, 1995) of As. A biomarker can be primarily operational if it can be detected in the field. However, PCs in the deep-rooted samples may be hard to find (Grill et al., 1988). At the same concentration of exposure, there is a decreasing order of PCs response: Hg>Cd>As>Te>Ag>Cu>Ni>Sb>Au>Sn>Se>Bi>Pb>W>Zn (Tukendrof & Rauser, 1990). In contrast to root, the shoot may be easy to sample, but PCs in leaves might represent a late response. Albeit, foliar PCs have been designated as a biochemical indicator of forest decline (McLaughlin & Wang, 1996). In As-hyperaccumulator *Pteris vittata* has duplication of ACR3 and ACR3;1 gene, which code for vacuolar transporter for arsenite, make it a hyperaccumulator for As by sequestration of AsIII into the vacuole which is unique to fern (Indriolo et al., 2010). This gene is absent in flowering plants.

### 10.4.3 Metallothionein in detoxification of As in plant

Metallothioneins (MT) are low molecular weight proteins ranging from 4-8kDa. These are Cys-rich proteins that can bind with heavy metals via thiol groups, thus forming heavy metal-MT complexes (Hamer, 1986). MTs have various roles in the entire life of the plants during stressful as well as normal conditions (Tripathi et al., 2015). The important roles of MTs are homeostasis of the essential heavy metals, protecting against oxidative damage, and improving protection against various heavy metals by making heavy metal-MT complexes and sequestering inside vacuole (Gasic & Korban, 2007). Plant MTs are divided into three groups based on the arrangement of cysteine residues (Cobbett & Goldsbrough, 2002). Most of the plant MTs are class 1 proteins containing two smaller cysteine-rich domains and a large spacer region devoid of cysteine. Class II MTs are grouped in 3 cysteine-rich domains separated by 10–15 amino acid residues (Lane et al.,1987; Zhou et al., 2005; Zhou et al., 2006). Class III MTs belong to PCs (Cobbett & Goldsbrough, 2002). Class 1 MTs can be further grouped in different types (type I-IV) differing in cysteine residues distribution. The expression of MTs is modulated by various factors such as hormonal signals and the presence of heavy metals. There are reports available advocating the protective roles of MT in plants during As stress. The study on rice suggested the presence of 11 class 1 MT genes in the rice genome, which are differentially expressed in growth and development in different tissues. The study indicated that class1 MT proteins in rice differ in tissue localization and have metal coordination chemistry. The study showed that some MTs (OsMT-I-Id) are highly expressed in the root as compared to Pb or Cd exposure. Similarly, another MT (OsMT-I-Ib) was highly expressed in both roots and shoot compared to Pb and Cd exposure. These two MTs (OsMT-I-Id and OsMT-I-Ib) can be used as a biomarker of As exposure in rice (Gautam et al., 2012). The expressions profile of rice MT genes were studied in AsV tolerant rice cultivar Triguna and AsV sensitive cultivar, IET-4716 MTs expression (OsMT-I-IIa, OsMT-I-IId, OsMT-I-Id, OsMT-I-Ie, OsMT-I-IIc) were always higher in As tolerant Triguna than AsV sensitive IET-4716 rice cultivar when grown in 50μM AsV for 7 days (Gautam et al., 2012). Chakrabarty et al. (2009) had also shown modulation in the expression of some MT-like proteins. Gene expression analysis showed differential expression of MT gene in different tissues, at different developmental stages and differentially at different heavy metals and at different exposure concentrations and times of the stress (Castiglione et al., 2007; Ahn et al., 2012).

## 10.5 Markers related to protein metabolism during As stress

In the plant cells, the cellular damage is dependent on the oxidation state of As. As a result of the high affinity of AsIII to sulfhydryl groups in proteins, AsIII is found to be 100 fold more toxic than AsV (Abbas et al., 2018). Furthermore, by interacting with thiol-containing proteins, AsIII can convert functional proteins to nonfunctional proteins (Finnegan & Chen 2012). Until now, over 100 AsIII sensitive proteins were identified, including structural proteins, transcription factors, metabolic enzymes, proteolytic proteins and redox regulatory enzyme activities (Shen et al., 2013). Requejo & Tena (2005) identified As responsive proteins in the roots of maize. It has been observed that around 10% of the total proteins were considerably altered under the AsV exposure, including 7 enzymes i.e., three SODs, two GPXs, one PRX, and one pBQR involved in redox homeostasis. Additionally, four functionally heterogeneous proteins such as two enzymes involved in energy metabolism, i.e., succinyl-CoA synthetase and ATP synthase, cytochrome P450 enzymes, guanine nucleotide-binding protein β subunit (signaling protein) have been identified during As stress, thus, confirming a significant effect on protein metabolism. In *Vigna radiata* and *Oryza sativa* (Ismail 2012; Singh et al., 2009), protein content was substantially reduced upon As exposure. Proteins related to cellular homeostasis and metabolism are hydrolyzed by protease and peptidases to form distinct amino acids, proteins, and peptides (Jakubczyk et al., 2020). In *Oryza sativa*, reduced protein content was observed to be associated with As elicited suppression of nitrate and nitrite reductase activity (Singh et al., 2009). Though, modified free amino acids and proteins content due to proteases degradation was observed upon As treatment resulting in reduced the growth and development of crops such as mung bean and rice (Chandrakar et al., 2016). Disrupted nutrient metabolism has been observed by amino acid profiling, which leads to the amino acids and proteins damage by As-induced ROS, releasing reactive carbonyl species (Kumar et al., 2014). ROS induces the oxidation of amino acids such as Arg, His, Lys, Pro, Thr and Trp of protein into γ-glutamyl semialdehyde, amino ketobutyrate and amino adipic semialdehyde contributes to the disruption of protein activities (Yaduet al., 2016). As contamination leads to higher biosynthesis of an essential amino acid (Ile, Leu, Val, Phe, Lys, Thr, Met) in low As accumulating rice genotype (LAARG) in contrast to highly As accumulating rice genotype (HAARG) when grown in highly As contaminate regions of West Bengal (Dwivedi et al., 2012). Stimulation of protein oxidation or carbonylation by ROS under As exposure has yet to be explored. Therefore, it is assumed that protein carbonylation can be used as an oxidative stress biomarker of As toxicity since a close association of carbonyl derivatives formation with As was investigated in animals (Mandal, 2017).

## 10.6 Conclusion and future prospect

Plants modulate the genomic, proteomic, and metabolomic responses to alleviate the As stress. The mechanisms of As uptake, transport, and detoxification pathways are different in As-hyperaccumulator or nonhyperaccumulator plants including crops. The variation in metabolite, activities of certain antioxidant enzymes such as MDA, SOD, APX, CAT, and GR authenticate the manifestation of oxidative stress during the metalloid stress in plants, and some of them seem to perform as biomarkers. Transporters OsPT4, OsPT8, OsNIP3;1, OsPIPs, OsHAC1, OsNRAMP1, which transport AsIII or AsV appear to be biomarkers of As stress which is supported by the performance of transgenic plants. Glyceraldehyde 3-P dehydrogenase (PvGAPC1), organic cation transporter (PvOCT4) and GST (PvGSTF1, also arsenate reductase activity) are upregulated by As and required for As tolerance in the gametophyte of *Pteris vittata* and these genes were highly upregulated by As and may emerge as a biomarker of As. PvACR3gene from *Pteris vittata* seems to be unique, transporting AsIII from root to shoot, present in mosses, pteridophytes, and gymnosperms but missing in angiosperms. As such PvACR3 may be a reliable biomarker in ferns. Phytochelatins (PCs), the prevalent metal detoxification system in plants, appeared to be qualitative biomarkers of As stress. The synthesis of PCs is PC synthase dependent. Metallothioneins, the cysteine-rich proteins have both metal homeostatic and detoxification roles besides other functions. Eleven MTs have been identified from rice, out of which two MTs, OsMT1-1d, and OsMT1-1b, seem to be mainly expressed during As stress and may work as biomarkers of As stress. Molecular biology techniques have discovered significant prospects in emerging As-tolerant plants. Additional studies are essential for classifying genetic markers linked with the As-tolerant phenotype. Low As accumulating cultivars can be selected using these markers. Transgenic methodologies including these signalling components may be helpful in generating As-free plants. Gene editing tools for example CRISPR (cluster regularly interspaced short palindromic repeats)-associated nuclease (Cas) [CRISPR/Cas9]; ZFNs (zinc-finger nucleases) and TALENs (transcription activator like effector nucleases) [ZFN/ TALEN] can be utilized to reduce As accumulation in grain by targeting potential genes.

## Acknowledgments

RDT and PKS are grateful to the National Academy of Sciences, Prayagraj, India for the NASI-Senior Scientist Platinum Jubilee fellowship and RAship, respectively. In addition, SD acknowledges the financial assistance from SERB Project File No.YSS/2015/000742. The MS. No. CSIR-NBRI_MS/2022/01/04 was allocated by the institutional ethics committee after subjecting the MS through ethical and plagiarism check.

## References

Abbas, G., Murtaza, B., Bibi, I., Shahid, M., Niazi, N.K., Khan, M.I., …, Hussain, M., 2018. Arsenic uptake, toxicity, detoxification, and speciation in plants: physiological, biochemical, and molecular aspects. Int. J. Environ. Res. Public Health 15 (1), 59.

Abercrombie, J.M., Halfhill, M.D., Ranjan, P., Rao, M.R., Saxton, A.M., Yuan, J.S., Stewart, C.N., 2008. Transcriptional responses of *Arabidopsis thaliana* plants to As (V) stress. BMC Plant Biol. 8 (1), 1–15.

Ahmad, A., Khan, W.U., Shah, A.A., Yasin, N.A., Naz, S., Ali, A., …, Batool, A.I., 2021. Synergistic effects of nitric oxide and silicon on promoting plant growth, oxidative stress tolerance and reduction of arsenic uptake in *Brassica juncea*. Chemosphere 262, 128384.

Ahmad, M.A., Gupta, M., 2013. Exposure of *Brassica juncea* (L) to arsenic species in hydroponic medium: comparative analysis in accumulation and biochemical and transcriptional alterations. Environ. Sci. Pollut. Res. 20 (11), 8141–8150.

Ahn, Y.O., Kim, S.H., Lee, J., Kim, H., Lee, H.S., Kwak, S.S., 2012. Three *Brassica rapa* metallothionein genes are differentially regulated under various stress conditions. Mol. Biol. Rep. 39 (3), 2059–2067.

Álvarez-Robles, M.J., Bernal, M.P., Sánchez-Guerrero, A., Sevilla, F., Clemente, R., 2020. Major As species, lipid peroxidation and protein carbonylation in rice plants exposed to increasing As (V) concentrations. Heliyon 6 (8), e04703.

Amna, S., Qamar, S., Naqvi, A.A.T., Al-Huqail, A.A., Qureshi, M.I., 2020. Role of sulfur in combating arsenic stress through upregulation of important proteins, and in-silico analysis to study the interaction between phosphate transporter (PHO1), arsenic and phosphate in spinach. Plant Physiol. Biochem. 157, 348–358.

Anjum, N.A., Sharma, P., Gill, S.S., Hasanuzzaman, M., Khan, E.A., Kachhap, K., …, Tuteja, N., 2016. Catalase and ascorbate peroxidase representative $H_2O_2$ detoxifying heme enzymes in plants. Environ. Sci. Pollut. Res. 23 (19), 19002–19029.

Arnetoli, M., Vooijs, R., ten Bookum, W., Galardi, F., Gonnelli, C., Gabbrielli, R., …, Verkleij, J.A., 2008. Arsenate tolerance in *Silene paradoxa* does not rely on phytochelatin-dependent sequestration. Environ. Pollut. 152 (3), 585–591.

Awasthi, S., Chauhan, R., Srivastava, S., Tripathi, R.D., 2017. The journey of arsenic from soil to grain in rice. Front. Plant Sci. 8, 1007.

Bebianno, M.J., Geret, F., Hoarau, P., Serafim, M.A., Coelho, M.R., Gnassia-Barelli, M., Romeo, M., 2004. Biomarkers in *Ruditapes decussatus*: a potential bioindicator species. Biomarkers 9 (4-5), 305–330.

Begum, M.C., Islam, M.S., Islam, M., Amin, R., Parvez, M.S., Kabir, A.H., 2016. Biochemical and molecular responses underlying differential arsenic tolerance in rice (*Oryza sativa* L.). Plant Physiol. Biochem. 104, 266–277.

Bianucci, E., Furlan, A., del Carmen Tordable, M., Hernández, L.E., Carpena-Ruiz, R.O., Castro, S., 2017. Antioxidant responses of peanut roots exposed to realistic groundwater doses of arsenate: identification of glutathione S-transferase as a suitable biomarker for metalloid toxicity. Chemosphere 181, 551–561.

Bleeker, P.M., Hakvoort, H.W., Bliek, M., Souer, E., Schat, H., 2006. Enhanced arsenate reduction by a CDC25-like tyrosine phosphatase explains increased phytochelatin accumulation in arsenate-tolerant *Holcus lanatus*. Plant J. 45 (6), 917–929.

Cai, C., Lanman, N.A., Withers, K.A., DeLeon, A.M., Wu, Q., Gribskov, M., …, Banks, J.A., 2019. Three genes define a bacterial-like arsenic tolerance mechanism in the arsenic hyperaccumulating fern *Pteris vittata*. Curr. Biol. 29 (10), 1625–1633.

Cao, X., Ma, L.Q., Tu, C., 2004. Antioxidative responses to arsenic in the arsenic-hyperaccumulator Chinese brake fern (*Pteris vittata* L.). Environ. Pollut. 128 (3), 317–325.

Castiglione, S., Franchin, C., Fossati, T., Lingua, G., Torrigiani, P., Biondi, S., 2007. High zinc concentrations reduce rooting capacity and alter metallothionein gene expression in white poplar (*Populus alba* L. cv. Villafranca). Chemosphere 67 (6), 1117–1126.

Catarecha, P., Segura, M.D., Franco-Zorrilla, J.M., García-Ponce, B., Lanza, M., Solano, R., …, Leyva, A., 2007. A mutant of the *Arabidopsis* phosphate transporter PHT1; 1 displays enhanced arsenic accumulation. Plant Cell 19 (3), 1123–1133.

Chakrabarty, D., Trivedi, P.K., Misra, P., Tiwari, M., Shri, M., Shukla, D., …, Tuli, R., 2009. Comparative transcriptome analysis of arsenate and arsenite stresses in rice seedlings. Chemosphere 74 (5), 688–702.

Chandrakar, V., Naithani, S.C., Keshavkant, S., 2016. Arsenic-induced metabolic disturbances and their mitigation mechanisms in crop plants: a review. Biologia (Bratisl) 71 (4), 367–377.

Chandrakar, V., Yadu, B., Meena, R.K., Dubey, A., Keshavkant, S., 2017. Arsenic-induced genotoxic responses and their amelioration by diphenylene iodonium, 24-epibrassinolide and proline in *Glycine max* L. Plant Physiol. Biochem. 112, 74–86.

Chao, D.Y., Chen, Y., Chen, J., Shi, S., Chen, Z., Wang, C., …, Salt, D.E., 2014. Genome-wide association mapping identifies a new arsenate reductase enzyme critical for limiting arsenic accumulation in plants. PLoS Biol. 12 (12), e1002009.

Chen, Y., Xu, W., Shen, H., Yan, H., Xu, W., He, Z., Ma, M., 2013. Engineering arsenic tolerance and hyperaccumulation in plants for phytoremediation by a PvACR3 transgenic approach. Environ. Sci. Technol. 47 (16), 9355–9362.

Chen, Y., Han, Y.H., Cao, Y., Zhu, Y.G., Rathinasabapathi, B., Ma, L.Q., 2017. Arsenic transport in rice and biological solutions to reduce arsenic risk from rice. Front. Plant Sci. 8, 268.

Cobbett, C., Goldsbrough, P., 2002. Phytochelatins and metallothioneins: roles in heavy metal detoxification and homeostasis. Annu. Rev. Plant Biol. 53 (1), 159–182.

Danh, L.T., Truong, P., Mammucari, R., Foster, N., 2014. A critical review of the arsenic uptake mechanisms and phytoremediation potential of *Pteris vittata*. Int. J. Phytorem. 16 (5), 429–453.

Degola, F., Fattorini, L., Bona, E., Sprimuto, C.T., Argese, E., Berta, G., di Toppi, L.S., 2015. The symbiosis between *Nicotiana tabacum* and the endomycorrhizal fungus *Funneliformis mosseae* increases the plant glutathione level and decreases leaf cadmium and root arsenic contents. Plant Physiol. Biochem. 92, 11–18.

Demircan, N., Cucun, G., Uzilday, B., 2020. Mitochondrial alternative oxidase (AOX1a) is required for the mitigation of arsenic-induced oxidative stress in *Arabidopsis thaliana*. Plant Biotechnol. Rep. 14 (2), 235–245.

Deshmukh, R.K., Vivancos, J., Ramakrishnan, G., Guérin, V., Carpentier, G., Sonah, H., …, Bélanger, R.R., 2015. A precise spacing between the NPA domains of aquaporins is essential for silicon permeability in plants. Plant J. 83 (3), 489–500.

Du, Y.Y., Wang, P.C., Chen, J., Song, C.P., 2008. Comprehensive functional analysis of the catalase gene family in *Arabidopsis thaliana*. J. Integr. Plant Biol. 50 (10), 1318–1326.

Dubey, A.K., Kumar, N., Sahu, N., Verma, P.K., Chakrabarty, D., Behera, S.K., Mallick, S., 2016. Response of two rice cultivars differing in their sensitivity towards arsenic, differs in their expression of glutaredoxin and glutathione S transferase genes and antioxidant usage. Ecotoxicol. Environ. Saf. 124, 393–405.

Dvorak, P., Krasylenko, Y., Zeiner, A., Samaj, J., Takac, T., 2020. Signaling toward ROS-scavenging enzymes in plants. Front. Plant Sci. 11, 2178.

Dwivedi, S., Mishra, A., Tripathi, P., Dave, R., Kumar, A., Srivastava, S., …, Nautiyal, C.S., 2012. Arsenic affects essential and non-essential amino acids differentially in rice grains: inadequacy of amino acids in rice based diet. Environ. Int. 46, 16–22.

Ellis, D.R., Gumaelius, L., Indriolo, E., Pickering, I.J., Banks, J.A., Salt, D.E., 2006. A novel arsenate reductase from the arsenic hyperaccumulating fern *Pteris vittata*. Plant Physiol. 141 (4), 1544–1554.

Ernst, W.H.O., 1995. Sampling of plant material for chemical analysis. Sci. Total Environ. 176 (1-3), 15–24.

Farooq, M.A., Gill, R.A., Islam, F., Ali, B., Liu, H., Xu, J., …, Zhou, W., 2016. Methyl jasmonate regulates antioxidant defense and suppresses arsenic uptake in *Brassica napus* L. Front. Plant Sci. 7, 468.

Fatima, R.A., Ahmad, M., 2005. Certain antioxidant enzymes of Allium cepa as biomarkers for the detection of toxic heavy metals in wastewater. Sci. Total Environ. 346 (1-3), 256–273.

Finnegan, P., Chen, W., 2012. Arsenic toxicity: the effects on plant metabolism. Front. Physiol. 3, 182.

Frugoli, J.A., Zhong, H.H., Nuccio, M.L., McCourt, P., McPeek, M.A., Thomas, T.L., McClung, C.R., 1996. Catalase is encoded by a multigene family in *Arabidopsis thaliana* (L.) Heynh. Plant Physiol. 112 (1), 327–336.

Gailer, J., Lindner, W., 1998. On-column formation of arsenic-glutathione species detected by size-exclusion chromatography in conjunction with arsenic-specific detectors. J. Chromatogr. B Biomed. Sci. Appl. 716 (1-2), 83–93.

Garbinski, L.D., Rosen, B.P., Chen, J., 2019. Pathways of arsenic uptake and efflux. Environ. Int. 126, 585–597.

Gasic, K., Korban, S.S., 2007. Transgenic Indian mustard (*Brassica juncea*) plants expressing an Arabidopsis phytochelatin synthase (AtPCS1) exhibit enhanced As and Cd tolerance. Plant Mol. Biol. 64 (4), 361–369.

Gautam, N., Verma, P.K., Verma, S., Tripathi, R.D., Trivedi, P.K., Adhikari, B., Chakrabarty, D., 2012. Genome-wide identification of rice class I metallothionein gene: tissue expression patterns and induction in response to heavy metal stress. Funct. Integr. Genomics 12 (4), 635–647.

Geret, F., Serafim, A., Bebianno, M.J., 2003. Antioxidant enzyme activities, metallothioneins and lipid peroxidation as biomarkers in *Ruditapes decussatus*? Ecotoxicology 12 (5), 417–426.

Grill, E., Winnacker, E.L., Zenk, M.H., 1987. Phytochelatins, a class of heavy-metal-binding peptides from plants, are functionally analogous to metallothioneins. Proc. Natl. Acad. Sci. 84 (2), 439–443.

Grill, E., Winnacker, E.L., Zenk, M.H., 1988. Occurrence of heavy metal binding phytochelatins in plants growing in a mining refuse area. Experientia 44 (6), 539–540.

Gupta, D.K., Srivastava, S., Huang, H.G., Romero-Puertas, M.C., Sandalio, L.M., 2011. Arsenic tolerance and detoxification mechanisms in plantsDetoxification of Heavy Metals. Springer, Berlin, Heidelberg, pp. 169–179.

Hamer, D.H., 1986. Metallothionein. Annu. Rev. Biochem. 55 (1), 913–951.

Hasanuzzaman, M., Bhuyan, M.H.M., Zulfiqar, F., Raza, A., Mohsin, S.M., Mahmud, J.A., …, Fotopoulos, V., 2020. Reactive oxygen species and antioxidant defense in plants under abiotic stress: revisiting the crucial role of a universal defense regulator. Antioxidants 9 (8), 681.

Hayashi, S., Kuramata, M., Abe, T., Takagi, H., Ozawa, K., Ishikawa, S., 2017. Phytochelatin synthase OsPCS1 plays a crucial role in reducing arsenic levels in rice grains. Plant J. 91 (5), 840–848.

Hughes, M.F., Beck, B.D., Chen, Y., Lewis, A.S., Thomas, D.J., 2011. Arsenic exposure and toxicology: a historical perspective. Toxicol. Sci. 123 (2), 305–332.

Indriolo, E., Na, G., Ellis, D., Salt, D.E., Banks, J.A., 2010. A vacuolar arsenite transporter necessary for arsenic tolerance in the arsenic hyperaccumulating fern *Pteris vittata* is missing in flowering plants. Plant Cell 22 (6), 2045–2057.

Ismail, G.S.M., 2012. Protective role of nitric oxide against arsenic-induced damages in germinating mung bean seeds. Acta. Physiol. Plant. 34 (4), 1303–1311.

Jain, M., Ghanashyam, C., Bhattacharjee, A., 2010. Comprehensive expression analysis suggests overlapping and specific roles of rice glutathione S-transferase genes during development and stress responses. BMC Genom. 11 (1), 1–17.

Jakubczyk, A., Karaś, M., Rybczyńska-Tkaczyk, K., Zielińska, E., Zieliński, D., 2020. Current trends of bioactive peptides-New sources and therapeutic effect. Foods 9 (7), 846.

Jocelyn, P.C., 1972. The occurrence, chemical properties, metabolism and biological function of thiols and disulphidesBiochemistry of the SH Group10. Academic Press, London.

Kanwar, M.K., Poonam, Bhardwaj, R, 2015. Arsenic induced modulation of antioxidative defense system and brassinosteroids in *Brassica juncea* L. Ecotoxicol. Environ. Saf. 115, 119–125.

Karimi, N., Ghaderian, S.M., Raab, A., Feldmann, J., Meharg, A.A., 2009. An arsenic-accumulating, hypertolerant brassica, *Isatis capadocica*. New Phytol. 184 (1), 41–47.
Kato, S., Ueno, T., Fukuzumi, S., Watanabe, Y., 2004. Catalase reaction by myoglobin mutants and native catalase: mechanistic investigation by kinetic isotope effect. J. Biol. Chem. 279 (50), 52376–52381.
Kertulis-Tartar, G.M., Rathinasabapathi, B., Ma, L.Q., 2009. Characterization of glutathione reductase and catalase in the fronds of two *Pteris* ferns upon arsenic exposure. Plant Physiol. Biochem. 47 (10), 960–965.
Khabudaev, K.V., Petrova, D.P., Grachev, M.A., Likhoshway, Y.V., 2014. A new subfamily LIP of the major intrinsic proteins. BMC Genom. 15 (1), 1–7.
Khan, A.A., Ellis, D.R., Huang, X., Norton, G.J., Meharg, A.A., Salt, D.E., Csonka, L.N., 2018. Glutathione-S-transferase from the arsenic hyperaccumulator fern *Pteris vittata* can confer increased arsenate resistance in *Escherichia coli*. BioRxiv, 379511.
Khan, E., Gupta, M., 2018. Arsenic–silicon priming of rice (*Oryza sativa* L.) seeds influence mineral nutrient uptake and biochemical responses through modulation of Lsi-1, Lsi-2, Lsi-6 and nutrient transporter genes. Sci. Rep. 8 (1), 1–16.
Klepp, L.I., Eirin, M.E., Garbaccio, S., Soria, M., Bigi, F., Blanco, F.C., 2019. Identification of bovine tuberculosis biomarkers to detect tuberculin skin test and IFNγ release assay false negative cattle. Res. Vet. Sci. 122, 7–14.
Kumar, A., Dwivedi, S., Singh, R.P., Chakrabarty, D., Mallick, S., Trivedi, P.K., …, Tripathi, R.D., 2014. Evaluation of amino acid profile in contrasting arsenic accumulating rice genotypes under arsenic stress. Biol. Plant. 58 (4), 733–742.
Kumar, V., Dietz, K.J., Sharma, S.S., 2021. Differential arsenic-induced membrane damage and antioxidant defence in isolated pinna segments of selected fern species from Western Himalaya. J. Plant Growth Regul. 40, 855–866.
Lane, B., Kajioka, R., Kennedy, T., 1987. The wheat-germ Ec protein is a zinc-containing metallothionein. Biochem. Cell. Biol. 65 (11), 1001–1005.
Li, N., Wang, J., Song, W.Y., 2016. Arsenic uptake and translocation in plants. Plant Cell Physiol. 57 (1), 4–13.
Li, Y., Wang, X., Zhang, H., Wang, S., Ye, X., Shi, L., …, Ding, G., 2019. Molecular identification of the phosphate transporter family 1 (PHT1) genes and their expression profiles in response to phosphorus deprivation and other abiotic stresses in *Brassica napus*. PLoS One 14 (7), e0220374.
Ma, J.F., Yamaji, N., Mitani, N., Tamai, K., Konishi, S., Fujiwara, T., …, Yano, M., 2007. An efflux transporter of silicon in rice. Nature 448 (7150), 209–212.
Ma, J.F., Yamaji, N., Mitani, N., Xu, X.Y., Su, Y.H., McGrath, S.P., Zhao, F.J., 2008. Transporters of arsenite in rice and their role in arsenic accumulation in rice grain. Proc. Natl. Acad. Sci. 105 (29), 9931–9935.
Mandal, P., 2017. Molecular insight of arsenic-induced carcinogenesis and its prevention. Naunyn Schmiedeberg's Arch. Pharmacol. 390 (5), 443–455.
McLaughlin, S.B., Yichun, W., 1996. Foliar phytochelatin levels as indicators of forest decline. Trends Plant Sci. 1 (8), 249–250.
Meyer, J.M., Grobbelaar, N., Vleggaar, R., Louw, A.I., 1992. Fluoroacetyl-coenzyme A hydrolase-like activity in *Dichapetalum cymosum*. J. Plant Physiol. 139 (3), 369–372.
Mitani-Ueno, N., Yamaji, N., Zhao, F.J., Ma, J.F., 2011. The aromatic/arginine selectivity filter of NIP aquaporins plays a critical role in substrate selectivity for silicon, boron, and arsenic. J. Exp. Bot. 62 (12), 4391–4398.
Mosa, K.A., Kumar, K., Chhikara, S., Mcdermott, J., Liu, Z., Musante, C., …, Dhankher, O.P., 2012. Members of rice plasma membrane intrinsic proteins subfamily are involved in arsenite permeability and tolerance in plants. Transgenic Res. 21 (6), 1265–1277.
Mylona, P.V., Polidoros, A.N., Scandalios, J.G., 1998. Modulation of antioxidant responses by arsenic in maize. Free Radical Biol. Med. 25 (4-5), 576–585.

Nahar, N., Rahman, A., Ghosh, S., Nawani, N., Manual, A., 2017. Functional studies of AtACR2 gene putatively involved in accumulation, reduction and/or sequestration of arsenic species in plants. Biologia (Bratisl) 72 (5), 520–526.

National Research Council, 1989. Biologic Markers in Reproductive Toxicology. National Academy Press, Washington, DC.

Pawlik-Skowrońska, B., Pirszel, J., Kalinowska, R., Skowroński, T., 2004. Arsenic availability, toxicity and direct role of GSH and phytochelatins in As detoxification in the green alga *Stichococcus bacillaris*. Aquat. Toxicol. 70 (3), 201–212.

Perry, J.J.P., Shin, D.S., Getzoff, E.D., Tainer, J.A., 2010. The structural biochemistry of the superoxide dismutases. Biochim. Biophys. Acta (BBA) Proteins Proteom. 1804 (2), 245–262.

Potdukhe, R.M., Bedi, P., Sarangi, B.K., Pandey, R.A., Thul, S.T., 2018. Root transcripts associated with arsenic accumulation in hyperaccumulator *Pteris vittata*. J. Biosci. 43 (1), 105–115.

Raab, A., Ferreira, K., Meharg, A.A., Feldmann, J., 2007. Can arsenic–phytochelatin complex formation be used as an indicator for toxicity in *Helianthus annuus*? J. Exp. Bot. 58 (6), 1333–1338.

Rahman, A., Mostofa, M.G., Alam, M., Nahar, K., Hasanuzzaman, M., Fujita, M., 2015. Calcium mitigates arsenic toxicity in rice seedlings by reducing arsenic uptake and modulating the antioxidant defense and glyoxalase systems and stress markers. Biomed. Res. Int. 2015, 340812, 12 pages. http://dx.doi.org/10.1155/2015/340812.

Rai, A., Tripathi, P., Dwivedi, S., Dubey, S., Shri, M., Kumar, S., …, Tuli, R., 2011. Arsenic tolerances in rice (*Oryza sativa*) have a predominant role in transcriptional regulation of a set of genes including sulphur assimilation pathway and antioxidant system. Chemosphere 82 (7), 986–995.

Raza, A., Su, W., Gao, A., Mehmood, S.S., Hussain, M.A., Nie, W., …, Zhang, X., 2021. Catalase (CAT) gene family in rapeseed (*Brassica napus* L.): Genome-wide analysis, identification, and expression pattern in response to multiple hormones and abiotic stress conditions. Int. J. Mol. Sci. 22 (8), 4281.

Requejo, R., Tena, M., 2005. Proteome analysis of maize roots reveals that oxidative stress is a main contributing factor to plant arsenic toxicity. Phytochemistry 66 (13), 1519–1528.

Sabir, F., Di Pizio, A., Loureiro-Dias, M.C., Casini, A., Soveral, G., Prista, C., 2020a. Insights into the selectivity mechanisms of grapevine NIP aquaporins. Int. J. Mol. Sci. 21 (18), 6697.

Sabir, F., Gomes, S., Loureiro-Dias, M.C., Soveral, G., Prista, C., 2020b. Molecular and functional characterization of grapevine NIPs through heterologous expression in aqy-null *Saccharomyces cerevisiae*. Int. J. Mol. Sci. 21 (2), 663.

Saidi, I., Yousfi, N., Borgi, M.A., 2017. Salicylic acid improves the antioxidant ability against arsenic-induced oxidative stress in sunflower (*Helianthus annuus*) seedling. J. Plant Nutr. 40 (16), 2326–2335.

Sánchez-Bermejo, E., Castrillo, G., Del Llano, B., Navarro, C., Zarco-Fernández, S., Martinez-Herrera, D.J., …, Leyva, A., 2014. Natural variation in arsenate tolerance identifies an arsenate reductase in *Arabidopsis thaliana*. Nat. Commun. 5 (1), 1–9.

Schmöger, M.E., Oven, M., Grill, E., 2000. Detoxification of arsenic by phytochelatins in plants. Plant Physiol. 122 (3), 793–802.

Scott, N., Hatlelid, K.M., MacKenzie, N.E., Carter, D.E., 1993. Reactions of arsenic (III) and arsenic (V) species with glutathione. Chem. Res. Toxicol. 6 (1), 102–106.

Sharma, S., Anand, G., Singh, N., Kapoor, R., 2017. Arbuscular mycorrhiza augments arsenic tolerance in wheat (*Triticum aestivum* L.) by strengthening antioxidant defense system and thiol metabolism. Front. Plant Sci. 8, 906.

Shen, S., Li, X.F., Cullen, W.R., Weinfeld, M., Le, X.C., 2013. Arsenic binding to proteins. Chem. Rev. 113 (10), 7769–7792.

Shi, S., Wang, T., Chen, Z., Tang, Z., Wu, Z., Salt, D.E., …, Zhao, F.J., 2016. OsHAC1;1 and OsHAC1;2 function as arsenate reductases and regulate arsenic accumulation. Plant Physiol. 172 (3), 1708–1719.

Shri, M., Kumar, S., Chakrabarty, D., Trivedi, P.K., Mallick, S., Misra, P., …, Tuli, R., 2009. Effect of arsenic on growth, oxidative stress, and antioxidant system in rice seedlings. Ecotoxicol. Environ. Saf. 72 (4), 1102–1110.

Shukla, D., Kesari, R., Mishra, S., Dwivedi, S., Tripathi, R.D., Nath, P., Trivedi, P.K., 2012. Expression of phytochelatin synthase from aquatic macrophyte *Ceratophyllum demersum* L. enhances cadmium and arsenic accumulation in tobacco. Plant Cell Rep. 31 (9), 1687–1699.

Siddiqui, F., Tandon, P.K., Srivastava, S., 2015. Arsenite and arsenate impact the oxidative status and antioxidant responses in *Ocimum tenuiflorum* L. Physiol. Mol. Biol. Plants 21 (3), 453–458.

Singh, A.P., Dixit, G., Kumar, A., Mishra, S., Kumar, N., Dixit, S., …, Tripathi, R.D., 2017. A protective role for nitric oxide and salicylic acid for arsenite phytotoxicity in rice (*Oryza sativa* L.). Plant Physiol. Biochem. 115, 163–173.

Singh, A.P., Dixit, G., Kumar, A., Mishra, S., Singh, P.K., Dwivedi, S., …, Tripathi, R.D., 2016. Nitric oxide alleviated arsenic toxicity by modulation of antioxidants and thiol metabolism in rice (*Oryza sativa* L.). Front. Plant Sci. 6, 1272.

Singh, H.P., Kaur, S., Batish, D.R., Sharma, V.P., Sharma, N., Kohli, R.K., 2009. Nitric oxide alleviates arsenic toxicity by reducing oxidative damage in the roots of *Oryza sativa* (rice). Nitric Oxide 20 (4), 289–297.

Singh, N., Ma, L.Q., Srivastava, M., Rathinasabapathi, B., 2006. Metabolic adaptations to arsenic-induced oxidative stress in *Pteris vittata* L and *Pteris ensiformis* L. Plant Sci. 170 (2), 274–282.

Song, W.Y., Mendoza-Cózatl, D.G., Lee, Y., Schroeder, J.I., Ahn, S.N., Lee, H.S., …, Martinoia, E., 2014a. Phytochelatin–metal (loid) transport into vacuoles shows different substrate preferences in barley and *Arabidopsis*. Plant Cell Environ. 37 (5), 1192–1201.

Song, W.Y., Park, J., Mendoza-Cózatl, D.G., Suter-Grotemeyer, M., Shim, D., Hörtensteiner, S., …, Martinoia, E., 2010. Arsenic tolerance in *Arabidopsis* is mediated by two ABCC-type phytochelatin transporters. Proc. Natl. Acad. Sci. 107 (49), 21187–21192.

Song, W.Y., Yamaki, T., Yamaji, N., Ko, D., Jung, K.H., Fujii-Kashino, M., …, Ma, J.F., 2014b. A rice ABC transporter, OsABCC1, reduces arsenic accumulation in the grain. Proc. Natl. Acad. Sci. 111 (44), 15699–15704.

Souri, Z., Karimi, N., de Oliveira, L.M., 2018. Antioxidant enzymes responses in shoots of arsenic hyperaccumulator, *Isatis cappadocica* Desv., under interaction of arsenate and phosphate. Environ. Technol. 39 (10), 1316–1327.

Srivastava, D., Verma, G., Chauhan, A.S., Pande, V., Chakrabarty, D., 2019. Rice (*Oryza sativa* L.) tau class glutathione S-transferase (OsGSTU30) overexpression in *Arabidopsis thaliana* modulates a regulatory network leading to heavy metal and drought stress tolerance. Metallomics 11 (2), 375–389.

Srivastava, M., Ma, L.Q., Singh, N., Singh, S., 2005. Antioxidant responses of hyper-accumulator and sensitive fern species to arsenic. J. Exp. Bot. 56 (415), 1335–1342.

Srivastava, S., Srivastava, A.K., Sablok, G., Deshpande, T.U., Suprasanna, P., 2015. Transcriptomics profiling of Indian mustard (*Brassica juncea*) under arsenate stress identifies key candidate genes and regulatory pathways. Front. Plant Sci. 6, 646.

Surgun-Acar, Y., Zemheri-Navruz, F., 2019. 24-Epibrassinolide promotes arsenic tolerance in *Arabidopsis thaliana* L. by altering stress responses at biochemical and molecular level. J. Plant Physiol. 238, 12–19.

Talukdar, D., 2013. Arsenic-induced changes in growth and antioxidant metabolism of fenugreek. Russian J. Plant Physiol. 60 (5), 652–660.

Tiwari, M., Sharma, D., Dwivedi, S., Singh, M., Tripathi, R.D., Trivedi, P.K., 2014. Expression in *Arabidopsis* and cellular localization reveal involvement of rice NRAMP, OsNRAMP 1, in arsenic transport and tolerance. Plant Cell Environ. 37 (1), 140–152.

Tiwari, S., Sarangi, B.K., 2017. Comparative analysis of antioxidant response by *Pteris vittata* and *Vetiveria zizanioides* towards arsenic stress. Ecol. Eng. 100, 211–218.

Tiwari, M., Kidwai, M., Dutta, P., Narayan, S., Gautam, N., Chawda, K., Shirke, P.A., Mishra, A.K., Chakrabarty, D., 2022. A tau class glutathione-S-transferase (OsGSTU5) confers tolerance against arsenic toxicity in rice by accumulating more arsenic in root. J. Hazard. Mater. 426, 128100.

Tripathi, R.D., Srivastava, S., Mishra, S., Singh, N., Tuli, R., Gupta, D.K., Maathuis, F.J., 2007. Arsenic hazards: strategies for tolerance and remediation by plants. Trends Biotechnol. 25 (4), 158–165.

Tripathi, P., Singh, P.K., Mishra, S., Gautam, N., Dwivedi, S., Chakrabarty, D., Tripathi, R.D., 2015. Recent advances in the expression and regulation of plant metallothioneins for metal homeostasis and tolerance. Environmental waste management, CRC Press, Boca Raton, pp. 551–564.

Trukhachev, V.I., Chmykhalo, V.K., Belanova, A.A., Beseda, D.K., Chikindas, M.L., Bren, A.B., …, Zolotukhin, P.V., 2021. Probiotic biomarkers and models upside down: from humans to animals. Vet. Microbiol., 109156.

Tukendorf, A., Rauser, W.E., 1990. Changes in glutathione and phytochelatins in roots of maize seedlings exposed to cadmium. Plant Sci. 70 (2), 155–166.

Upadhyay, A.K., Singh, N.K., Singh, R., Rai, U.N., 2016. Amelioration of arsenic toxicity in rice: comparative effect of inoculation of *Chlorella vulgaris* and *Nannochloropsis* sp. on growth, biochemical changes and arsenic uptake. Ecotoxicol. Environ. Saf. 124, 68–73.

Venisse, J.S., Õunapuu-Pikas, E., Dupont, M., Gousset-Dupont, A., Saadaoui, M., Faize, M., …, Roeckel-Drevet, P., 2021. Genome-wide identification, structure characterization, and expression pattern profiling of the aquaporin gene family in *Betula pendula*. Int. J. Mol. Sci. 22 (14), 7269.

Verma, P.K., Verma, S., Pande, V., Mallick, S., Deo Tripathi, R., Dhankher, O.P., Chakrabarty, D., 2016. Overexpression of rice glutaredoxin OsGrx_C7 and OsGrx_C2. 1 reduces intracellular arsenic accumulation and increases tolerance in *Arabidopsis thaliana*. Front. Plant Sci. 7, 1048.

Wang, C., Na, G., Bermejo, E.S., Chen, Y., Banks, J.A., Salt, D.E., Zhao, F.J., 2018. Dissecting the components controlling root-to-shoot arsenic translocation in *Arabidopsis thaliana*. New Phytol. 217 (1), 206–218.

Wanke, D., Üner Kolukisaoglu, H., 2010. An update on the ABCC transporter family in plants: many genes, many proteins, but how many functions? Plant Biol. 12, 15–25.

Wojas, S., Clemens, S., SkŁodowska, A., Antosiewicz, D.M., 2010. Arsenic response of AtPCS1-and CePCS-expressing plants–effects of external As (V) concentration on As-accumulation pattern and NPT metabolism. J. Plant Physiol. 167 (3), 169–175.

Wu, Z., Ren, H., McGrath, S.P., Wu, P., Zhao, F.J., 2011. Investigating the contribution of the phosphate transport pathway to arsenic accumulation in rice. Plant Physiol. 157 (1), 498–508.

Xu, W., Dai, W., Yan, H., Li, S., Shen, H., Chen, Y., …, Ma, M., 2015. *Arabidopsis* NIP3; 1 plays an important role in arsenic uptake and root-to-shoot translocation under arsenite stress conditions. Mol. Plant 8 (5), 722–733.

Yadu, B., Chandrakar, V., Keshavkant, S., 2016. Responses of plants to fluoride: an overview of oxidative stress and defense mechanisms. Fluoride 49 (3), 293.

Yamazaki, S., Ueda, Y., Mukai, A., Ochiai, K., Matoh, T., 2018. Rice phytochelatin synthases OsPCS1 and OsPCS2 make different contributions to cadmium and arsenic tolerance. Plant Direct 2 (1), e00034.

Yan, H., Gao, Y., Wu, L., Wang, L., Zhang, T., Dai, C., …, He, Z., 2019. Potential use of the *Pteris vittata* arsenic hyperaccumulation-regulation network for phytoremediation. J. Hazard. Mater. 368, 386–396.

Yang, Z., Mhamdi, A., Noctor, G., 2019. Analysis of catalase mutants underscores the essential role of $CATALASE_2$ for plant growth and day length-dependent oxidative signalling. Plant Cell Environ. 42 (2), 688–700.

Ye, Y., Li, P., Xu, T., Zeng, L., Cheng, D., Yang, M., …, Lian, X., 2017. OsPT4 contributes to arsenate uptake and transport in rice. Front. Plant Sci. 8, 2197.

Zamocky, M., Furtmüller, P.G., Obinger, C., 2008. Evolution of catalases from bacteria to humans. Antioxid. Redox Signaling 10 (9), 1527–1548.

Zhao, F.J., Ago, Y., Mitani, N., Li, R.Y., Su, Y.H., Yamaji, N., …, Ma, J.F., 2010. The role of the rice aquaporin Lsi1 in arsenite efflux from roots. New Phytol. 186 (2), 392–399.

Zhao, F.J., Dunham, S.J., McGrath, S.P., 2002. Arsenic hyperaccumulation by different fern species. New Phytol. 156 (1), 27–31.

Zhou, G.K., Xu, Y.F., Liu, J.Y., 2005. Characterization of a rice class II metallothionein gene: tissue expression patterns and induction in response to abiotic factors. J. Plant Physiol. 162 (6), 686–696.

Zhou, G., Xu, Y., Li, J., Yang, L., Liu, J.Y., 2006. Molecular analyses of the metallothionein gene family in rice (*Oryza sativa* L.). BMB Rep. 39 (5), 595–606.

CHAPTER 11

# Bioindicators of soil contaminated with organic and inorganic pollutants

**Meenu Gautam, Srishti Mishra, Madhoolika Agrawal**
Laboratory of Air Pollution and Global Climate Change, Department of Botany, Institute of Science, Banaras Hindu University, Varanasi, Uttar Pradesh, India

## 11.1 Introduction

Soils are composite of air, water, organic matter, minerals, countless organisms and decayed organisms. It is the "skin of the earth," capable of supporting plant life and is vital to life on the earth. Hence, it is crucial to keep the soil, pollution free and nourishing. However, due to various natural (volcanic eruption, sea salt spray, earthquakes, landslides, hurricanes, flood, etc.) and anthropogenic sources (industrial, commercial and agricultural activities, waste generations from municipal sectors, etc.), contamination of soil has become a vital problem in most of the countries specifically at places where no proper management and importance of natural balancing occur. Varieties of compounds/contaminants in soil due to natural and anthropogenic sources include both inorganic and organic pollutants (Gautam et al., 2021). Organic pollutants consist of detergents, disinfectants, agrochemicals, petroleum hydrocarbons, polycyclic aromatic hydrocarbons (PAHs), solvents and fatty acids (Gautam et al., 2021). Inorganic pollutants include metals (Fe, Al, Cd, Zn, Pb, Cr, Cu, Cd, As, Sb, Ni Co, etc.) and inorganic salts (carbonates, phosphates, sulfates, nitrates, etc.) (Gautam et al., 2016).

### 11.1.1 Sources of organic pollutants in soil

Contaminants can be called pollutants at a level when it impairs the regulation and functioning of balanced natural mechanisms and cause risk to organisms. Soil contamination is not only limited to the earth surface, but is variably or invariably related to other components of ecosystem, i.e.,

*New Paradigms in Environmental Biomonitoring Using Plants*
DOI: https://doi.org/10.1016/B978-0-12-824351-0.00001-8

water, air, flora and fauna (Snousy, 2016). Pollutants are mainly produced by anthropogenic sources where improper balancing mechanism exist (Fig. 11.1). Such sources include agricultural activities, industrial processes, medical and domestic materials used in day to day life which are sources of many organic pollutants in the soil (Pignatello et al., 2010). Persistent organic compounds include all those which have capacity to persist in nature for long time including polychlorinated biphenyls (PCBs), pesticides, Polychlorinated dibenzo-p-dioxins and polychlorinated dibenzo-p-furans (PCDD/Fs) and Polycyclic aromatic hydrocarbons (PAHs), etc., generated from various man-made sources (Breivik et al., 2004). Other anthropogenic sources such as rubber sealants, transformer, capacitors, plasticizers in paints, chemicals used in controlling pests, materials used in paper and dye industries, printing materials, fossil-fuel combustion, urban runoff, wastes, industrial by-products, emissions from traffic and roadways are also important in generating organic pollutants into the soil (Fabiańska et al., 2016). Improper handling and application of pesticides, improper disposal of wastes, leakage and accidents of oils and chemicals are also other major sources of organic pollutants (Snousy, 2016).

Ever increasing population leads to food demand which further triggers the use of agrochemicals and pesticides in the field, directly or indirectly contaminating the soil, water and air; which in turn alter the quality of food, which is related to food safety and human health (Carvalho 2017). Trend of using pesticides has been increasing now-a-days and has now reached up to in billions dollars per year where herbicides became more concerned contaminants as compared to insecticides in most of the agricultural practices in USA (US Geological Survey 2021). Thousands of pesticide's formulations have been used in the world where application of these for agricultural purpose has directly or indirectly affected the soil, ground water and surface water composition due to its persistent and toxic nature in the environment (Horsak et al., 1964). Exhaustive use of resources, intensive agricultural practices and use of high concentrations of organic chemicals have severe toxic effects on soil properties and related environment (Chowdhury et al., 2008).

### 11.1.2 Sources of metals in soil

Metals occur naturally in the soil through weathering of parent materials, volcanic eruptions, sea salt spray, landslides and storms, etc. Air borne particles, solid wastes and wastewater from anthropogenic sources such as mining, metallurgy, disposal of metallic wastes in inappropriate way

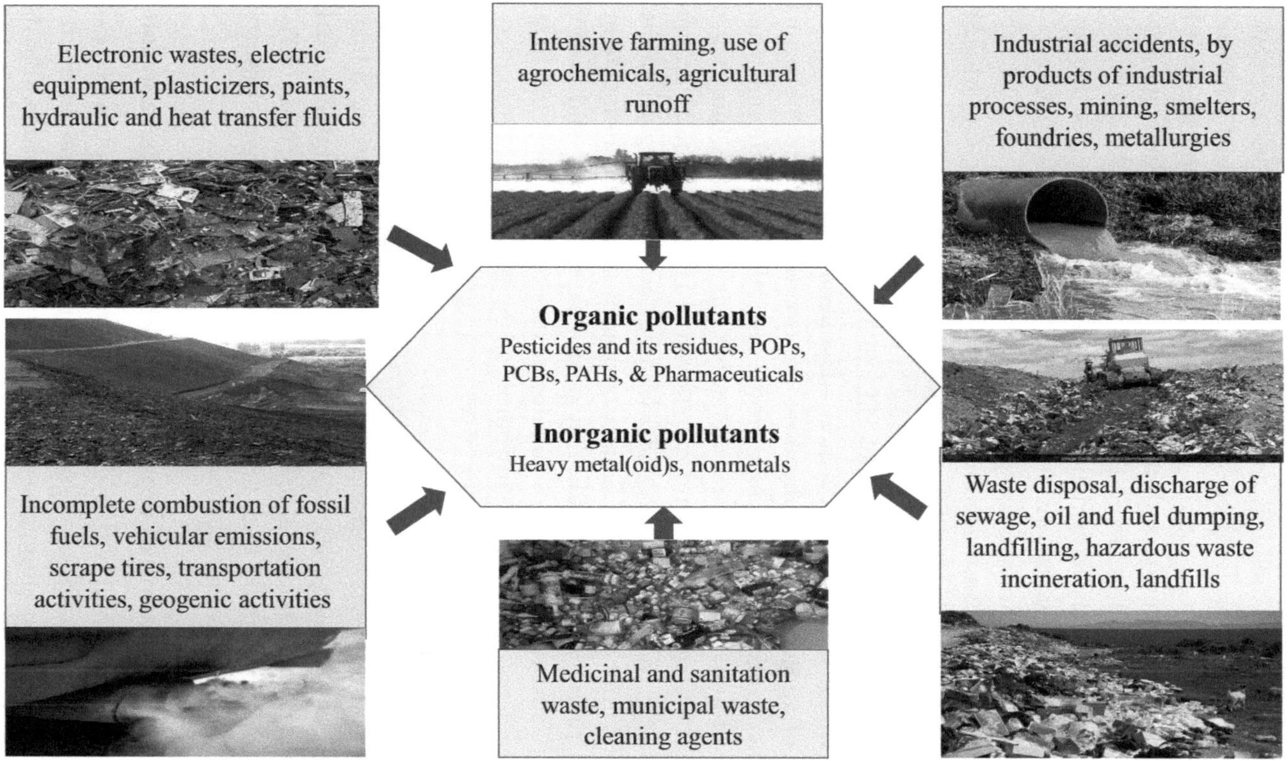


**Fig. 11.1** ***Sources of organic and inorganic (metals) pollutants from different sectors.***

to landfills, leaded gasoline and lead-based paints, land application of fertilizer, animal manures, biosolids, compost, pesticides, coal combustion residues, petrochemicals and atmospheric deposition accelerate the metal contamination level of soil and cause high risks to human health, plants, animals and ecosystems (Gautam et al., 2016). Application of huge quantities of fertilizers to soils in an intensive farming system also supplies toxic elements (F, Hg, Cd and Pb) as impurities (Rai et al., 2019). Uses of pesticides extensively in agriculture and horticulture are also the sources of substantial concentrations of soil metals (Rai et al., 2019). Use of biosolids (livestock manures, composts and sewage sludge) as land application involuntarily cause accumulation of metals i.e. Zn, Cu, Mo, Se, Tl, Ni, Sb, Pb, Cd, Cr, As, and Hg, etc. in soil (Abdullahi, 2015). Irrigation of land using wastewater from municipal and industrial sectors is commonly practiced in several parts of the world, thus enhancing the pollutants' load in soil (Khalid et al., 2018). Farmers are generally less bothered about the impact of agrochemical applications on environment and are primarily concerned on maximizing their profits on yields. Even though concentrations of metals in wastewater effluents are relatively less, but long-term irrigation of agricultural land can however results in metal accumulation in soil (Khalid et al., 2018). Mining and metallurgical industries play pre-eminent role in wide distribution of metal contaminants in soil beyond their toxic thresholds that pose risk to human and ecological health (Gautam et al., 2016). Besides, textiles, tanneries, petrochemicals from accidental oil spills or utilization of petroleum-based products and pharmaceutical facilities are additives to toxic metals in soil (Wuana and Okieimen, 2011).

Metals originating from airborne sources including duct or stack emissions of gas, air or vapor streams as well as fugitive emissions including dust emerging from waste piles or storage areas are usually released as particulates (Wuana and Okieimen 2011). Besides, some metals such as Cd, As, and Pb can volatilize under high temperature processing and get converted to oxides, condense as fine particulates and reach the soil.

### 11.1.3 Biomonitoring of organic and inorganic pollutants in soil

The concentrations of both organic and inorganic pollutants when exceed their critical thresholds in soil may cause alteration in soil properties resulting in adversities in biodiversity and food chain contamination (Fig. 11.2). Biomonitoring of pollutants' levels in soil must be the foremost step to assess the soil health in particular area prior to adopting any management

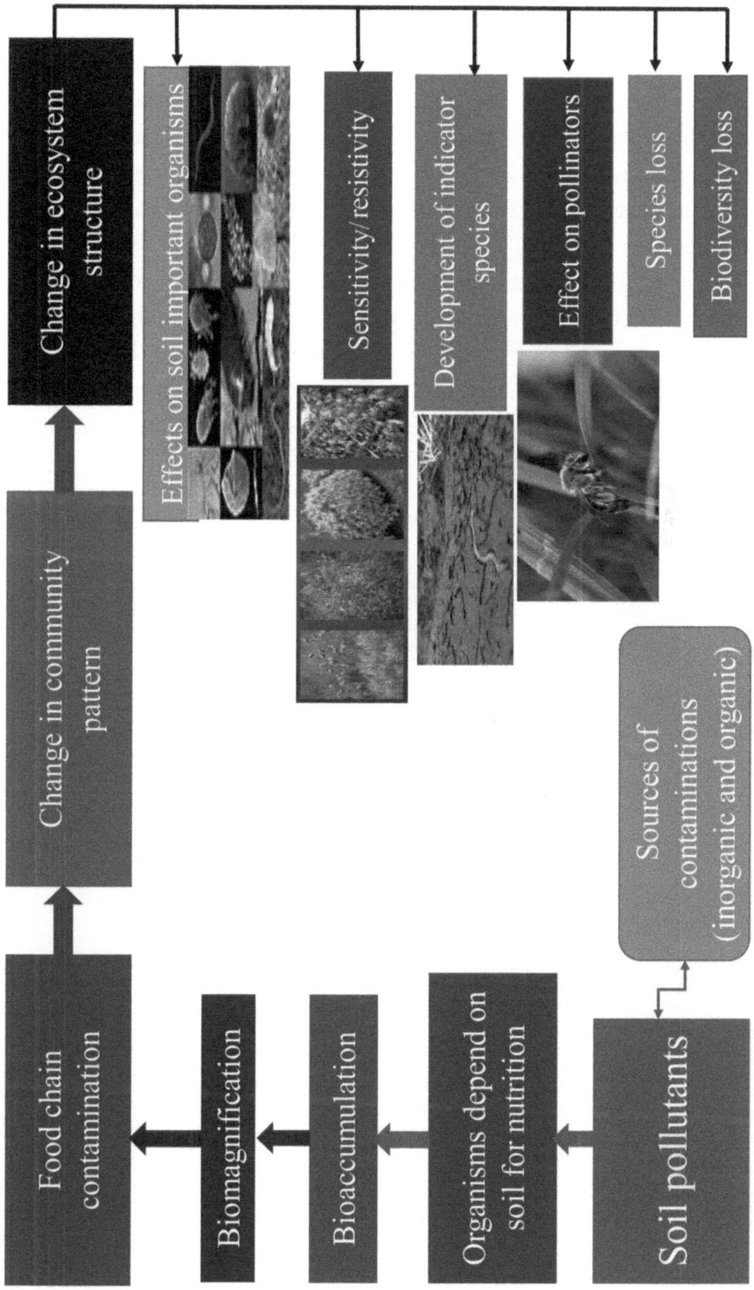


**Fig. 11.2** ***Responses of bioindicator towards organic and inorganic (metals) pollutants in soil.***

techniques. This include preparation of a soil monitoring proposal, execution of the approved soil monitoring plan, interpretation and reporting of results of the soil monitoring and then preparation and execution of soil management plan indicated by the results of soil monitoring. This guideline provides background and description of soil management and soil monitoring programs. There are several biological organisms including plants, animals, fungi and algae, etc. which are the indicators of soil health. Such indicator organisms are sensitive, while some are tolerant towards soil pollutants.

The chapter aims to explicate the information based on the literature in using organisms as bioindicators in the analysis of soil contaminated with organic and metal pollutants at different levels of the biological. The study also emphasizes upon the future prospects of utilizing bioindicators to monitor the soil pollution including both organic and inorganic pollutants (Fig. 11.3).

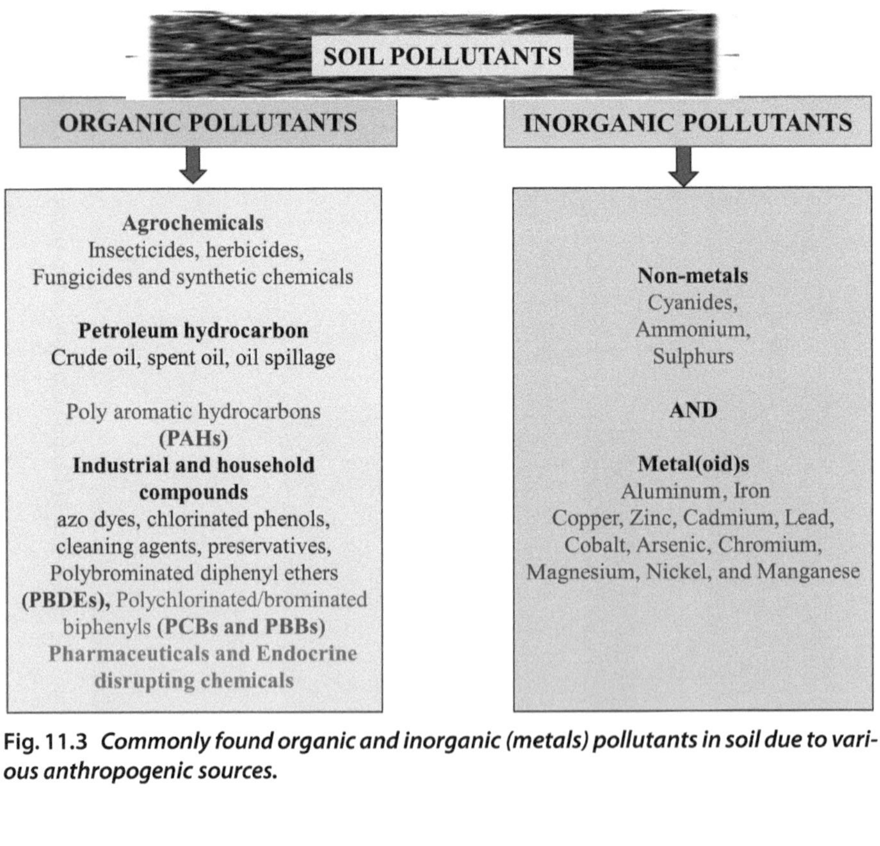


**Fig. 11.3** ***Commonly found organic and inorganic (metals) pollutants in soil due to various anthropogenic sources.***

## 11.2 Literature search

Peer reviewed literatures available at the Web of Science (from 1964 till the end of May 26, 2021) and documents from official sites of US EPA, Geological Survey, OECD, and ISO were referred to review on bioindicators of organic and inorganic pollutants.

## 11.3 Methods to monitor/assess soil organic and inorganic pollutants

Soil contaminants include a wide range of inorganic and organic pollutants, but here we will discuss inorganic pollutants such as metalloids and organic pollutants including pesticides, PAHs, petroleum hydrocarbons, PCBs, personal care products, pharmaceuticals and endocrine disrupting chemicals arising from different sources. Many analytical methods are used in analyses of different organic compounds from soil such as types of pesticides and residues, PAHs, PCBs, PCDD/Fs, Polybrominated diphenyl ethers (PBDEs), Organochlorine pesticides (OCPs) and petroleum hydrocarbons by using liquid chromatography-mass spectrometry (Belmonte Vega et al., 2005), gas chromatography- mass spectrophotometry (Kuo et al., 2015), high performance liquid chromatography, fluorescence detector and gas chromatography FID after extraction and clean up through different processes (Harmsen and Rietra, 2018). For inorganic components such as metalloids (Ni, Zn, Mn, Fe, Al, Cr, Cd, Cu, As, Pb and Hg, etc.), quantification is done through atomic absorption spectrophotometer (AAS), cold vapor atomic absorption spectrometry (CV-AAS), electro-thermal atomic absorption spectrometry (ET-AAS), hydride generation atomic fluorescence spectrometry (HG-AFS), hydride generation atomic absorption spectrometry (HG-AAS), inductively coupled plasma mass spectrometry (ICP-MS) and inductively coupled plasma optical emission spectrometry (ICP-OES) (Ferreira et al., 2010).

## 11.4 Types and importance of bioindicators

Ecological indicators used for monitoring and/or indexing any parameters or pollutants using biological organisms either at individual level or at landscape level are termed as bioindicator (Parmar et al., 2016). Use of bioindicators for monitoring and remediation of pollutants is best way to measure, analyze and manage the pollution level in soil. Bioindicators are single or group of organisms which can be used to monitor the presence/absence of toxic substances and their beneficial/hazardous effects on growth and survival, defense and sensitive behavior with respect to their outside environment (Gerhardt 2002). Toxicological bioindicators are mainly included for monitoring soil contaminants either inorganic or organic (Gerhardt 2002). Even though certain plant and animal bioindicators are adversely affected by soil contaminants, but biomonitoring through such bioindicators is also a propitiatory method towards sustainable approach.

Many plants, microorganism and animals acted as bioindicator for detecting toxic chemicals and their effects as well as natural or artificial methods accessible for testing the mechanisms of action (Zaghloul et al., 2020). Pollinators act as important bioindicators for healthy and unhealthy ecosystems to understand the interactions between plants and ecological sustainability for conserving biological diversity (Rollin et al., 2016). Increasing pollution levels in the environment have high risk and many indicator species including higher and lower plants, microorganisms, algal, fungal and bacterial species, lichens, planktons, pollinator insects, invertebrates and soil dwelling organisms helped in detecting and remediating contaminants in soil (Zaghloul et al., 2020). Generally bioindicators include three categories based on organism's types- microbes, animals and plants bioindicators (Fig. 11.4) for detecting positive and negative changes in the environment (Parmar et al., 2016); whereas based on the area of monitoring, they are categorized into pollution, ecological, environmental and biodiversity bioindicators.

## 11.5 Bioindicators of organic pollutants in soil

Plants and animals from microscopic to higher structural levels play preeminent role in ecological balance through various physiological and biochemical mechanisms wherein biomonitoring is one of the essential parts in ecological role (Zaghloul et al., 2020). Not only plants and animals but their products for instance, honey from honeybees and powder of fruits of carob and chestnut are used as indicator of pollutants present in the environment (Naccari et al., 2014).

Use of pesticides in agricultural practices causes severe impact on ecosystem and diversities of invertebrates, and are used as important bioindicators of pesticide pollution (Knillmann et al., 2018). Living organisms are variably or invariably related to each other; hence, presence of organic contaminants in any organism across food chain causes their sensitivity by lowering the capacity to tolerate and mark them differently crucial for life on the earth.

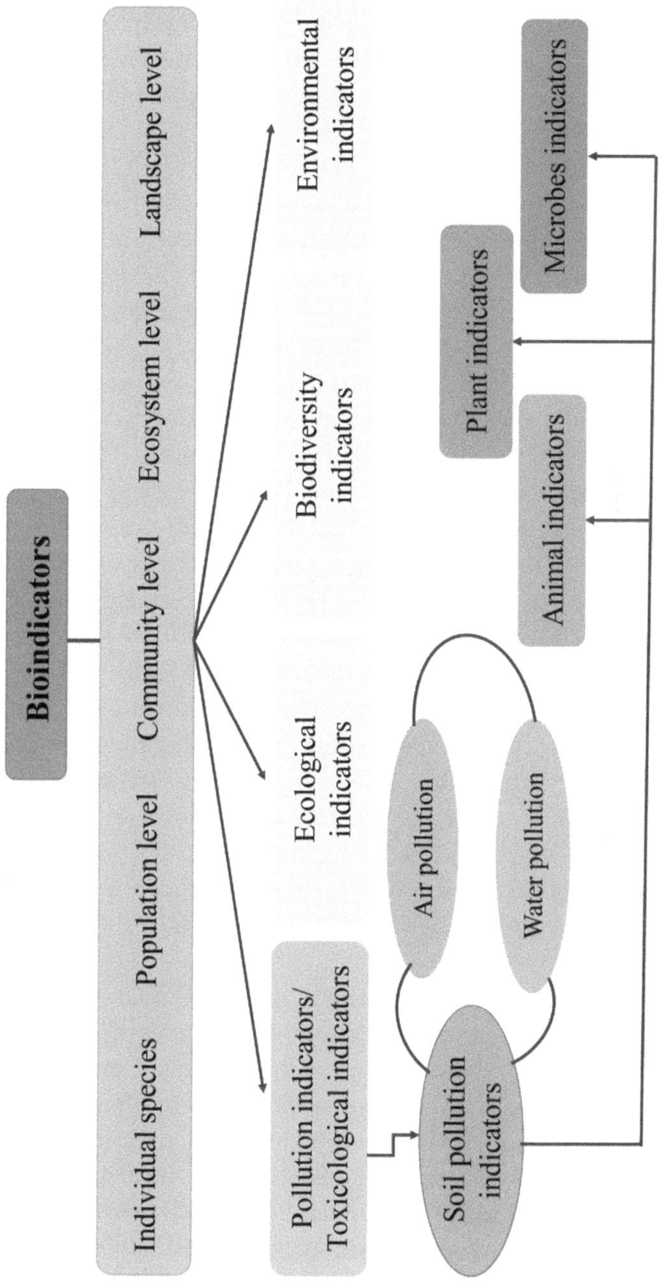


**Fig. 11.4** *Categorization of bioindicators and their relation with monitoring area.*

### 11.5.1 Terrestrial invertebrates

Terrestrial invertebrates play important role in maintaining the soil health through gaseous exchange and nutrient cycling but in recent decades frequent use of agro-chemicals for pest management has severely affected the diversity and functioning of such organisms which in turn have unfavorably affected the soil micro-flora and fauna, enzymatic activities and nutrient cycling (Ali and Naaz, 2013). The Organization for Economic Cooperation and Development (OECD) and International Organization for Standardization (ISO) have mentioned *Eisenia fetida* (earthworm) as an important bioindicator for soil health for evaluating the toxicity of organic pollutants in soil (OECD, 1984; ISO, 1993). In a National Research/Bioindicator Programme conducted in France in year 2006–2012, earthworm was considered as an important bioindicator tool for monitoring, characterizing and for risk assessment of soil health with respect to PAHs and metal pollution (Pérès et al., 2011). Moreover, biomass, abundance, species and functional structures, and ecological traits of earthworm were used to assess the soil quality under the program. Pesticides and other organic pollutants have many toxic and latent effects on annelids, i.e., earthworms which directly or indirectly affect the food chain (Paoletti 1999).

Pollinators are very crucial in ecological functioning and sustainability along with monitoring environmental contaminants (Kevan, 1999). Honeybees are very crucial pollinators in the environment and their products, i.e., honey which they collect from plants play important role as bioindicator for the presence of pesticides and its residues in the environment (de Oliveira et al., 2016). Many organophosphate pesticides used in agricultural practices were found in bee-wax and honey samples collected from apiaries nearby agriculture areas of Greece (Balayiannis and Balayiannis, 2008).

Exposure of glyphosate pesticide on the African giant snail showed genetic changes and suggested to monitor pesticides in soil and water (Udebuani et al., 2018). Zebra mussel indicates the presence of dichloro-diphenyl-trichloroethane (DDT) with significant negative effects on endocrine system (Binelli et al., 2001). In agricultural practices, use of insecticide Fipronil showed significant damaging effects on the crayfish (Schlenk et al., 2001). Molluscs showed higher accumulation of OCPs and polybrominated diphenyl ethers (PBDEs) from soil and suggested as suitable bioindicator of OCPs and PBDEs in soil (Wang et al., 2008).

### 11.5.2 Higher and lower plants'

Similar to invertebrates, many plant species are widely used as bioindicators of organic pollutants in soil (Table 11.1). Tomato, lettuce, and wheat

**Table 11.1** Plant and animal bioindicators of organic pollutants in soils.

| S. No. | Bioindicators | Organic pollutants | References |
|---|---|---|---|
| **A.** | **Higher plants** | | |
| 1. | *Terminalia catappa* | Oil spillage | Punwong et al. (2017) |
| 2. | *Hypericum perforatum*, *Agrostemma githago*, *Lychnis flos-cuculi*, *Allium cepa* and *Vicia faba* | Herbicides | de Souza et al. (2016); Franzaring et al. (2001) |
| 3. | *Solanum lycopersicum*, *Lactuca sativa* and *Triticum aestivun* | Carbamazepine, dihydroxyl-CBZ and epoxide-CBZ | Ben Mordechay et al. (2018) |
| 4. | *Allium cepa* and *Zea mays* | Polycyclic aromatic compounds | Odutayo et al. (2016) |
| 5. | *Coronopus dydimus* | Chlorpyrifos and Malathion insecticides | Mishra et al. (2020) |
| 6. | *Pinus sylvestris* | POPs, PCBs | Holoubek (2000); Ockenden et al. (1998) |
| 7. | *Triticum sativa*, *Bromous mermis*, *Linum usitatissimum* and *Agropyron deserterum* | Spent oil | Sharifi et al. (2007) |
| 8. | *Medicago sativa* and *Vigna ungiculata* | Crude oil | Shahriari et al. (2007); Olutayo et al. (2007) |
| 9. | *Lectuca sativa* and *Daucus carota* | Pharmaceutical contaminants | Azanu et al. (2016) |
| 10. | *Melilotus officinalis* and *Festuca arundinacea* | PAHs | Parrish et al. (2005) |
| **B.** | **Ferns** | | |
| | *Azolla filiculoidws* | Mono-azo dye | Khataee et al. (2013) |
| **C.** | **Mosses** | | |
| 1. | *Hypnum cupressiforme*, *Hylocomium splendens* and *Sphagnum* sp. | PAHs | Holoubek (2000); Wegener (1992) |
| 2. | mosses | PCBs, HCHs, OCPs, PAHs and herbicides | Khaldi (2014); Lim et al. (2006) |
| **D.** | **Fungi** | | |
| | *Aspergillus nidulans* | Glyphosate herbicide | Mesnage et al. (2020) |

*(Continued)*

**Table 11.1** Cont'd

| S. No. | Bioindicators | Organic pollutants | References |
|---|---|---|---|
| E. | **Algae** | | |
| 1. | Indigenous soil algae | Pesticides | Bérard et al. (2004) |
| 2. | *Scenedesmus obliquus* | PAHs, oil pollutants | Tang et al. (2010) |
| 3. | *Monoraphidium braunii* | Endocrine disrupting chemicals | Gattullo et al. (2012) |
| 4. | *Chlamydomonas reinhardtii*, *Scenedesmus quadricauda* | PAHs, PCBs and petroleum hydrocarbons | Ben Chekroun et al. (2014) |
| 5. | *Chlamydomonas reinhardtii* | Fluroxypyr pesticide | Zhang et al., 2011 |
| F. | **Lichens** | | |
| | *Hypogymnia physodes* | POPs and PCBs | Ockenden et al. (1998) |
| G. | **Microbes** | | |
| 1. | *Eschirichia* and *Morticella* spp. | OCPs | Egbe et al. (2021) |
| 2. | *Sphingomonas* GY2B, *Burkholderia cepacia*, *Psedomonas* GP3A and *Pandoraea pnomenusa* | PAHs, oil pollutants | Tang et al. (2010) |
| H. | **Invertebrates** | | |
| 1. | Honeybees | Organophosphate pesticides, pesticide and pesticide residues | De Oliveira et al. (2016); Balayiannis-Balayiannis (2008) |
| 2. | Earthworms | Pesticides and PAHs | Pérès et al. (2011) |
| 3. | *Archachatina marginata* | Glyphosate pesticide | Udebuani et al. (2018) |
| 4. | *Dreissena polymorpha* | DDT insecticide | Binelli et al. (2001) |
| 5. | *Procambarus clarkii* And *Procambarus zonangulus* | Fipronil insecticide | Schlenk et al. (2001) |
| 6. | *Mytilus edulis* and *Crassostrea talienwhanensis* | OCPs and PBDEs | Wang et al. (2008) |

CBZ, Carbamazepine; POPs, Persistent Organic Pollutants; PCBs, Polychlorinated Biphenyls; PAHs, Polyaromatic Hydrocarbons; HCHs, Hexachlorocyclohexanes; OCPs, Organochlorine pesticides; PBDEs, Polybrominated diphenyl ethers.

were used as indicators for presence of pharmaceutical and personal care products such as carbamazepine, dihydroxyl-CBZ and epoxide-CBZ in soil of wastewater treated area (Ben Mordechay et al., 2018). *Allium cepa* and *Zea mays* have been reported to monitor the toxic effects of PAHs in the soil (Odutayo et al., 2016). Sensitivity and toxicity of *A. cepa*, *Tradescantia* sp. and *Vicia faba* against herbicides were assessed as cell abrasion and mutation (de Souza et al., 2016). *Lectuca sativa* and *Daucus carota* are reported as tolerant plants against. tetracycline and amoxicillin (Azanu et al., 2016).

Biomonitoring of some organophosphate pesticides such as methyl parathion and profenfos indicated radical changes in chromosomal behavior and growth of *Lathyrus sativus* (Ganguly et al., 2010). *Melilotus officinalis* and *Festuca arundinacea* due to their sensitivity in soil contaminated by PAHs are known bioindicators (Parrish et al., 2005). Presence of PAHs in *Hypnum cupressiforme* and *Pinus sylvestris* is also used in biomonitoring studies (Holoubek et al., 2000). *Pinus sylvestris* and *Hypogymnia physodes* are known to monitor high load of POPs and PCBs in soil as a part of intra and interspecies monitoring study of vegetation (Ockenden et al., 1998). *Azolla filiculoides* fern is reported as tolerant plant by attaining defensive mechanism under the exposure of mono-azo dye (Khataee et al., 2013). In addition, *Azolla filiculoides* and *Xanthoria parietina* showed sensitivity towards diclofenac application (Vannini et al., 2018).

Franzaring et al. (2001) reported high concentrations of chlorpropham and ethofumesate herbicides in soil using agriophyte *Agrostemma githago*. Punwong et al. (2017) elucidated the leaf anatomy of *Terminalia catappa* as an indication of oil pollution along the coastline of Rayong province in Thailand. Spent oil contamination in soil showed phytotoxic effects on germination rate of *Triticum sativa*, *Bromous mermis*, *Linum usitatissimum*, and *Agropyron deserterum* (Sharifi et al., 2007). *Vigna ungiculata* showed significant tolerance in soil contaminated by crude oil (Olutayo et al., 2007) whereas *Medicago sativa* showed sensitivity (Shahriari et al., 2007).

Mosses were identified as bioindicator of PCBs, HCH and OCPs due to their accumulation in plant parts (Lim et al., 2006). Peat moss (*Sphagnum* sp.) has been known to be used in active biomonitoring of carcinogenic PAHs and metals in soil (Wegener et al., 1992). Change in biodiversity pattern of bryophytes reflects their sensitivity towards application of herbicides in the environment (Khaldi, 2014). Fungus *Aspergillus nidulans* is used as an important bioindicator for soil health, having unique tolerance towards glyphosate herbicide (Mesnage et al., 2020). Accumulation of organic pollutants can be seen in algal species which can be used in monitoring such as *Chlamydomonas reinhardtii*, and *Scenedesmus quadricauda*, etc. (Ben Chekroun et al., 2014). Community size and photosynthetic processes in soil algae show sensitivity towards atrazine pesticide (Bérard et al., 2004). *Chlamydomonas reinhardtii* acted as model organism for fluroxypyr pesticide (Zhang et al., 2011). In a study of Jin et al. (2012), green algae *Chlamydomonas reinhardtii* and *Monoraphidium braunii* showed high capacity to bioaccumulate and tolerate the herbicides prometryne and bisphenol, respectively from the soil (Gattullo et al., 2012).

### 11.5.3 Case studies

Selected case studies were attempted to assess the population of bioindicator under the effects of organic chemicals and their consequent impacts on human beings. Usage of wood preservatives, i.e., creosote, pentachlorophenol, chromated copper arsenate and furadan without proper management in residential area of Southern United States resulted in a reduction in crayfish population commonly used as food product and cause serious illness such as cancer, respiratory issues, skin rashes and gastrointestinal illness in local people (Horsak et al., 1964). In South East Spain, an integrated pest management case study was conducted on the use of PAHs, DDTs, PCBs and many phenolic compounds in soil with intense agricultural practices for commercial purposes where higher concentrations of the contaminants were found in soil as well as in crops (Plaza-Bolaños et al., 2012). In an another study conducted at Vikuge state farm, Tanzania showed that leakage of obsolete pesticides by improper handling and management practices resulted in higher concentration of DDT and HCHs in nearby area of farm which further caused their accumulation in maize, rice and hay like fodders (Kishimba and Mihale, 2004).

## 11.6 Bioindication of metals in soil

Presence of metals in soil at toxic levels may lead to changes in the growth and development pattern of soil biota because metals are commonly considered as stimulators or inhibitors of life processes (Szyczewski et al., 2009). Soil health under contamination of toxic metals can be assessed by using suitable biological indicators. Terrestrial plants as well as soil inhabiting organisms such as microflora and fauna (insects, earthworms, nematodes, fungi, and protozoa) have been utilized as an preeminent bioindicators of soil toxicity (Park et al., 2011).

Uptake of metals from soil into plants and animals occur via different routes. In terrestrial plants, metal uptake is accomplished via roots from soil and transported to aboveground plant parts. Metal's accumulation and distribution in plants depend upon the type of species, concentrations of metals in soil, the element species, pH, cation exchange capacity, bioavailability and vegetation period, etc. (Filipović-Trajković et al., 2012). Metals beyond their phytotoxic thresholds show symptoms such as stunted growth, change in leaf color, necrosis, chlorosis, curling and bronzing of leaves, reduction in yield and plants' death (Gautam et al., 2016). On the other side, there are many plant species which are highly metal tolerant and thrive under high bioaccumulation of metals (Gautam and Agrawal, 2019; Chowdhury and Maiti, 2016). Uptake of metals by soil fauna takes place through diffusion or ingestion of contaminated plants. Metals above the critical concentrations for animals inhibit their normal physiological functioning, metabolic activities, growth and reproductive pattern. Metals stored in the intracellular matrices of the organisms are excreted in feces and passed to fetus/eggs (Stankovic and Stankovic, 2013). Unlike plants, animals up to certain extent can avoid anthropogenic stressors by their mobility.

### 11.6.1 Plant bioindicators

Algae, fungi, and plant communities play a fundamental role in nutrient cycling and life on the earth. The pre-eminent sources of elements in them are their growing media, i.e., soil. They are nonmotile organisms always exposed to pollutants at their sites of growth which ultimately cause alteration in their normal physiological and biochemical processes. Algae, fungi, lower and higher plants growing at metal polluted sites exhibit metal bioaccumulation leading to growth reduction, altered physiology and metabolism, lower yield and biomass production (Stankovic et al., 2014).

Metals' accumulation and distribution in the plant not merely depends on the contents of the metals, physical and chemical compositions of soil, but also varies in different parts of the plant (Nagajyoti et al., 2010). Biomaterials such as micro- and macroalgae, fungi, lichens, mosses, tree bark and leaves of higher plants have been used to detect the deposition, accumulation and distribution of metals in soil (Stankovic et al., 2014). These organisms play the essential role in the biochemical cycling of metals and are sensitive bioindicators toward metal pollution in soil (Stankovic et al., 2014). Significant reductions in algal biomass of *Scenedesmus dimorphus* have often been distinguished in metal polluted soils (Efroymson et al., 2020). Mosses have been known to assess the trend and levels of metals within and around urban and industrial areas (Suchara et al., 2011). Mosses and higher fungi serving as ecological indicators have developed mechanisms to accumulate metals from contaminated soil (Stankovic and Stankovic 2013). General reductions in diversity of microfungal have often been reported in polluted soils containing metals (Zn, Cu, Pb, Cd, and As) (Stankovic et al., 2014).

Higher plants have been utilized as bioindicators in areas with significantly higher level of pollution load. Roots, bark, leaves, and flowers from higher plants are possible bioindicators of metal pollution in soil (Filipović-Trajković et al., 2012). Serbula et al. (2012) stated that most commonly used parts of higher plants include bark, leaves and needles of varying trees. Distribution of metals in plants is often unequal where bark accumulates higher metals than roots, leaves, and fruits (Filipović-Trajković et al., 2012). Sensitive to moderate metal tolerant plants are characterized by following symptoms such as necrosis, chlorosis, bronzing, reduced seed sprouting, stunted plant growth, and reduction in number, size of leaves, biomass, and yield (Gautam et al., 2016). Plants are directly exposed to the contaminant, without any dilution or filtration of the sample (Xiao et al., 2020). Wang and Zhang (2020) showed that the metal contents in studied plants at Shangdong abandoned karst bauxite mines, southwestern China were relatively high, particularly in *Pteris vittata* which is suggested to tolerate and accumulate high levels of metals and can be a suitable bioindicator of soil metal contamination caused by mining activities.

*Allium cepa* is the most commonly used plant to determine the mutagenic, genotoxic and cytotoxic effects of noxious elements in polluted soil (Grant, 1994). Its cellular kinetics favors the speedy growth of roots accredited to increase in cellular division. Therefore in cellular division of meristematic cells, the record of mitotic activity and anomalies can be easily monitored (Grant, 1994). Leme and Marin-Morales (2009) affirmed that the *A. cepa* test is a fast and sensitive technique to detect genotoxic and mutagenic substances dispersed in the environment. *Vicia faba* is a popular plant, widely used to evaluate the impact of metals on physiological and cytological studies (Kanaya et al., 1994). Radiobiological test was conducted in *V. faba* grown in metal contaminated soil to investigate the chromosomal aberrations, and since then has been widely used for genotoxicity studies for evaluation of sister chromatid exchange (Kihlman and Andersson, 1984).

#### 11.6.1.1 Case studies

Phyto-sociological study conducted at abandoned red mud dumpsite (RMD) in comparison to residential (RS) and forest sites (FS) at HINDALCO, Renukoot, India showed that total number of herbaceous species at FS, RS, and RMD sites were 56, 43, and 37, respectively. Whereas based on the total number of tree species, study sites were in the order of FS (47) > RS (29) > RMD (8). Moreover, total number of shrub species was 26 at FS, 11 at RS, and 9 at RMD sites. The aggregate number of woody, shrub and herbaceous species at RMD site were low (Gautam et al., 2018). Importance value index of metal sensitive species was low while that of tolerant species was higher at RMD site. Presence of invasive species such as *Urena lobata, Stylosanthes scabra, Caesalpinia bonduc*, and *Acacia nilotica* at only RMD site was suggested as an indication of high metal contamination along with high salinity, alkalinity as well as poor nutrient status (Gautam et al., 2018).

A plant community structure study by Pandey et al. (2014) revealed minimum numbers of herbaceous species (19) at both Raniganj and Jharia coalfields compared to the reference site (Central Institute of Mining and Fuel Research (CIMFR)). Both the coalfields were dominated by *Cyanodon dactylon* and *Alternanthra paronychoides* which acted as bioindicator, whereas *Achyranthes aspera, Convolvulus* sp., *Dicanthium* sp. *Eclipta alba*, and *Solanum* sp. were the sensitive species found only at reference site and were completely vanished from coal mining areas. *Eragrsotis* sp., *Cynosuroides* sp., and *Setaria glauca* were identified as indicator species only found at coal fields. A significant reduction in number of tree species was found in coal fields while indicator tree species like *Butea monosperma, Ficus benghalensis, Ficus religiosa*, and *Psidium guajava* were dominant at coal fields (Pandey et al., 2014). Dutta and Agrawal (2003) observed significant reductions in aboveground biomass of *Cassia siamea* grown on the wasteland of coal mining area. Important Value Index of indicator species such as *Achyranthus aspera, Cassia tora* and *Eclipta alba* decreased and those of tolerant species such as *Alternanthra* sp., *Cynodon* sp., *Cyprus* sp., and *Sida* sp. increased with decrease in distance from HINDALCO Industries Ltd., Renukoot, an important aluminum smelter in India (Narayan et al., 1994).

Jeddi et al. (2021) studied on selected tree species (*Eucalyptus occidentalis* Endl., *Acacia salicina* Lindl., *Cupressus sempervirens* L., *Casuarina equisetifolia* L., and *Tamarix aphylla* (L.) Karst.) to monitor metal pollution in soil of arid Mediterranean region in southern Tunisia. They have found higher mean concentrations of Pb and Cd in bark than in leaves of trees; whereas, leaves showed relatively higher mean concentrations of Zn and Cu than bark. Based on metal accumulation index, *E. occidentalis* and *T. aphylla* accumulated more metals compared to other tree species. Therefore, the study suggested utilization of both the tree species to monitor the pollution of Zn, Cu, Pb, and Zn in soil of arid Mediterranean region.

### 11.6.2 Animal bioindicators

Terrestrial animals mostly used as soil bioindicators are invertebrates such as arachnids, springtails, isopods, oligochaetes (earthworms), gastropods and nematodes (Stankovic and Stankovic 2013). Soil nematodes are the most copious metazoan at all the trophic levels on the earth (Ferris 2010). The advantages of nematodes commonly used as bioindicator are short lifespan, simple and inexpensive cultivation (300–350 eggs per nematode), availability of complete genome sequence and genetically modified strains, and a rapid response towards environmental change (Bouyanfif et al., 2019). Nematodes are commonly served as bioindicators and the ecological indices to reveal the natural and human-induced changes in soil ecosystems (Chauvin et al., 2020). Community structure, diversity and function of nematodes are important indices of metal contamination in soil (Park et al., 2011). Based on previous studies, soil contamination by metals has an overall negative effect on nematode's abundance, diversity and community structure (Chauvin et al., 2020). A meta-analysis by Chauvin et al. (2020) showed a decline in the abundance and diversity of nematode community in soil contaminated with metals (As, Cd, Cr, Co, Cu, Hg, Ni, Mo, Pb, Zn, and Tl). The concentrations of metalloids such as Cr, Cd, Pb, Zn, and Ni influence the soil nematode community structure (Park et al., 2011). In the United States, a guide booklet was prepared to describe the use of nematode *Caenorhabditis elegans* in soil toxicity tests (Boyle and Kakouli-Duarte, 2008).

The importance of earthworms in metal pollution monitoring is widely recognized in terrestrial ecosystems. They have been extensively studied, and are approved for use in toxicity testing by the United States Environmental Protection Agency (US EPA) and the European Economic Community and the Organization for Economic Cooperation and Development (OECD, 2004). Earthworms in soil enhance organic matter content and help in formation of soil aggregate structure which enables their utilization for the rehabilitation of contaminated soil. Earthworms are considered as a good candidate for soil bioindicator because they are better represented in soil system with respect to density; respond to variety of ecological and environmental factors (Pérès et al., 2011). There are several factors which make earthworms excellent bioindicators of metal toxicity in soil, such as their habit, habitat and significant trophic position in terrestrial food webs thereby serving as food for animals at higher trophic level (de Andréa, 2010).

Due to the great importance in the soil and wide distribution, earthworms specifically *Eisenia andrei* and *Eisenia fetida* are widely chosen for soil toxicity tests by many regulatory agencies of several countries (de Andréa, 2010). Other species such as *Lumbricus rubellus* and *Lumbricus terrestris* are important in studies regarding metal bioaccumulation (Veltman et al., 2007). Suther et al. (2008) stated significant positive correlation amid contents of metals (Zn, Fe, Cu, Pb, and Cd) in tissues of earthworm (*Metaphire posthuma* and *Lampito mauritii*) and surrounding soil. *E. foetida* is capable of accumulating Hg, Cd, Cu, Pb, and Zn; and significant positive correlations were found between metal contents within the earthworm and contaminated soil (Olayinka et al., 2011; Hirano and Tamae 2011). A research carried out in the dry lands to investigate the ability of *Namalycastis indica* to absorb metals in its tissues ranging from 5 to 250 ppm found that after 14 days of exposure, it displayed high capacity of their tissues to accumulate Cd, Ni, Cu, and Co even under higher temperature reflecting a sign of tolerance (Al-Maliki et al., 2021).

Gastropods play important ecological roles in terrestrial ecosystems due to their ubiquitous distribution and enormous species number. Gastropods represent the only molluscs class in terrestrial ecosystem which can be used for bioindication and biomonitoring of metal pollution. Fritsch et al. (2011) found that both *Cantareus aspersus* and *Cepaea nemoralis* found at "Metaleurop-Nord" smelter in Northern France are capable of concentrating Cd, Cu, Pb, and Zn in their tissues which resulted in their reduced growth and diversity. *Deroceras reticulatum* snails found across much of North Africa and America, Europe, and Atlantic islands are good bioindicators of soil pollution (Douglas and Tooker, 2012). Snails being in the diet of many species such as toads, birds, beetles and snakes, hold a significant position in food web.

Terrestrial arthropods such as isopoda, diplopoda and collembola are amongst most suitable organisms to evaluate contamination of metals in soil due to their direct contact with contaminants present (Hopkin et al., 1989). Isopoda is one of the largest orders of crustaceans and terrestrial isopoda are extensively used in soil toxicity tests where the key essential parameters of evaluation include abundance of individuals, reproduction rates and survival (Niemeyer et al., 2009). Metals are the main toxic agents in soil evaluated through Isopoda, since these invertebrates bioaccumulate these elements. In this regard, several studies have been conducted to study metal bioaccumulation, cytotoxic effect and the detoxification mechanisms (Odendaal and Reinecke, 2003). Diplopods behold the prime importance in recycling of nutrients, aeration and fertilization of soil (Dangerfield and Telford, 1989). Due to the habitat of diplopods, they are greatly influenced by the deposition of metals, organic compounds and many complex compounds in the soil (Dangerfield and Telford, 1989).

## 11.7 Genetically modified organisms as bioindicators

Transgenic plants/animals or genetically modified organisms (GMOs) are produced for many agriculture and valuable purposes such as to increase crop production, resistant to diseases, quality and quantity enhancement (Verhoog, 1992). GMOs are described as organisms manipulated through genetic engineering acquiring novel phenotypic and genotypic traits for beneficial uses of human (Zhu et al., 2013). The GMOs are known to be adapted under adverse environmental conditions and hence could be a suitable bioindicator for soil quality assessment (Nazir and Iqbal, 2019). In addition, there are many controversies related to ecological risk and biocontaminants arose with GMOs in the environment (Zhu et al., 2013).

Some pesticides are banned in developed country but are still used in some developing nations; therefore, genetic engineering is involved in production of GMOs for agricultural productions which are resistant towards pesticides (López et al., 2012). Biosafety and risk related issue restricts the use of GMOs as bioindicator (Herring, 2015). For instance, a study by Candolfi et al. (2004) showed adverse effects on nontarget invertebrates' community in cultivation of genetically modified *Zea mays* and synthetic insecticides. The GMOs are introduced to reduce the use of pesticides and insecticides indirectly as we know that pesticides have many deleterious effects on nontargeted organisms in addition to some toxicological effects on applied environment (Zhu et al., 2013). Superbug which is GMO patented by Dr. Anand Mohan Chakrabarty is important for monitoring crude oil contaminants in the environment (Pandey and Arora, 2020). Compared to impact of pesticides on honey bees, GMO crops have lower impact on colony collapse of honey bees but narrowing genetic variability has severe impacts (Bekić et al., 2014).

Many transgenic plants were developed to monitor organic pollutants such as *Arabidopsis thaliana*, *Nicotiana tabacum*, *Sesbania gradiflora* and *Medicago sativa* taking strain from some microbes but need regulatory approval for commercialized use (Rylott et al., 2015). Genetically modified species of *Allium cepa* has potential to screen the effect of diclofop-methyl and lindane showing genotoxic and cytotoxic effects (Mesi and Kopliku, 2013). Hur et al. (2011) showed that genetically modified popular (*Populus alba* × *Populus tremular* var. glandulosa) was relatively more efficient in accumulating metals from contaminated soil compared to wild type popular and suggesting the former one as an efficient bioindicator of metal pollution in soil. Adaptation and rejection occur with GMOs but naturally growing plants have no such controversy in commercial use to monitor any pollutants.

## 11.8 Advantages and disadvantages of bioindicators

Numerous advantages of bioindicators have spurred legislative mandates regarding their use across the world and their inclusion in several international accords. Yet bioindicators have certain limitations in their utilization to monitor pollution level in soil. We cannot discriminate natural variability from human made changes which limits the applicability of bioindicators in heterogeneous soil environment. Consequently, populations of indicator species are influenced by multiple factors such as disease, parasitism, competition and predation (Holt and Miller, 2010). Another disadvantage of bioindicators is that their indicator ability is scale-dependent. Besides, bioindicator species variably and invariably have differing habitat requirements and suitable living environment. Bioindicators can be employed at a range of scales, i.e., from the cellular to the ecosystem level to evaluate the health of a particular ecosystem (Holt and Miller, 2010). They cumulatively bring information from the physical, chemical and biological components of the ecosystem (Holt and Miller, 2010). From management viewpoint, bioindicators inform our actions as to what is and is not biologically sustainable.

## 11.9 Futuristic approach in use of bioindicators to monitor soil pollution

Bioindicators can be used to predict the natural state of a certain region and the degree of contamination (Khatri and Tyagi, 2015). Hence, there are many fold advantages of utilizing biological organisms as ecological indicators-

- Synergetic and antagonistic impacts of pollutants on plants and animals can be monitored
- Primary stage diagnosis and harmful effects of pollutants to human beings can be monitored.
- Can be easily counted up due to their prevalence.
- Economically viable option when compared with other specialized measuring techniques.

## 11.10 Conclusions

Constant release of harmful substances into the terrestrial environment due to anthropogenic processes has triggered the worrisome regarding their effects on living organisms. Organic and metal pollutants induce significant damage to physicochemical and biological processes in soil. Use of biological organisms as ecological indicators is reliable and cost-effective approach to assess the toxicological analyses in soil. Morphological, physiological, biochemical, and genotoxic evaluations of bioindicators provide complete understanding regarding the action of contaminants in soil. It is important to highlight that the choice of the bioindicator according to the habitat condition is essential for successful monitoring of soil environment. Bioindicators such as plants and animals either show antagonistic or synergistic effects in presence of organic and metal pollutants in soil. Sensitive species show reduced abundance, diversity, growth, and development while tolerant species sustain the stressed soil environment. Bioindicators pose both advantages and disadvantages but due to their existence in natural environment, it is the most suitable and cost effective approach to assess the soil health as preliminary steps. Hence, accepting the importance of using bioindicators in assessment of toxicity of substrates in soil will be extremely beneficial with respect to future perspective.

## Acknowledgments

Authors are thankful to the Head and the Coordinators, CAS and DST-FIST in Botany, and Coordinators, IoE and ISLS, Banaras Hindu University, Varanasi for basic facilities. Meenu Gautam is thankful to the Council of Scientific and Industrial Research (File No. 09/013(0857)/2018-EMR-I), New Delhi, for research associateship. Srishti Mishra is grateful to Banaras Hindu University for proving financial support as University Research Fellowship of University Grants Commission, New Delhi, India.

## References

Abdullahi, M.S., 2015. Soil contamination, remediation and plants: prospects and challenges. In: Khalid, R.H., Muhammad, S., Munir, O., Ahmet, R.M. (Eds.), Soil Remediation and Plants. Elsevier, New York, pp. 1–525.

Ali, A., Naaz, I., 2013. Earthworm biomarkers: the new tools of environmental impact assessment. Biosci. Biotechnol. Res. Commun. 6 (2), 163–169.

Al-Maliki, S., Al-Taey, D.K., Al-Mammori, H.Z., 2021. Earworms and eco-consequences: considerations to soil biological indicators and plant function: a review. Acta Ecol. Sinica 41, 512–523.

Azanu, D., Mortey, C., Darko, G., Weisser, J.J., Styrishave, B., Abaidoo, R.C., 2016. Uptake of antibiotics from irrigation water by plants. Chemosphere 157, 107–114.

Balayiannis, G., Balayiannis, P., 2008. Bee honey as an environmental bioindicator of pesticides' occurrence in six agricultural areas of Greece. Arch. Environ. Contam. Toxicol. 55 (3), 462–470.

Bekić, B., Jeločnik, M., Subić, J., 2014. Honey Bee Colony Collapse Disorder (*Apis mellifera* L.) possible causes. Sci. Papers Series Manage. Econ. Eng. Agric. Rural Dev. 14 (2), 13–18.

Belmonte Vega, A., Garrido Frenich, A., Martínez Vidal, J.L., 2005. Monitoring of pesticides in agricultural water and soil samples from Andalusia by liquid chromatography coupled to mass spectrometry. Anal. Chim. Acta 538 (1–2), 117–127.

Ben Chekroun, K., Sánchez, E., Baghour, M., 2014. The role of algae in bioremediation of organic pollutants. Int. Res. J. Public Environ. Health 1 (2), 19–32.
Ben Mordechay, E., Tarchitzky, J., Chen, Y., Shenker, M., Chefetz, B., 2018. Composted biosolids and treated wastewater as sources of pharmaceuticals and personal care products for plant uptake: a case study with carbamazepine. Environ. Pollut. 232, 164–172.
Bérard, A., Rimet, F., Capowiez, Y., Leboulanger, C., 2004. Procedures for determining the pesticide sensitivity of indigenous soil algae: a possible bioindicator of soil contamination?. Arch. Environ. Contam. Toxicol. 46 (1), 24–31.
Binelli, A., Bacchetta, R., Vailati, G., Galassi, S., Provini, A., 2001. DDT contamination in Lake Maggiore (N. Italy) and effects on zebra mussel spawning. Chemosphere 45 (4–5), 409–415.
Bouyanfif, A., Jayarathne, S., Koboziev, I., Moustaid-Moussa, N., 2019. The nematode *Caenorhabditis elegans* as a model organism to study metabolic effects of ω-3 polyunsaturated fatty acids in obesity. Adv. Nutr. 10 (1), 165–178.
Boyle, S., Kakouli-Duarte, T., 2008. The effects of chromium VI on the fitness and on the β-tubulin genes during in vivo development of the nematode *Steinernema feltiae*. Sci. Total Environ. 404 (1), 56–67.
Breivik, K., Alcock, R., Li, Y.F., Bailey, R.E., Fiedler, H., Pacyna, J.M., 2004. Primary sources of selected POPs: regional and global scale emission inventories. Environ. Pollut. 128 (1–2), 3–16.
Candolfi, M.P., Brown, K., Grimm, C., Reber, B., Schmidli, H., 2004. A faunistic approach to assess potential side-effects of genetically modified Bt-corn on non-target arthropods under field conditions. Biocontrol Sci. Technol. 14 (2), 129–170.
Carvalho, F.P., 2017. Pesticides, environment, and food safety. Food Energy Secur. 6 (2), 48–60.
Chauvin, C., Trambolho, M., Hedde, M., Makowski, D., Cérémonie, H., Jimenez, A., Villenave, C., 2020. Soil nematodes as indicators of heavy metal pollution: a meta-analysis. Open J. Soil Sci. 10 (12), 579–601.
Chowdhury, A., Maiti, S.K., 2016. Identification of metal tolerant plant species in mangrove ecosystem by using community study and multivariate analysis: a case study from Indian Sunderban. Environ. Earth Sci. 75 (744), 1–21.
Chowdhury, A., Pradhan, S., Saha, M., Sanyal, N., 2008. Impact of pesticides on soil microbiological parameters and possible bioremediation strategies. Indian J. Microbiol. 48 (1), 114–127.
Dangerfield, J.M., Telford, S.R., 1989. Are millipedes important for soil fertility. Zimbabwe Sci. News 23 (7/9), 66–68.
de Andréa, M.M., 2010. Use of earthworms as bioindicators of soil contamination. Acta Zool. Mexicana 26 (Especial 2), 95–107.
de Oliveira, R.C., do N Queiroz, S.C., da Luz, C.F.P., Porto, R.S., Rath, S., 2016. Bee pollen as a bioindicator of environmental pesticide contamination. Chemosphere 163, 525–534.
de Souza, C.P., Guedes, T.A., Fontanetti, C.S, 2016. Evaluation of herbicides action on plant bioindicators by genetic biomarkers: a review. Environ. Monit. Assess. 188 (12), 1–12.
Douglas, M.R., Tooker, J.F., 2012. Slug (Mollusca: Agriolimacidae, Arionidae) ecology and management in no-till field crops, with an emphasis on the mid-Atlantic region. J. Integr. Pest Manage. 3 (1), C1–C9.
Dutta, R.K., Agrawal, M., 2003. Restoration of opencast coal mine spoil by planting exotic tree species: a case study in dry tropical region. Ecol. Eng. 21 (2-3), 143–151.
Efroymson, R.A., Pattullo, M.B., Mayes, M.A., Mathews, T.J., Mandal, S., Schoenung, S., 2020. Exploring the sustainability and sealing mechanisms of unlined ponds for growing algae for fuel and other commodity-scale products. Renew. Sustain. Energy Rev. 121, 109708.

Egbe, C.C., Oyetibo, G.O., Ilori, M.O., 2021. Ecological impact of organochlorine pesticides consortium on autochthonous microbial community in agricultural soil. Ecotoxicol. Environ. Saf. 207 (2020), 111319.
Fabiańska, M.J., Kozielska, B., Konieczyński, J., Kowalski, A., 2016. Sources of organic pollution in particulate matter and soil of Silesian Agglomeration (Poland): evidence from geochemical markers. Environ. Geochem. Health 38 (3), 821–842.
Ferreira, S.L.C., Miró, M., Silva, Da, E., G.P., Matos, G.D., Reis, D., P., S., Brandao, G.C., Santos, D., W., N.L., Duarte, A.T., Vale, M.G.R., Araujo, R.G.O, 2010. Slurry sampling-an analytical strategy for the determination of metals and metalloids by spectroanalytical techniques. Appl. Spectrosc. Rev. 45 (1), 44–62.
Ferris, H., 2010. Form and function: metabolic footprints of nematodes in the soil food web. Eur. J. Soil Biol. 46 (2), 97–104.
Filipović-Trajković, R., Ilić, Z.S., Šunić, L., Andjelković, S., 2012. The potential of different plant species for heavy metals accumulation and distribution. J. Food Agric. Environ. 10, 959–964.
Franzaring, J., Kempenaar, C., Van der Eerden, L.J.M, 2001. Effects of vapours of chlorpropham and ethofumesate on wild plant species. Environ. Pollut. 114 (1), 21–28.
Fritsch, C., Coeurdassier, M., Gimbert, F., Crini, N., Scheifler, R., De Vaufleury, A., 2011. Investigations of responses to metal pollution in land snail populations (*Cantareus aspersus* and *Cepaea nemoralis*) from a smelter-impacted area. Ecotoxicology 20 (4), 739–759.
Ganguly, S., Bhattacharya, S., Mandi, S., Tarafdar, J., 2010. Biological detection and analysis of toxicity of organophosphate- and azadirachtin-based insecticides in Lathyrus sativus L. Ecotoxicology 19 (1), 85–95.
Gattullo, C.E., Bährs, H., Steinberg, C.E.W., Loffredo, E., 2012. Removal of bisphenol A by the freshwater green alga Monoraphidium braunii and the role of natural organic matter. Sci. Total Environ. 416, 501–506.
Gautam, M., Agrawal, M., 2019. Identification of metal tolerant plant species for sustainable phytomanagement of abandoned red mud dumps. Appl. Geochem. 104, 83–92.
Gautam, M., Mishra, S., Agrawal, M., 2021. Cause, effects and sustainable approach to remediate contaminated soil. In: Prasad, R. (Ed.), Environmental Pollution and Remediation. Springer Nature, Singapore ISBN: 978-981-15-5498-8.
Gautam, M., Pandey, B., Agrawal, M., 2018. Identification of indicator species at abandoned red mud dumps in comparison to residential and forest sites, accredited to soil properties. Ecol. Indic. 88, 88–102.
Gautam, M., Pandey, D., Agrawal, S.B., Agrawal, M., 2016. Metals from mining and metallurgical industries and their toxicological impacts on plants. In: Singh, A., Prasad, S.M., Singh, R.P. (Eds.), Plants Response to Xenobiotics. Springer, Singapore, pp. 231–272 ISBN: 978-981-10-2859-5.
Gerhardt, A., 2002. Bioindicator species and their use in biomonitoring. Environ. Monit. 1, 77–123 file:///C:/Users/HP%20PC/Downloads/Gerhardt_Biomonitoringspeciesl.PDF. Assessed 06.06.2021.
Grant, W.F., 1994. The present status of higher plant bioassays for the detection of environmental mutagens. Mut. Res. Fundam. Mol. Mech. Mutagen. 310 (2), 175–185.
Harmsen, J., Rietra, R.P.J.J., 2018. 25 years monitoring of PAHs and petroleum hydrocarbons biodegradation in soil. Chemosphere 207, 229–238.
Herring, R.J., 2015. State science, risk and agricultural biotechnology: Bt cotton to Bt Brinjal in India. J. Peasant Stud. 42 (1), 159–186.
Hirano, T., Tamae, K., 2011. Earthworms and soil pollutants. Sensors 11, 11157–11167.
Holoubek, I., Kořínek, P., Šeda, Z., Schneiderová, E., Holoubková, I., Pacl, A., Tříska, J., Cudlín, P., Čáslavský, J., 2000. The use of mosses and pine needles to detect persistent organic pollutants at local and regional scales. Environ. Pollut. 109 (2), 283–292.
Holt, E.A., Miller, S.W., 2010. Bioindicators: using organisms to measure environmental impacts. Nat. Educ. Knowl. 3 (10), 8.

Hopkin, S.P., Hames, C.A.C., Bragg, S., 1989. Terrestrial isopods as biological indicators of zinc pollution in the reading area, south-east England. In: Ferrara, F., Argano, R., Manicastri, C., Schmalfus, H., Taiti, S. (Eds.), Proceedings of the Second Symposium on the Biology of Terrestrial Isopods. Monitore Zoologico Italiano, 477–488.

Horsak, R.D., Bedient, P.B., Hamilton, M.C., Thomas, F.B, 1964. Pesticides. In: Horsak, R.D., Bedient, P.B., Hamilton, M.C., Thomas, F.B. (Eds.), Environmental Forensics: Contaminant Specific Guide. Elsevier Inc, pp. 143–165.

Hur, M., Kim, Y., Song, H.R., Kim, J.M., Im Choi, Y., Yi, H., 2011. Effect of genetically modified poplars on soil microbial communities during the phytoremediation of waste mine tailings. Appl. Environ. Microbiol. 77 (21), 7611.

ISO. International Standards Organization, 1993. Soil-Quality-Effects of Pollutants on Earthworms (Eisenia fetida). Part 1 Determination of Acute Toxicity Using Artificial Soil Substrate. International Standardization Organization, Geneva, Switzerland ISO11268–1.

Jeddi, K., Fatnassi, M., Chaieb, M., Siddique, K.H., 2021. Tree species as a biomonitor of metal pollution in arid Mediterranean environments: case for arid southern Tunisia. Environ. Sci. Pollut. Res. 28 (22), 28598–28605.

Jin, Z.P., Luo, K., Zhang, S., Zheng, Q., Yang, H., 2012. Bioaccumulation and catabolism of prometryne in green algae. Chemosphere 87 (3), 278–284.

Kanaya, N., Gill, B.S., Grover, I.S., Murin, A., Osiecka, R., Sandhu, S.S., Andersson, H.C., 1994. *Vicia faba* chromosomal aberration assay. Mut. Res. Fundam. Mol. Mech. Mutagen. 310 (2), 231–247.

Kevan, P.G., 1999. Pollinators as bioindicators of the state of the environment: species, activity and diversityInvertebrate Biodiversity as Bioindicators of Sustainable Landscapes74. Woodhead Publishing Limited.

Khaldi, F., 2014. Study of changes in physiological and enzymatic metabolism of a bryophyte “rhytidium rugosum” under the effect of a herbicide “TREFLAN”. Int. J. Res. Agric. Sci. 1 (1), 16–20.

Khalid, S., Shahid, M., Bibi, I., Sarwar, T., Shah, A.H., Niazi, N.K., 2018. A review of environmental contamination and health risk assessment of wastewater use for crop irrigation with a focus on low and high-income countries. Int. J. Environ. Res. Public Health 15 (5), 895.

Khataee, A.R., Movafeghi, A., Vafaei, F., Salehi Lisar, S.Y., Zarei, M., 2013. Potential of the aquatic fern azolla filiculoides in biodegradation of an azo dye: modeling of experimental results by artificial neural networks. Int. J. Phytorem. 15 (8), 729–742.

Khatri, N., Tyagi, S., 2015. Influences of natural and anthropogenic factors on surface and groundwater quality in rural and urban areas. Front. Life Sci. 8 (1), 23–39.

Kihlman, B.A., Andersson, H.C., 1984. Root tips of *Vicia faba* for the study of the induction of chromosomal aberrations and sister chromatid exchanges. In: Kilbey, B.J., Legator, M., Nichols, W., Ramel, C. (Eds.), Handbook of Mutagenicity Procedures2nd ed. Elsevier, Amsterdam, pp. 531–554.

Kishimba, M.A., Mihale, M.J., 2004. Levels of pesticide residues and metabolites in soil at Vikuge farm, Kibaha district, Tanzania–a classic case of soil contamination by obsolete pesticides. Tanz. J. Sci. 30 (2), 77–86.

Knillmann, S., Orlinskiy, P., Kaske, O., Foit, K., Liess, M., 2018. Indication of pesticide effects and recolonization in streams. Sci. Total Environ. 630, 1619–1627.

Kuo, Y.C., Chen, Y.C., Lin, M.Y., Young, L.H., Hsu, H.T., Liou, S.H., Wu, T.N., Wang, L.C., Tsai, P.J., 2015. Ambient air concentrations of PCDD/Fs, coplanar PCBs, PBDD/Fs, and PBDEs and their impacts on vegetation and soil. Int. J. Environ. Sci. Technol. 12 (9), 2997–3008.

Leme, D.M., Marin-Morales, M.A., 2009. *Allium cepa* test in environmental monitoring: a review on its application. Mut. Res. Rev. Mut. Res. 682 (1), 71–81.

Lim, T.B., Xu, R., Tan, B., Obbard, J.P., 2006. Persistent organic pollutants in moss as bioindicators of atmospheric pollution in Singapore. Chemosphere 64 (4), 596–602.

López, S.L., Aiassa, D., Benítez-Leite, S., Lajmanovich, R., Mañas, F., Poletta, G., Śnchez, N., Simoniello, M.F., Carrasco, A.E., 2012. Pesticides used in South American GMO-based agriculture. A review of their effects on humans and animal models. Adv. Mol. Toxicol. 6, 41–75.

Mesi, A., Kopliku, D., 2013. Cytotoxic and genotoxic potency screening of two pesticides on *Allium cepa* L. Proc. Technol. 8, 19–26.

Mesnage, R., Oestreicher, N., Poirier, F., Nicolas, V., Boursier, C., Vélot, C., 2020. Transcriptome profiling of the fungus *Aspergillus nidulans* exposed to a commercial glyphosate-based herbicide under conditions of apparent herbicide tolerance. Environ. Res. 182 (January), 109116.

Mishra, S., Mukherjee, A., Agrawal, M., 2020. Changes in herbaceous community pattern, growth and biomass accumulation under single and combined treatment of chlorpyrifos and malathion insecticides. Plant Arch. 20 (2), 3503–3509.

Naccari, C., Macaluso, A., Giangrosso, G., Naccari, F., Ferrantelli, V., 2014. Risk assessment of heavy metals and pesticides in honey from Sicily (Italy). J. Food Res. 3 (2), 107.

Nagajyoti, P.C., Lee, K.D., Sreekanth, T.V.M., 2010. Heavy metals, occurrence and toxicity for plants: a review. Environ. Chem. Lett. 8 (3), 199–216.

Narayan, D., Agrawal, M., Pandey, J., Singh, J., 1994. Changes in vegetation characteristics downwind of an aluminium factory in India. Ann. Bot. 73 (5), 557–565.

Nazir, S., Iqbal, M.Z, 2019. Molecular identification of genetically modified crops for biosafety and legitimacy of transgenes. In: Yuan-Chuan, C, Shiu-Jau, C (Eds.), Gene Editing - Technologies and Applications. IntechOpen.

Niemeyer, J.C., Santos, V.C., Araújo, P.B.D., Silva, Da, E., M., 2009. Reproduction of Cubaris murina (Crustacea: Isopoda) under laboratory conditions and its use in ecotoxicity tests. Braz. J. Biol. 69 (1), 137–142.

Ockenden, W.A., Steinnes, E., Parker, C., Jones, K.C., 1998. Observations on persistent organic pollutants in plants: implications for their use as passive air samplers and for POP cycling. Environ. Sci. Technol. 32 (18), 2721–2726.

Odendaal, J.P., Reinecke, A.J., 2003. Quantifying histopathological alterations in the hepatopancreas of the woodlouse *Porcellio laevis* (Isopoda) as a biomarker of cadmium exposure. Ecotoxicol. Environ. Saf. 56 (2), 319–325.

Odutayo, F.O.I., Oyetade, O.B., Esan, E.B., Adaramola, F.B., 2016. Effects of industrial effluent on the environment using *Allium cepa*. Int. J. Environ. Pollut. Res. 4 (3), 28–39.

OECD, Organization for Economic Cooperation and Development, 2004. Earthworm Reproduction Test. Guideline for Testing Chemicals No. 222. OECD, Paris, France.

OECD, 1984. Guideline for Testing of Chemicals, No. 207, Earthworm Acute Toxicity. Organization for Economic Cooperation and Development, Paris, France.

Olayinka, O.T., Idowu, A.B., Dedeke, G.A., Akinloye, O.A., Ademolu, K.O., Bamgbola, A.A, 2011. Earthworm as bio-indicator of heavy metal pollution around Lafarge, Wapco Cement Factory, Ewekoro, Nigeria, Proceedings of the Environmental Management Conference. Abeokuta, Nigeria. Federal University of Agriculture, pp. 489–496.

Olutayo, M., Science, P., Harcourt, P., Harcourt, P., State, R., 2007. Effects of amendments and bioaugumentation of soil polluted with crude oil, automotive gasoline oil, and spent engine oil on the growth of cowpea (*Vigna ungiculata* L. Walp). Sci. Res. Essays 2 (5), 147–149.

Pandey, B., Agrawal, M., Singh, S., 2014. Assessment of air pollution around coal mining area: emphasizing on spatial distributions, seasonal variations and heavy metals, using cluster and principal component analysis. Atmos. Pollut. Res. 5 (1), 79–86.

Pandey, P., Arora, N.K., 2020. Prof. Ananda Mohan Chakrabarty: the superbug superhero!. Environ. Sustain. 3 (3), 333–335.

Park, B.Y., Lee, J.K., Ro, H.M., Kim, Y.H., 2011. Effects of heavy metal contamination from an abandoned mine on nematode community structure as an indicator of soil ecosystem health. Appl. Soil Ecology 51, 17–24.

Parmar, T.K., Rawtani, D., Agrawal, Y.K., 2016. Bioindicators: the natural indicator of environmental pollution. Front. Life Sci. 9 (2), 110–118.

Parrish, Z.D., Banks, M.K., Schwab, A.P., 2005. Effect of root death and decay on dissipation of polycyclic aromatic hydrocarbons in the rhizosphere of yellow sweet clover and tall fescue. J. Environ. Qual. 34 (1), 207–216.

Paoletti, M.G., 1999. The role of earthworms for assessment of sustainability and as bioindicators. Agric. Ecosyst. Environ. 74 (1–3), 137–155.

Pérès, G., Vandenbulcke, F., Guernion, M., Hedde, M., Beguiristain, T., Douay, F., Cluzeau, D., 2011. Earthworm indicators as tools for soil monitoring, characterization and risk assessment. An example from the national Bioindicator programme (France). Pedobiologia 54, S77–S87.

Pignatello, J.J., Katz, B.G., Li, H., 2010. Sources, interactions, and ecological impacts of organic contaminants in water, soil, and sediment: an introduction to the special series. J. Environ. Qual. 39 (4), 1133–1138.

Plaza-Bolaños, P., Padilla-Sánchez, J.A., Garrido-Frenich, A., Romero-González, R., Martínez-Vidal, J.L., 2012. Evaluation of soil contamination in intensive agricultural areas by pesticides and organic pollutants: South-eastern Spain as a case study. J. Environ. Monit. 14 (4), 1182–1189.

Punwong, P., Juprasong, Y., Traiperm, P., 2017. Effects of an oil spill on the leaf anatomical characteristics of a beach plant (*Terminalia catappa* L.). Environ. Sci. Pollut. Res. 24 (27), 21821–21828.

Rai, P.K., Lee, S.S., Zhang, M., Tsang, Y.F., Kim, K.H., 2019. Heavy metals in food crops: health risks, fate, mechanisms, and management. Environ. Int. 125, 365–385.

Rollin, O., Benelli, G., Benvenuti, S., Decourtye, A., Wratten, S.D., Canale, A., Desneux, N., 2016. Weed-insect pollinator networks as bio-indicators of ecological sustainability in agriculture. A review. Agron. Sustain. Dev. 36 (1), 1–22.

Rylott, E.L., Johnston, E.J., Bruce, N.C., 2015. Harnessing microbial gene pools to remediate persistent organic pollutants using genetically modified plants—a viable technology? J. Exp. Bot. 66 (21), 6519–6533.

Schlenk, D., Huggett, D.B., Allgood, J., Bennett, E., Rimoldi, J., Beeler, A.B., Block, D., Holder, A.W., Hovinga, R., Bedient, P., 2001. Toxicity of fipronil and its degradation products to Procambarus sp.: field and laboratory studies. Arch. Environ. Contam. Toxicol. 41 (3), 325–332.

Serbula, S.M., Miljkovic, D.D., Kovacevic, R.M., Ilic, A.A., 2012. Assessment of airborne heavy metal pollution using plant parts and topsoil. Ecotoxicol. Environ. Saf. 76, 209–214.

Shahriari, M.H., Firozabadi, G.S., Azizi, M., Kalantari, F., Minai-Tehrani, D., 2007. Study of growth and germination of *Medicago sativa* (Alfalfa) in light crude oil-contaminated soil. Res. J. Agric. Biol. Sci. 3 (1), 46–51. Retrieved 10 February 2021, from. https://www.researchgate.net/publication/266499510.

Sharifi, M., Sadeghi, Y., Akbarpour, M., 2007. Germination and growth of six plant species on contaminated soil with spent oil. Int. J. Environ. Sci. Technol. 4 (4), 463–470.

Snousy, M.G., 2016. An overview on the organic contaminants. SDRP J. Earth Sci. Environ. Stud. 1 (2), 1–9.

Stankovic, S., Stankovic, A.R., 2013. Bioindicators of toxic metalsGreen Materials for Energy, Products and Depollution. Springer, Dordrecht, pp. 151–228.

Stankovic, S., Kalaba, P., Stankovic, A.R., 2014. Biota as toxic metal indicators. Environ. Chem. Lett. 12 (1), 63–84.

Suchara, I., Sucharova, J., Hola, M., Reimann, C., Boyd, R., Filzmoser, P., Englmaier, P., 2011. The performance of moss, grass, and 1-and 2-year old spruce needles as bioindicators of contamination: a comparative study at the scale of the Czech Republic. Sci. Total Environ. 409 (11), 2281–2297.

Suther, S., Singh, S., Dhawan, S., 2008. Earthworms as bioindicator of metals (Zn, Fe, Mn, Cu, Pb and Cd) in soils: Is metal bioaccumulation affected by their ecological category? Ecol. Eng. 32 (2), 99–107. doi:https://doi.org/10.1016/j.ecoleng.2007.10.003.

Szyczewski, P., Siepak, J., Niedzielski, P., Sobczyński, T., 2009. Research on heavy metals in Poland. Poland J. Environ. Stud. 18 (5), 755.

Tang, X., He, L.Y., Tao, X.Q., Dang, Z., Guo, C.L., Lu, G.N., Yi, X.Y., 2010. Construction of an artificial microalgal-bacterial consortium that efficiently degrades crude oil. J. Hazard. Mater. 181 (1–3), 1158–1162.

Udebuani, A., Ezeji, E., Agu, R., Onwurah, I., Abara, P., 2018. Bioavailability and Genotoxicity of glyphosate treated soil on *Archachatina marginata*. Nat. Sci. 16 (7), 119–125.

US Geological Survey: Science for a changing world, https://pubs.usgs.gov/circ/circ1225/html/sources.html. Assessed January. 24, 2021.

Vannini, A., Paoli, L., Vichi, M., Bačkor, M., Bačkorová, M., Loppi, S., 2018. Toxicity of Diclofenac in the Fern *Azolla filiculoides* and the Lichen *Xanthoria parietina*. Bull. Environ. Contam. Toxicol. 100 (3), 430–437. https://doi.org/10.1007/s00128-017-2266-4.

Veltman, K., Huijbregts, M.A., Vijver, M.G., Peijnenburg, W.J., Hobbelen, P.H., Koolhaas, J.E., Hendriks, A.J., 2007. Metal accumulation in the earthworm *Lumbricus rubellus*. Model predictions compared to field data. Environ. Pollut. 146 (2), 428–436.

Verhoog, H., 1992. The concept of intrinsic value and transgenic animals. J. Agric. Environ. Ethics 5 (2), 147–160.

Wang, D.F., Zhang, Z.H., 2020. Higher plants as bioindicators of metal contamination from Shangdong abandoned karst bauxite, southwestern China. Plant Biol. 22 (2), 323–330.

Wang, Y., Wang, T., Li, A., Fu, J., Wang, P., Zhang, Q., Jiang, G., 2008. Selection of bioindicators of polybrominated diphenyl ethers, polychlorinated biphenyls, and organochlorine pesticides in mollusks in the Chinese Bohai Sea. Environ. Sci. Technol. 42 (19), 7159–7165.

Wegener, J.W.M., van Schaik, M.J.M., Aiking, H., 1992. Active biomonitoring of polycyclic aromatic hydrocarbons by means of mosses. Environ. Pollut. 76 (1), 15–18.

Wuana, R.A., Okieimen, F.E., 2011. Heavy metals in contaminated soils: a review of sources, chemistry, risks and best available strategies for remediation. Int. Sch. Res. Notices 2011, 1–20.

Xiao, C., Wang, L., Zhou, Q., Huang, X., 2020. Hazards of bisphenol A (BPA) exposure: a systematic review of plant toxicology studies. J. Hazard. Mater. 384, 121488.

Zaghloul, A., Saber, M., Gadow, S., Awad, F., 2020. Biological indicators for pollution detection in terrestrial and aquatic ecosystems. Bull. Natl. Res. Centre 44 (1), 1–11.

Zhang, S., Qiu, C.B, Zhou, Y., Jin, Z.P., Yang, H., 2011. Bioaccumulation and degradation of pesticide fluroxypyr are associated with toxic tolerance in green alga *Chlamydomonas reinhardtii*. Ecotoxicology 20, 337–347. https://doi.org/10.1007/s10646-010-0583-z.

Zhu, B., Douville, M., Gagné, F., 2013. Genetically modified organisms in the environmentEncyclopedia of Aquatic Ecotoxicology. Springer, Netherlands, pp. 539–546.

CHAPTER 12

# Responses of plants to metallic nanoparticles under coexposure to metals and metalloids

**Banita Kumari Paswan, Subodh Kumar, Archana Dwivedi, Poornima Vajpayee**
Ecotoxicology, Bioremediation and Nanotoxicology Laboratory, Department of Botany University of Lucknow, Lucknow, U.P., India

## 12.1 Introduction

Nanotechnology based on engineered nanoparticles (carbon based or metal based), ENPs having at least one dimension between 1 and 100 nm (American Society for Testing and Materials (ASTM) 2012) has been evolved in order to provide an improved clean environmental system, reducing hazardous effects of environmental pollutants, combating nutrient deficiencies, enhanced growth and crop productivity in agri- environment, new clinical interventions in early detection of deadly diseases and their treatments. Engineered NPs can be carbon-based or metal-based (Peralta-Videa et al., 2011).

Metal-based are grouped in metals, metal oxides, and quantum dots (Peralta-Videa et al., 2011). More specific term metallic nanoparticles are the nanomaterials which have at least one metallic elemental component in the single Nano molecule and are classified in metals, metal oxides, and quantum dots. Titanium nanoparticles ($TiO_2$), iron oxide nanoparticles ($Fe_2O_3$), gold nanoparticles (nAu), silver nanoparticles (nAg), etc. are example of some commercially produced engineered metallic nanoparticles (EMNPs) and are extensively used in bio sciences, agri-environments, reducing environmental pollutants engineering and in number of commercial products formulations such as paint, paper, ink and plastic, sunscreens, antimicrobial creams, etc. (Rico et al., 2015). The rapid advancement of nanotechnology significantly increases the manufacturing, application and release of engineered metallic nanoparticles into the environment, leading to a potential increase in human and environmental exposure to EMNPs

*New Paradigms in Environmental Biomonitoring Using Plants*
DOI: https://doi.org/10.1016/B978-0-12-824351-0.00012-2

(Schwabe et al., 2015; Goswami et al., 2017). These applications have led to the frequent occurrence of metallic NPs in the environment, with agricultural soil as an important sink (Majumdar et al., 2016). Metallic nanoagrichemicals significantly modify the soil chemical properties (Wang et al., 2018) which may result into reduced crop vigor, quality and productivity. Nanoparticles with their unique surface properties can alter the physiological, biochemical attributes of plants after internalization and upward movement of these particles to shoots affecting metabolomes and transcriptomes modulating various vital activities of plants.

EMNPs are new born contaminants in the environment which may effects plant growth, reproduction and production by their phytotoxicological effects. The coexposure of EMNPs with other environmental pollutants including metal and metalloids is global thrust area to study the synergistic impact of ENMPs on plants. Metallic Nanoparticles (NPs) have recently emerged as potential agents for plants to ameliorate heavy metal and metalloid stress in plants. Various studies has been surfaced to investigate the role of metallic and non-metallic nanoparticles in elucidating their role in ameliorating the phytotoxicity plants caused by toxic metals and reducing their accumulation in plant tissues globally (You et al., 2021; Rizwan et al., 2021; Chen et al., 2021). Both positive and negative effects of coexposure of metallic Nanoparticles with metals metalloids has been reported in plants which vary with plant species, size of the metallic nanoparticles, exposure routs, toxic metal/metalloid as well as soil types (Rizwan et al., 2021, Zhou et al., 2021). Starch-stabilized nanoscale zerovalent iron has been used by (Gong et al., 2017) for nanophytoremediation of heavy metal contaminated sites. Starch-stabilized nanoscale zerovalent iron decrement the growth of ramie (*Boehmeria nivea* (L.) Gaudich) than that of control under Cd stress but significantly enhanced the cadmium (Cd) and iron (Fe) uptake in the phytoremediator plant grown in Cd-contaminated sediments, as compared with respective control. The group further concluded that nanoparticles at low concentration can improve the efficiency of phytoremediation. Similarly, titanium dioxide ($TiO_2$) NPs upregulated Cd uptake in Cd-stressed soybean has been documented (Singh and Lee, 2016). Radish seeds were exposed to n-$TiO_2$ and in combination using a seed germination and seedling growth toxicity test OECD 208. $TiO_2$ NPs (high doses) were not able to alleviate the phytotoxic responses in Raphanus sativus L. Parvus (Manesh et al., 2018). Here, an attempt has been done to review the responses of various plants under the coexposure of ENMPs with heavy metals and metalloids.

## 12.2 Coexposure of EMNPs with metalloids

Arsenic (As), a ubiquitous metalloid is highly toxic to all forms of life including plants. It is categorized as a potent class-one non-threshold carcinogen in humans (Li et al., 2017, Smith et al., 2002, Awasthi et al., 2017, Tripathi et al., 2007). High concentrations of Arsenic as environmental pollutant in global environment is present due to geogenic and anthropogenic activities (Awasthi et al., 2017, Patel et al., 2005, Mawia et al., 2021). The most common Inorganic forms of Arsenic in environment are arsenate (AsV) and arsenite, (AsIII). Both inorganic forms are toxic to plants. Arsenate, an analogue of phosphate enters inside plants through high-affinity phosphate transporters (Meharg and Hartley-Whitaker, 2002, Abedin et al., 2002) and disrupts various phosphate-dependent aspects of metabolism inside plant cells due to inadequate phosphate supply. Being an analogue of arsenic it competes with phosphate during phosphorylation reactions, resulting in the formation of unstable and short-lived AsV adducts viz. Oxidative phosphorylation, ATP synthesis, (Tripathi et al., 2007, Finnegan and Chen, 2012). Arsenite makes its route to inside plants via aquaglyceroporins and interferes with sulfhydryl groups of enzymes and proteins in cytosol (Meharg and Jardine, 2003) which causes detrimental effects on functioning of proteins.

Groundwater pollution by arsenic has been documented globally particularly in Bangladesh al., China, Vietnam, Nepal and India (Chauhan et al., 2009, Khanam et al., 2021) It is reported that parts of all the states and countries surveyed in the Ganga–Meghna–Brahmaputra (GMB) plain, which has an area of approximately 500 000 $km^2$ and a population over 500 million in parts of all the states and countries surveyed in the Ganga–Meghna–Brahmaputra (GMB) plain are at risk from ground water arsenic contamination (Chakraborti et al., 2004). Agricultural crops accumulate high concentrations of arsenic and consumers of these crops are at high risk of arsenic poisoning. Rice is the major crop in several highly arsenic contaminated areas. Rice could accumulate up to 2 mg $kg^1$ As in grains (Islam et al., 2004) and up to 92 mg $kg^1$ in straw (Abedin et al., 2002). Arsenic accumulation in rice and other food crops can be attain through removal of arsenic from agricultural fields or by introducing low arsenic accumulating tolerant species. Recently, nanotechnology based on EMNPs attracted worldwide scientists to reduce arsenic accumulation in rice and other plants of commercial importance.

A pot experiment has been conducted by Liu et al (Liu et al., 2014) to study the impact of nanosized silica sols (10-30 nm) on rice. It has been observed by them that foliar spray of these particles increased grain yield and reduced arsenic accumulation in grain and stalk after 21 days of exposure. Same group of scientists reported that foliar spray of same nanoparticles produce similar results when sprayed to the foliage of the rice plants grown hydroponically (Liu et al., 2014). Further, field studies conducted by Liu et al. (2014) reported that nanosized silica sols (10–30 nm) enhanced the number of tillers in rice and reduced arsenic contents in grains (Table 12.1)

$SiO_2$ nanoparticles (Size: 170 ± 10 nm) when exposed in the presence of arsenic (25 μm and 50 μM) in hydroponics for 7 days reduced arsenic accumulation in plant tissues of maize (Tripathi et al., 2016). Further, an increment in photosynthesis was recorded in maize plants. ROS mediated toxic effects of arsenic was ameliorated by the $SiO_2$ nanoparticles in maize (Tripathi et al., 2016).

$\alpha$-$Fe_2O_3$ (Size: <30 nm), mixed with soil (Dose: 0, 0.05, 0.1, 0.2% w/w) along with arsenic (0, 6, 12, 24, 48, 98 mg $kg^{-1}$) and filled in pots. Maize plants were grown in these pots for 75 days (Gil-Diaz et al., 2016). An increase in dry weights of roots and aerial parts of maize was observed. Arsenic accumulation in aerial plant tissues and roots interestingly reduced by the $\alpha$-$Fe_2O_3$ (Gil-Diaz et al., 2016).

Gil-Diaz et al. (Mansouri et al., 2017) documented that exposure of nZVI (nanoscale zerovalent iron, size: ~ 50 nm, nZVI, Dose: 0, 1.0, 10%) to Barley in arsenic contaminated soils reduced As levels in shoot and roots,. incremented growth (plant height, dry weight) and K and Na levels in shoot. Further, it has been reported that NPs enhanced the pH of postharvest soil Gil-Diaz et al. (Mansouri et al., 2017).

Wetland plants could be used as suitable biofilters for the phytoremediation of metal(loid)s and other environmental contaminants from waters and wastewaters due to their ability to accumulate and retain the contaminants in their roots and other plant tissues (Pardo et al., 2016). A low-resistance internal pathway consisting of root aerenchyma (a porous tissue) for the exchange of gases between aerial and submerged tissues is present in some wet land plants (*Typha latifolia, Phragmites australis* etc.) for respiration under anaerobic conditions (Visser et al., 2000). The oxygenated rhizosphere of these wetland plants promotes formation of a natural iron plaque (mixture of crystalline and amorphous iron (oxyhydr)oxides, composed mainly of Fe $^{3++}$ minerals: lepidocrocite, goethite, or ferrihydrite) on the surface of their roots due to precipitation of iron oxides (Pardo et al., 2016, Liang et al., 2006). The iron plaques exhibit an ability of the meta(lloid) adsorption

**Table 12.1** Amelioration of metalloid accumulation and phytotoxicity in plants by EMNPs.

| Plant | Coexposure of EMNPs | Metalloid | Treatment type, duration | Effects | References |
|---|---|---|---|---|---|
| Rice | Nano-sized silica sols Size: 10–30 nm Dose: 200 mL (5 mM | As 0, 50, 100 μM | Hydroponic 20 days old plants for 10 days exposed by Foliar spray | NPs increased plant biomass. Decreased As in shoots and roots and increased its percentage in cell walls of shoot and roots. Reduced MDA and Electrolyte leakage and increased antioxidant enzyme activities | (Liu et al., 2014) |
| Rice | Nano-sized silica sols Size: 10–30 nm Dose: 100 mL/pot (5 mM) | As 93.6 mg $Kg^{-1}$ | pot experiment, foliar spray of NPs, 120 days | NPs increased yield of grains and stalks while decreased As content in these parts | (Liu et al., 2014) |
| Rice | Nano-sized silica sols Size: 10–30 nm Dose: 3000 mL/ha (5 mM) | As 93.6 mg $Kg^{-1}$ | Field experiment for 100days | NPs increased number of tillers, thousand grains weight, and yield while decreased As contents in grains | (Liu et al., 2014) |
| Maize | $SiO_2$ Size: 170 ±10 nm Dose: 10 μM | As 25, 50 μM | Hydroponic, 7d | NPs enhanced plant growth, photosynthesis and reduced oxidative stress and As accumulation in plants | (Tripathi et al., 2016) |
| Maize | α-Fe2O3 Size: <30 nm Dose: 0, 0.05, 0.1, 0.2% w/w | As 0, 6, 12, 24, 48, 98 mg kg | soil in pots,75 days | NPs increased dry weights of roots and aerial parts and decreased As concentrations in these parts. Decreased available As in the soil | (Gil-Diaz et al., 2016) |
| Barley | nZVI (nanoscale zervalent iron) Size: ~ 50 nm Dose: 0, 1.0, 10% | As 5800 ±450 mg $Kg^{-1}$ | soil in pots, 33 days | NPs incremented growth (plant height, dry weight) and K and Na in shoot while reduced As levels in shoot and roots. NPs enhanced the pH of post- harvest soil. | (Mansouri et al., 2017) |

*(Continued)*

**Table 12.1** Cont'd

| Plant | Coexposure of EMNPs | Metalloid | Treatment type, duration | Effects | References |
|---|---|---|---|---|---|
| *Phragmites australis* | $\gamma Fe_2O_3$ (size 20-100 nm) alone or in combination with $FeSO_4$ | As 100 mg/ per L (added as $Na_2HAsO_4$) | hydroponic, exposed for 5 d | Combination of $nFe_2O_3$ and $Fe^{2+}$ salts boosted iron plaque formation in plant roots. Increased iron plaque formation enhanced *P. australis* As immobilization capacity. | (Pardo et al., 2016) |
| *Brassica juncea* | $Fe_3O_4$ Size: ~ 80 nm Dose: 500 mg L | As 150 μM | Hydroponic, 14-D-old seedlings exposed for 4 days | NPs increased seed germination, root and shoot lengths, and chlorophyll contents. Reduced SOD, CAT, APX activities, and MDA and $H_2O_2$ contents in plant | (Praveen et al., 2017) |
| Rice | CuO NPs, spherical. Size: 9 to 22 nm in diameter, with an average diameter of 14.3 nm | As 1 mgL-1 of As(III) or As(V), or 100 mgL-1 of CuO NPs or Cu(II), or different combinations of As and CuO NPs or Cu(II) | Hydroponic, 35 days old seedlings exposed to for 6 days | Reduction of copper accumulation and As speciation in rice seedlings was copper form (Cu++ /CuO NPs) regulated. Both forms significantly decreased the accumulation of total As in rice tissues, with Cu(II) exhibiting significantly greater effect than CuO NPs. CuO NPs and Cu++ decreased 45% and 47% of As(V) in rice roots, respectively. | (Wang et al., 2019) |

| Plant | Coexposure of EMNPs | Metalloid | Treatment type, duration | Effects | References |
|---|---|---|---|---|---|
| *Vigna radiata* | $Fe_2O_3$ NPs<br>Size: 10–50 nm (hexagon)<br>Dose:100,200 and 400 | As $AsO_4$ 3– equivalent to 0.5, 1 and 2 μM | hydroponic, exposed for 8 days | Reduction of As accumulation achieved through adsorption on Fe2O3 NPs, antagonise arsenate induced oxidative stress and restricted to root surface limit transport of toxic iron levels. | (Nisha Shabnam et al., 2019) |
| *Glycine max* | ZnO<br>size: 20 nm<br>Dose:25 μM | As | | Levels of reduced glutathione content, ascorbic acid, and various photosynthesis- and antioxidant-relevant enzymes increased, decreased MDA content, higher expression of defense- and detoxification-related genes compared to control during coexposure of NPs | (Zeeshan et al., 2021) |
| Rice | Biogenic NPS of magnesium oxide nanoparticles (MgO-NPs, size 38-57 nm) | As | | significantly enhanced the plant biomass, antioxidant enzymatic contents, and decreased reactive oxygen species and upward As translocation as compared with control treatment | (Wu et al., 2021) |

(*Continued*)

Table 12.1 Cont'd

| Plant | Coexposure of EMNPs | Metalloid | Treatment type, duration | Effects | References |
|---|---|---|---|---|---|
| Rice | Two types of nano-$TiO_2$ with anatase and rutile structures Dose: 0, 10, 100, 1000 mg $L^{-1}$ | 1 mg As(V)/L $Na_3AsO4$ | Hydroponic, 7d | nano-$TiO_2$ significantly reduced arsenic bioaccumulation in rice seedlings by 40%–90% via strong sorption process, growth was not affected, notably alleviated oxidative stress, caused by arsenic exposure. nano-$TiO_2$ amendment significantly decreased the iron plaque amount by 50%–63%, weakened the arsenic retention in this barrier by 47%–99%, Rutile nano-$TiO_2$ (1000 mg/L) reduced As accumulation with no significant oxidative stress. As bioaccumulation in rice seedlings and the response varied greatly with dose and type of NPs | Wu et al 2021 |
| Soyabean | $SiO_2$ (30 nm and 50nm | Hg three Hg exposure levels (0, 4.95 μmol/L and 24.75 μmol/L) | hydroponics for 30 d | reduce the accumulation of Hg in plants and increase plant biomass ameliorated toxicity to chlorophyll | Li et al (2020) |

and in plants their formation reached up to 10% of root dry weight and extended for near 20 mm in the rhizosphere (Hansel and Fendorf, 2001, Pan et al., 2014). An hydroponic study evaluated the iron plaque formation after external iron addition (nanomaghemite, $nFe_2O_3$ and $FeSO_4$, alone or in combination), as well as their impact on arsenic immobilization capacity (Pardo et al., 2016). The combination of $FeSO_4$ and $nFe_2O_3$ was proved to be most effective treatment for enhancing the As removal capacity *of P. Australis* (Pardo et al., 2016).

*Brassica juncea* plants were coexposed to $Fe_3O_4$ NPs (Size: ~ 80 nm, Dose: 500 mg $L^{-1}$) and arsenic (150 μM) in hydroponics for 4 days (Praveen et al., 2017). It has been observed that $Fe_3O_4$ NPs ameliorated the toxic effects of arsenic in *B. juncea* by reducing the arsenic accumulation in plant tissues (Table 12.1).

A recent study reported the reciprocal effect of coexisting CuO NPs or $Cu^{++}$ and As in rice seedlings and documented the unique "nano-effect" of CuO NPs. Hydroponic culture of rice seedlings were exposed with a combinations of 1 mg/L of arsenite (As(III)) or arsenate (As(V)) and 100 mg/ L of CuO NPs or Cu(II) for 6 days (Wang et al., 2019). The study concluded that Cu++ and spherical CuO NPs (size 9–22 nm) reduced the accumulation of total arsenic in rice. However, arsenic accumulation and arsenic speciation was copper form dependent as coexposure of As(V) with CuO NPs decrement 45% and coexistence of As(V) with Cu(II) caused a 47% increase in As(V) in rice roots, respectively. Further, It has been documented by them that coexisting As(III) or As(V) with CuO NPs lowered Cu levels in rice roots by 40% and 50%, respectively. However, both inorganic forms of arsenic did not affect Cu content in rice roots coexposed to $Cu^{++}$ (Wang et al., 2019).

Iron nanoparticles (NPs) are extensively used for the reduction of toxic levels of arsenic in water. It has been documented that the growth of *Vigna radiata* seedlings are slightly promoted by $Fe_2O_3$ NPs while arsenate reduced growth Shabnam et al. (Nisha Shabnam et al., 2019). However, arsenate induced decline in the growth of *V. radiata* was ameliorated by $Fe_2O_3$ NPs. $Fe^{++}$ alone and together with arsenate inhibited seed germination completely. They have concluded that $Fe_2O_3$-NPs could effectively ameliorate arsenic toxicity in *V. radiata* by restricting As availability to plant by adsorption of arsenic on its surface.

The arsenic (As V) uptake by soya bean plants exposed to 25 μM (As V) imposed phytotoxicity in plants by displaying reduction in photosynthetic efficiency, increased proline and glycine betaine accumulation in tissues, and altered antioxidant activity compared to an untreated control. The application of zinc oxide (25 μM ZnO) decreased arsenic imposed stress in tissues (root and shoot) of soybean plants and rescued plant health and vigor by intervening antioxidant defense system, enzymes regulating photosynthesis, increasing thiols and ascorbic acid (Zeeshan et al., 2021). Besides, increased expression of genes involved in various pathways of defense- and detoxification of arsenic toxicity in soybean plants also contribute to arsenic tolerance in nanoparticles treated soybean plants (Zeeshan et al., 2021). All these observations indicate that zinc nanoparticles may provide a powerful solution to arsenic toxicity exerted in *Glycine max* crops in agricultural soils.

The application(200 mg $Kg^{-1}$) og biogenic magnesium oxide nanoparticles (MgO-NPs, size 38-57 nm) from a native Enterobactor sp. Strain RTN2, genetically identified through 16S rRNA gene sequence analysis in arsenic contaminated soils significantly enhanced the plant biomass, antioxidant enzymatic contents, decreased reactive oxygen species and upward arsenic translocation as compared with control treatment (Wu et al., 2021). Further, biogenic MgO-NPs promoted the morpho-physiological profiles of arsenic stressed rice plants via decrementing arsenic translocation and reactive oxygen species in plants. The authors have concluded that biogenic MgO-NPs could be used as potent nanofertilizer for sustainable rice production in arsenic contaminated agrifields (Wu et al., 2021).

The possible application of $TiO_2$ nanoparticles (nano-$TiO_2$) to alleviate arsenic bioaccumulation in rice seedlings and such a functioning with their crystalline structure were investigated. Specifically, nano-$TiO_2$

$TiO_2$ nanoparticles (nano-$TiO_2$) with anatase and rutile structures and the bulk $TiO_2$ at 0, 10, 100, and 1000 mg/L were exposed to rice seedlings in the presence of arsenic (1 mg/L) hydroponically for 7 d to investigate their role in arsenic level reduction in rice (Ahmed et al., 2021). The study documented that nano-$TiO_2$ significantly reduced arsenic bioaccumulation in rice seedlings by 40%–90% via strong sorption process. The nano-$TiO_2$ has no adverse impact on growth of rice. They have suggested that this was probably achieved by the mitigation of oxidative stress imposed by toxic levels of arsenic in rice plants. Further, presence of nano-$TiO_2$ amendment significantly decreased the iron plaque amount by 50%–63%, weakened the arsenic retention in this barrier by 47%–99%, As bioaccumulation in rice seedlings and their response varied significantly with dose and type of NPs (Ahmed et al., 2021).

Mercury (Hg) pollution is a worldwide concern Due to long distance transportation and high bioaccumulation rate of methylmercury (MeHg) in food webs mercury pollution raised concern globally (Li et al., 2020). The impact of two sizes (30 nm and 50 nm) of engineered Nano-$SiO_2$ (two sizes:30 nm and 50 nm) on Hg accumulation in soybean plants has been studied to explore distribution of Hg in plant tissues and its impact on chlorophyll and antioxidant enzymes. Both $SiO_2$ NPs reduce accumulation of Hg in in various plant tissues and upward translocation of mercury. The study also report that toxic effects of Hg to chlorophyll content in soybean plants was also ameliorated by $SiO_2$ NPs (Li et al., 2020).

## 12.3 Coexposure of EMNPs with Heavy metals

Rizwan et al (Rizwan et al., 2019) reported that coexposure of ZnO NPs in the presence of arsenic enhanced root and shoot dry biomass, chlorophyll and reduced MDA, $H_2O_2$ and Cd concentration in maize plants.

Cadmium (Cd) transportation and root uptake in rice is highly affected by iron plaque formation on root surface in presence of iron containing materials in the rhizosphere A hydroponic study revealed that Nano-$Fe_3O_4$-modified biochar (BCeFe) induce iron plaque formation on surface of rice roots. The proportion of crystallized iron plaque (31.8%–35.9%) formed by BC-Fe was much higher than that formed by chelated iron (EDTA-Fe) and $FeSO_4$. The Cd translocation factor from root to and the Cd translocation factor from stem to leaf with BC-Fe treatment reduced by 84.7% and 80.0%, respectively. The study suggested that application of BC-Fe supported the formation of iron plaque in rice and increased the sequestration of Cd and Fe in roots, thus reduced the transportation and accumulation of Cd in aerial rice tissues (Hussain et al., 2020).

Paddy fields of southern China are sink of heavy metal pollutants due to direct discharge of untreated industrial waste in water bodies which are source of irrigation in these areas. It has been documented that foliage application of selenium nanoparticles ameliorates cadmium and Lead phytotoxicity in *Oryza sativa* through enhancing the grain yield, rice biomass, and Se contents in brown rice (Zhang et al., 2021).

Cadmium and drought are the most destructive of the abiotic stresses with negative consequences in terms of impaired metabolism, restricted nutrient use efficiency and disruptive photosynthesis of plants. A recent study documented the application of bioengineered iron oxide NPs from *Bacillus* strain RNT1 and hydrogel nanoparticles (spherical in shape with size 18-94 nm of both particles) for their potential combined use as nanofertilizer for modulation of growth and productivity of rice in cadmium polluted soils affected by drought (Ahmed et al., 2021). Further, They have documented that bioengineered iron oxide NPs exposed rice plants exhibit an increment in biomass, antioxidant enzyme activity, photosynthesis efficiency, nutrient acquisition together with reduced reactive oxygen species, upward cadmium transport via curtailed in rice plants. Another study on explore the impacts of single and combined use of ZnO NPs, Fe NPs and Si NPs on wheat growth and Cd intake by plants in a Cd-contaminated field revealed increment in straw and grain yield in wheat plants. The study documented reduced cadmium uptake in straw (84%) and grains (98%) during combined foliar application in comparison to untreated wheat plants (Ahmed et al., 2021).

The application of 100 mg $kg^{-1}$ bioengineered iron oxide NPs (FeO-NPs, SEM size 19–40 nm) from a native bacterial strain, Pantoea ananatis strain RNT4, in saline soils increased wheat plant growth, attaining 36.7% of additional length as compared with the control through mitigating the detrimental effects of abiotic stresses and thereby reprogramming the morpho-physiological state of wheat plants (Manzoor et al., 2021). In addition, the presence of bioengineered FeO-NPs in saline soil significantly enhanced the nutrient concentrations (N, P, and $K^+$) and reduced the $Na^+$ and $Cl^-$ components in the wheat grain. Interestingly, uptake of cadmium in wheat plants grown in cadmium polluted soils significantly (72%) decreased by bioengineered the FeO-NPs. It has been suggested that large surface area adsorption of cadmium by bioengineered FeO-NPs reduced its bioavailability to the plants and could be an ideal technology for saline agricultural soils polluted by cadmium for safe agriculture practice (Manzoor et al., 2021).

An attempt has been done to explore the role of ZnO-NPs (foliar) and 24-epibrassinolide (EBL; root dipping) individually or in combined form in to enhance the copper tolerance in tomato (Solanum lycopersicum) plants. It was apparent from the results of the study that Cu-stressed tomato (Solanum lycopersicum) plants reduced photosynthesis (17.3%), stomatal conductance (18.1%), plant height (19.7%), and nitrate reductase (NR) activity (19.2%) due to enhanced reactive oxygen species generated oxidative stress (superoxide radical $O_2^-$: 22.3%; hydrogen peroxide, $H_2O_2$; 26.2%; malondialdehyde, MDA; 29.4%) S. Lycopersicum (Faizan et al 2021). ZnO-NPs and/or EBL implemented via different modes improved photosynthetic activity, stomatal aperture, growth, cell viability and activity of antioxidant enzymes and proline contents augmenting copper tolerance in S. Lycopersicum (Faizan et al., 2021).

## 12.4 Role of Casparian strips (CSs) and suberin lamellae (SL) in reducing heavy metal uptake during coexposure of metallic Nanoparticles

Nutrients and other soluble metals enter in plants via roots through combination of symplastic pathway, apoplastic pathway or the coupled transcellular pathway (Barberon, 2017) from soil. Nutrients and other cations travel across all external cell layers into the central vasculature and upward to the shoots. (Barberon, 2017, Liu et al., 2021). Both the Casparian strips (CSs) and suberin lamellae (SL) are considered as vital barriers to extracellular ion movements. It has been documented that Casparian strips (CSs) and suberin lamellae (SL) play a critical role in checking the movement of cadmium to the xylem via apoplast (Lux et al., 2011). Recently, it has been demonstrated that $La_2O_3$ NPs exposure produced shorten and thicken roots in Zea mays and this was explained as $La_2O_3$ NPs induced early development of apoplastic barriers (Yue et al., 2017). Plant roots develop morphological plasticity e.g. shorten and thickened roots) to reduce acclimatize themselves under various abiotic stresses including heavy metal stress (Paez-Garcia et al., 2015). Roots often shorten and thicken under mineral and physical stress (de Freitas, 1988, Bennie, 1991). Enlarged root diameter is suggestive of altered cell expansion/division usually resulted by the isotropic growth of epidermal cells and apoplastic barriers, as a stress response to reduce water loss from roots to the environment (Veen, 1982, Wang et al., 2011). Rossi et al. (Rossi et al., 2017) documented that the coexistence of $CeO_2$NPs and Cd changed the endodermal suberization in Glycine max (L.) Merr. Cadmium uptake via the apoplastic pathway in the root has been proven significant for Cd accumulations in the hyperaccumulator *Sedum alfredii* (Tao et al., 2016).

The effects of $CeO_2$NPs on Cd uptake, transport and the consequent Cd accumulation in Sedum alfredii has been documented by Liu et al (Barberon, 2017). The study reveals that exposure of 50 or 500 mg $L^{-1}$ $CeO_2$NPs alone had no apparent damaging effects on plant growth. However, coexposure with Cd, the consistent $CeO_2$NPs decrease Cd accumulation in the roots and shoots up to 37% through apoplastic pathway. The elevated expression levels of genes (SaCASP, SaGPAT5, SaKCS20 and SaCYP86A1)associated with formation of Casparian strips (CSs) and suberi*n by* $CeO_2$*NPs in* Sedum alfredii under coexposure of cadmium has been observed. Simultaneously Cd decreased the Ce levels in the roots and affected its translocation from root to shoot. Uptake of both $CeO_2$NPs and Cd together influence the interactional uptake of both Ce and cadmium in roots of S. alfredii mainly via the apoplastic pathway which is primarily regulated by the development of Casparian strips and suberin (Barberon, 2017).

## 12.5 Conclusion and future perspectives

It could be concluded on the basis of studies reviewed here that amelioration of the toxicity of metals and metalloid and their accumulation in plants by metallic nanoparticles is metal and plant species specific and highly dependent on size of nanoparticles, mode of application and growth conditions. Metals and metalloid induce reactive oxygen species (ROS) mediated in phytotoxicity in plants. Metallic NPs in coexposure of metals and metalloids activate the cellular antioxidant enzymes and vital cellular processes to accumulate cellular antioxidants, osmoprotectants like proline which in turn decrement the phytotoxicity caused by ROS. Besides, accelerated formation of Casparian strips and suberin also reduce the accumulation of toxic metals through apoplastic pathway. These cellular features In future, metabolic engineering involving stress responsive genes transcript manipulation may provide new insights in the area of restoration of heavy metal/metalloid contaminated sights. These NP-induced cellular responses can give promising results, as far as the futuristic environmental biomonitoring studies are concerned.

The evidences documented in this chapter regarding reduction of metal//metalloid accumulation in crop plants might help in resolving the problem of excess metal/metalloids in crop plants leading improved food security with reduced risk to large population. There is need to conduct more planned robust studies involving various plants to understand the cellular mechanisms and processes going inside the plant cells to counteract phytotoxicity of metals/metalloids. We need to plan a few studies which authenticate the significant role of the NP induced cellular attributes in plants for biomonitoring studies.

## Acknowledgments

This review is the part of PhD dissertation of Banita Kumari Paswan. The financial support from UP. Higher Education in the form of Centre of Excellence: New emerging contaminant, metallic nanoparticles in agri environment: Toxicity, tolerance and metabolic adaptations in cereal crops (sankhya-360/sattar-4-2019-1035/2018) is gratefully acknowledged. Archana Dwivedi is highly indebted to Department of Science and Technology, New Delhi for providing financial assistance under WOS B scheme for women scientists (No. DST/WOS-B/2018/980(G).

## References

Abedin, M.J., Feldmann, J., Meharg, A.A., 2002. Uptake kinetics of arsenic species in rice plants. Plant Physiol. 128, 1120–1128. https://doi.org/10.1104/pp.010733.
Ahmed, T., Noman, M., Manzoor, N., Shahid, M., Abdullah, M., Ali, L., Wang, G., Hasheem, A., Abd_Allah, E.F., Li, B., Fahad Al-Arjani, Al–B, Alqarawi, A.A., 2021. Nanoparticle-based amelioration of drought stress and cadmium toxicity in rice via triggering the stress responsive genetic mechanisms. Ecotoxicol. Environ. Saf. 209, 111829. Article. https://doi.org/10.1016/j.ecoenv.2020.111829.
Ahmed, T., Noman, M., Manzoor, N., Shahid, M., Hussaini, K.M., Rizwan, M., Ali, S., Maqsood, A., Li, B., 2021. Green magnesium oxide nanoparticles-based modulation of cellular oxidative repair mechanisms to reduce arsenic uptake and translocation in rice (Oryza sativa L.) plants. Environ. Pollut. 288, 117785. doi:10.1016/j.envpol.2021.117785.
American Society for Testing and Materials (ASTM), 2012. Standard Terminology Relating to Nanotechnology. ASTM E2456-06. http://www.astm.org/Standards/E2456.htm.
Awasthi, S., Chauhan, R., Srivastava, S., Tripathi, R., 2017. The journey of arsenic from soil to grain in rice. Front. Plant Sci 8, 1007. https://doi.org/10.3389/fpls.2017.01007.
Barberon, M., 2017. The endodermis as a checkpoint for nutrients. New Phytol. 213, 1604–1610.
Bennie, ATP., 1991. Growth and mechanical impedance Editors. In: Waisel, Y, Eshel, A, Kafkafi, U (Eds.), Plant Roots: The Hidden Half. Marcel Dekker, New York, pp. 393–416.
Chakraborti, D., Sengupta, M.K., Rahman, M.M., Ahamad, S., Chowdhury, U.K., Hossain, M.D.A., Mukherjee, S.C., Pati, S., Saha, K.C., Dutta, R.N., Quamruzzaman, Q., 2004. Groundwater arsenic contamination and its health effects in the Ganga–Meghna–Brahmaputra plain. J. Environ. Monit. 6, 74N–83N.
Chauhan, V.S., Nickson, R.T., Chauhan, D., Iyengar, L., Sankararamakrishnan, N., 2009. Ground water geochemistry of Ballia district, Uttar Pradesh, India and mechanism of arsenic release. Chemosphere 75, 83–91. doi:10.1016/j.chemosphere.2008.11.065.
Chen, X., Wang, J., Hayat, K., Zhang, D., i Zhou, P., 2021. Small structures with big impact: Multi-walled carbon nanotubes enhanced remediation efficiency in hyperaccumulator *Solanum nigrum* L. under cadmium and arsenic stress. Chemosphere *276*, 130130. https://doi.org/10.1016/j.chemosphere.2021.130130.
de Freitas, PL., 1988. PhD dissertation. Department of Agronomy, Cornell University, Ithaca, N.Y.
Faizan, M., Bhat, J.A., Noureldeen, A., Ahmed, P., Yu, F., 2021. Zinc oxide nanoparticles and 24-epibrassinolide alleviates Cu toxicity in tomato by regulating ROS scavenging, stomatal movement and photosynthesis (2021). Ecotoxicol. Environ. Saf. 218, 112293. https://doi.org/10.1016/j.ecoenv.2021.112293.
Finnegan, P.M., Chen, W., 2012. Arsenic toxicity: the effects on plant metabolism. Front. Physiol. 3, 182. doi:10.3389/fphys.2012.00182.
Gil-Diaz, M., Diez-Pascual, S., Gonz´ alez, A., Alonso, J., Rodríguez-Vald´es, E., Gallego, J.R., Lobo, MC., 2016. A nanoremediation strategy for the recovery of an As-polluted soil. Chemosphere 149, 137–145. doi:10.1016/j.chemosphere.2016.01.106.
Gong, X., Huang, D., Liu, Y., Zeng, G., Wang, R., Wan, J., Zhang, C., Cheng, M., Qin, X., Xue, W., 2017. Stabilized nanoscale zerovalent iron mediated cadmium accumulation and oxidative damage of *Boehmeria nivea* (L.) Gaudich cultivated in cadmium contaminated sediments. Environ. Sci. Technol. 51, 11308–11316. doi:10.1021/acs.est.7b03164.
Goswami, L., Kim, K.-.H., Deep, A., Das, P., Bhattacharya, S.S., Kumar, S., Adelodun, AA., 2017. Engineered nanoparticles: nature, behaviour, and effect on the environment. J. Environ. Manage. 196, 297–315.

Hansel, C.M., Fendorf, S., 2001. Characterization of Fe plaque and associated metals on the roots of mine-waste impacted aquatic plants. Environ. Sci. Technol. 35, 3863–3868.

Hussain, B., Lin, Q., Hamid, Y., Sanaullah, M., Di, L., Hashmi, M.L.U.R., Khan, M.B., He, Z., Yang, X., 2020. Foliage application of selenium and silicon nanoparticles alleviates Cd and Pb toxicity in rice (Oryza sativa L.). Sci. Total Environ. 712, 136497. doi:10.1016/j.scitotenv.2020.136497.

Islam, M.R., Jahiruddin, M., Islam, S., 2004. Assessment of arsenic in the water–soil–plant systems in Gangetic flood plains of Bangladesh. Asian J. Plant Sci. 3, 489–493. doi:10.3923/ajps.2004.489.493.

Khanam, R., Hazra, G.C., Ghosh Bag, A., Kulsum, P.G.P.S., Chatterjee, N., Shukla, A.K., 2021. Risk assessment of arsenic toxicity through groundwater-soil-rice system in Maldah District, Bengal Delta Basin, India. Arch. Environ. Contam. Toxicol. 81, 438–448. doi:10.1007/s00244-021-00883-7.

Li, Y., Zhu, N., Liang, X., Bai, X., Zheng, L., Zhao, J., Li, Y.F., Zhang, Z., Gao, Y., 2020. Silica nanoparticles alleviate mercury toxicity *via* immobilization and inactivation of Hg(ii) in soybean (*Glycine max*). Environ. Sci. Nano 7, 1807–1817. doi:10.1039/d0en00091d.

Li, Z., Zhang, D., Xiong, X., Yan, B., Xie, W., Sheen, J., Li, J.F., 2017. A potent Cas9-derived gene activator for plant and mammalian cells. Nat. Plants 3, 930–936.

Liang, Y., Zhu, Y.G., Xia, Y., Li, Z., Ma, Y., 2006. Iron plaque enhances phosphorus uptake by rice (*Oryza sativa*) growing under varying phosphorus and iron concentrations. Ann. Appl. Biol. 149, 305–312.

Liu, C., Wei, L., Zhang, S., Xu, X., Li, F., 2014. Effects of nanoscale silica sol foliar application on arsenic uptake, distribution and oxidative damage defense in rice (*Oryza sativa* L.) under arsenic stress. RSC Adv. 4, 57227–57234. https://doi.org/10.1039/C4RA08496A.

Liu, Y., Persson, D.P., Li, J., Liang, Y., Li, T., 2021. Exposure of cerium oxide nanoparticles to the hyperaccumulator *Sedum alfredii* decreases the uptake of cadmium via the apoplastic pathway. J. Hazard. Mater. 417, 125955. https://doi.org/10.1016/j.jhazmat.2021.125955.

Lux, A., Martinka, M., Vaculik, M., White, P.J., 2011. Root responses to cadmium in the rhizosphere: a review. J. Exp. Bot. 62, 21–37.

Majumdar, S., Trujillo-Reyes, J., Hernandez-Viezcas, J.A., White, J.C., Peralta-Videa, J.R., Gardea-Torresdey, JL., 2016. Cerium biomagnification in a terrestrial food chain: influence of particle size and growth stage. Environ. Sci. Technol. 50, 6782–6792.

Manesh, R.R., Grassi, G., Bergami, E., Marques-Santos, L.F., Faleri, G., Liberatri, IC., 2018. Co-exposure to titanium dioxide nanoparticles does not affect cadmium toxicity in radish seeds (Raphanus sativus). Ecotoxicol. Environ. Saf. 148, 359–366. https://doi.org/10.1016/j.ecoenv.2017.10.051.

Mansouri, T., Golchin, A., Neyestani, MR., 2017. The effects of hematite nanoparticleson phytoavailability of arsenic and corn growth in contaminated soils. Int. J. Environ. Sci. Technol. 14, 1525–1534 2017.

Manzoor, N., Ahmed, T., Noman, M., Shahid, M., Nazir, M.M., Ali, L., Alnusaire, T.S., Li, B., Schulin, R., Wang, G., 2021. Iron oxide nanoparticles ameliorated the cadmium and salinity stresses in wheat plants, facilitating photosynthetic pigments and restricting cadmium uptake. Sci. Total Environ. 769, 145221. https://doi.org/10.1016/j.scitotenv.2021.145221.

Mawia, A.M., Hui, S., Zhou, L., Li, H., Tabassum, J., Lai, C., Wang, J., Shao, G., Wei, X., Tang, S., Luo, J., Hu, S., Hu, P., 2021. Inorganic arsenic toxicity rice and alleviation strategies in rice. J. Hazard. Mater. 408, 124751. https://doi.org/10.1016/j.jhazmat.2020.124751.

Meharg, A.A., Hartley-Whitaker, J., 2002. Arsenic uptake and metabolism in arsenic resistant and nonresistant plant species. New Phytol. 154, 29–43. https://doi.org/10.1046/j.1469-8137.2002.00363.x.

Meharg, A.A., Jardine, L., 2003. Arsenite transport into paddy rice (*Oryza sativa*) roots. New Phytol. 157, 39–44. doi:10.1046/j.1469-8137.2003.00655.x.

Nisha Shabnam, N., Kim, M., Kim, H., 2019. Iron (III) oxide nanoparticles alleviate arsenic induced stunting in *Vigna radiate*. Ecotoxicol. Environ. Saf. 183, 109496. https://doi.org/10.1016/j.ecoenv.2019.109496.

Paez-Garcia, A., Motes, C.M., Scheible, W.R., Chen, R.J., Blancaflor, E.B., Monteros, M.J., 2015. Root traits and phenotyping strategies for plant improvement. Plants 4, 334–355.

Pan, W., Wu, C., Xue, S., Hartley, W., 2014. Arsenic dynamics in the rhizosphere and its sequestration on rice roots as affected by root oxidation. J. Environ. Sci. 26, 892–899.

Pardo, T., Martínez-Fernández, D., de la Fuente, C., Clemente, R., Komárek, M., Bernal, M.P., 2016. Maghemite nanoparticles and ferrous sulfate for the stimulation of iron plaque formation and arsenic immobilization in Phragmites australis. Environ. Pollut. 219, 296–304. doi:10.1016/j.envpol.2016.10.014.

Patel, K.S., Shrivas, K., Brandt, R., et al., 2005. Arsenic contamination in water, soil, sediment and rice of central India. Environ. Geochem. Health 27, 131–145. https://doi.org/10.1007/s10653-005-0120-9.

Peralta-Videa, J.R., Zhao, L., Lopez-Moreno, M.L., de La Rosa, G., Hong, J., Gardea Torresday, J.L., 2011. Nanomaterials and the environment: a review for the biennium 2008–2010. J. Hazard. Mater. 186, 1–15. https://doi.org/10.1016/j.jhazmat.2010.11.020.

Praveen, A., Khan, E., Perwez, M., Sardar, M., Gupta, M., 2017. Iron oxide nanoparticles as nano-adsorbents: a possible way to reduce arsenic phytotoxicity in indian mustard plant (*Brassica juncea* L.). J. Plant Growth Regul. 37, 612–624. https://doi.org/10.1007/s00344-017-9760-0.

Rico, C.M., Peralta-Videa, J.R., Gardea Torresday, J.L., 2015. Chemistry, biochemistry of nanoparticles, and their role in antioxidant defense system in plants Editors. In: Siddiqui, MH, Al-Whaibi, M H, Mohammad, F (Eds.), Nanotechnology and Plant Sciences. Springer International Publishing, Switzerland. doi:10.1007/978-3-319-14502-0_1.

Rizwan, M., Ali, S., Rehman, M.Z., Adrees, M., Arshad, M., Qayyum, M.F., Ali, L., Hussain, A., Chatha, S.A.S., Imran, M., 2019. Alleviation of cadmium accumulation in maize (Zea mays L.) by foliar spray of zinc oxide nanoparticles and biochar to contaminated soil. Environ. Pollut. 248, 358–367.

Rizwan, M., Ali, S., Zia ur Rehman, M., Riaz, M., Adrees, M., Hussain, A., Zahir, Z.A., Rinklebe, A., 2021. Effects of nanoparticles on trace element uptake and toxicity in plants: A review. Ecotoxicol. Environ. Saf. *221*, 112437. https://doi.org/10.1016/j.ecoenv.2021.112437.

Rossi, L., Zhang, W., Schwab, A.P., Ma, X., 2017. Uptake, accumulation, and in planta distribution of coexisting cerium oxide nanoparticles and cadmium in *Glycine max* (L.) Merr. Environ. Sci. Technol. 51, 12815–12824. doi:10.1021/acs.est.7b03363.

Schwabe, F., Tanner, S., Schulin, R., Rotzetter, A., Stark, W., von Quadt, A., Nowack, B., 2015. Dissolved cerium contributes to uptake of Ce in the presence of differently sized CeO2-nanoparticles by three crop plants. Metallomics Integr. Biometal Sci. 7, 466–477.

Singh, J., Lee, B.K., 2016. Influence of nano-TiO2 particles on the bioaccumulation of Cd in soybean plants (*Glycine max*): a possible mechanism for the removal of Cd from the contaminated soil. J. Environ. Manage. 170, 88–96. doi:10.1016/j.jenvman.2016.01.015.

Smith, A.H., Lopipero, P.A., Bates, M.N., Steinmaus, CM., 2002. Public health. arsenic epidemiology and drinking water standards. Science 296, 2145–2146. doi:10.1126/science.1072896.

Tao, Q., Jupa, R., Luo, J., Lux, A., Kovac, J., Wen, Y., Li, T., 2016. The apoplasmic pathway via the root apex and lateral roots contributes to Cd hyperaccumulation in the hyperaccumulator *Sedum alfredii*. J. Exp. Bot. 68, 739–751.

Tripathi, D.K., Singh, S., Singh, V.P., Prasad, S.M., Chauhan, D.K., Dubey, NK., 2016. Silicon nanoparticles more efficiently alleviate arsenate toxicity than silicon in maize cultivar and hybrid differing in arsenate tolerance. Front Environ Sci 4, 1–14. https://doi.org/10.3389/fenvs.2016.00046.

Tripathi, R.D., Srivastava, S., Mishra, S., Singh, N., Tuli, R., Gupta, D.K., Maathuis, FJM., 2007. Arsenic hazards: strategies for tolerance and remediation by plants. Trends Biotechnol. 25, 158–165. https://doi.org/10.1016/j.tibtech.2007.02.003.

Veen, B.W., 1982. The influence of mechanical impedance on the growth of maize roots. Source. Plant Soil 66, 101–109.

Visser, E.J.W., Colmer, T.D., Blom, C.W.P.M., Voesenek, LACJ., 2000. Changes in growth, porosity, and radial oxygen loss from adventitious roots of selected mono- and icotyledonous wetland species with contrasting types of aerenchyma. Plant Cell Environ. 23, 1237–1245. https://doi.org/10.1046/j.1365-3040.2000.00628.x.

Wang, S.H., Kurepa, J., Smalle, JA., 2011. Ultra-small $TiO_2$ nanoparticles disrupt microtubular networks in Arabidopsis thaliana. Plant Cell Environ. 34, 811–820.

Wang, X., Sun, W., Ma, X., 2019. Differential impacts of copper oxide nanoparticles and Copper(II) ions on the uptake and accumulation of arsenic in rice (Oryza sativa). Environ. Pollut. 252, 967–973. https://doi.org/10.1016/j.envpol.2019.06.052.

Wang, X., Zhang, S., Sharifan, H., Ma, X., 2018. Elucidating the effects of cerium oxide nanoparticles and zinc oxide nanoparticles on arsenic uptake and speciation in rice (Oryza sativa) in a hydroponic system. Environ. Sci. Technol. 52, 10040–10047. doi:10.1021/acs.est.8b01664.

Wu, X., Hu, J., Wu, F., Zhang, X., Wang, B., Yang, Y., Shen, G., Liu, J., Tao, S., Wang, X., 2021. Application of $TiO_2$ nanoparticles to reduce bioaccumulation of arsenic in rice seedlings (*Oryza sativa* L.): a mechanistic study. J. Hazard. Mater. 405, 124047. https://doi.org/10.1016/j.jhazmat.2020.124047.

You, Y., Liu, L., Wang, Y., Li, J., Ying, Z., Hou, Z., Liu, H., Du, S., 2021. Graphene oxide decreases Cd concentration in rice seedlings but intensifies growth restriction. J. Hazard. Mater. *417*, 125958. https://doi.org/10.1016/j.jhazmat.2021.125958.

Yue, L., Ma, C., Zhan, X., White, J.C., Xing, B., 2017. Molecular mechanisms of maize seedling response to La2O3 NP exposure: water uptake, aquaporin gene expression and signal transduction. Environ. Sci. Nano 4, 843–855. https://doi.org/10.1039/C6EN00487C.

Zeeshan, M., Hu, Y.X., Iqbal, A., Salam, A., Liu, Y.X., Muhammad, I., Ahmed, S., Khan, A.H., Hale, B., Wu, H.Y., Zhou, X.B., 2021. Amelioration of AsV toxicity by concurrent application of ZnO-NPs and Se-NPs is associated with differential regulation of photosynthetic indexes, antioxidant pool and osmolytes content in soybean seedling. Ecotoxicol. Environ. Saf. 225, 112738. 738. https://doi.org/10.1016/j.ecoenv.2021.112738.

Zhang, J.Y., Zhou, H., Zeng, P., Wang, S.L., Yang, W.J., Huang, F., Liao, B.H., 2021. Nano-Fe3O4-modified biochar promotes the formation of iron plaque and cadmium immobilization in rice root. Chemosphere 276, 130212. doi:10.1016/j. chemosphere.2021.130212.

Zhou, P., Adeel, M., Shakoor, N., Guo, M., Hao, Y., Azeem, I., Li, M., Liu, M., Rui, Y., 2021. Application of nanoparticles alleviates heavy metals stress and promotes plant growth: an overview. Nanomaterials 11, 26. https://doi.org/10.3390/nano11010026.

CHAPTER 13

# Chromosomal and molecular indicators: A new insight in biomonitoring programs

**Priyanka Singh[a], Supriya Tiwari[b], Shashi Bhushan Agrawal[a]**
[a]Laboratory of Air Pollution and Climate Change, Department of Botany, Banaras Hindu University, Varanasi, Uttar Pradesh, India
[b]Laboratory of Ecotoxicology, Department of Botany, Banaras Hindu University, Varanasi, Uttar Pradesh, India

## 13.1 Introduction

Air pollution is a cocktail of various pollutants like ozone ($O_3$), sulfur dioxide ($SO_2$), carbon monoxide (CO), oxides of nitrogen ($NO_x$), volatile organic compounds (VOCs), dioxins, polycyclic aromatic hydrocarbons (PAHs), particulate matter (PM), and heavy metals, which directly or indirectly affects all kinds of life forms (Manisalidis et al., 2020; Bignal et al., 2007). Emission of air pollutants occurs not only by anthropogenic activities but by natural sources also, although anthropogenic sources like automobile exhaust and industrial activities are the main contributing factors, whose proportion has increased by several folds during the last century. The hazardous effects of air pollution in humans are well known as several epidemiological studies suggest that it may also be responsible for the increased rate of lung cancer and cardiovascular disorders (Hazari et al., 2018; Loomis et al., 2013).

In plants, air pollution has emerged as an important factor causing significant damage to their metabolic machinery, cumulating in a remarkable reduction in yield (Miao et al., 2017). Several studies indicate that plants have a high sensibility toward various air pollutants like $SO_2$, $NO_X$, ozone ($O_3$), formaldehyde, and other complex mixture of pollutants like cigarette smoke or fuel combustion (Gomez-Arroyo et al., 2018). Many of the air pollutants can act as potential mutagens and a few studies have also suggested that plant chromosomes are a sensitive indicator of air pollutants (Li et al., 2016; Agrawal and Chaudhary, 1992; Agrawal and Chaudhary, 1989; Ma, 1982). DNA is the primary target of most mutagens, causing

*New Paradigms in Environmental Biomonitoring Using Plants*
DOI: https://doi.org/10.1016/B978-0-12-824351-0.00010-9

chromosomal aberrations and genotoxicity (Agrawal et al., 1992). The damage in cell repair machinery causes mispairing which may lead to chromosomal aberrations and can also result in disturbances in the synthesis of DNA during the S-phase of the cell cycle, leading to aberrations. The common air pollution-induced DNA damages observed in plants are DNA fragmentation, strand- breaks, base loss (apurinic and apyrimidinic), transition mutation, and strand breaks with terminal sugar residues (Wagnor et al., 2021; Roldán-Arjona and Ariza, 2009; Wink et al., 1991). Likewise, the air pollutant-induced common cytological disorders are chromosomal stickiness, anaphase bridges, chromosomal rupture, ring formation, and laggards (Oladele et al., 2013). These cytological and molecular abnormalities induced by air pollution can act as efficient biomarkers and can be utilized in biomonitoring programs. The cytological and molecular indicators are reliable and convenient tools to detect the injuries induced by air pollutants.

## 13.2 Major air pollutants and their source of emission

Common ambient air pollutants are ozone ($O_3$), sulfur dioxide ($SO_2$), carbon monoxide (CO), nitrogen oxides (NOx), particulate matter, and lead. These air pollutants are generally classified as 'criteria' pollutants by the Environmental Protection Agency [US EPA, 2014]. Along with these criteria pollutants, other pollutants like polycyclic aromatic hydrocarbon (PAHs), volatile organic compounds (VOCs), and dioxins are of concern because of their carcinogenic and mutagenic effect on living beings (Fig. 13.1). Ozone, which is present in two layers of the atmosphere, one is at a height of 0-10 km in the troposphere (bad ozone) which acts as a common secondary air pollutant while the $O_3$ is present in the stratosphere at the height of 10-40 km, acts as a protective layer (good $O_3$) against sun's harmful radiation. The tropospheric $O_3$ (bad $O_3$) is a secondary pollutant, it is not directly emitted from any anthropogenic or natural sources, it is formed due to the reaction between $NO_2$ and VOCs in the presence of sunlight. $NO_x$ is commonly emitted from anthropogenic sources mainly vehicular pollution while sources of VOCs emission are both natural as well as anthropogenic.

Sulfur dioxide ($SO_2$) is a highly reactive colorless gas, about 95% of which in the atmosphere arises from anthropogenic sources, of which combustion of fossil fuel is responsible for nearly 80% while another minor contribution is from natural sources in the form of a volcanic eruption. Carbon monoxide (CO) is a colorless, odorless, and tasteless gas, it is toxic in nature and present in trace amounts in the atmosphere. The common

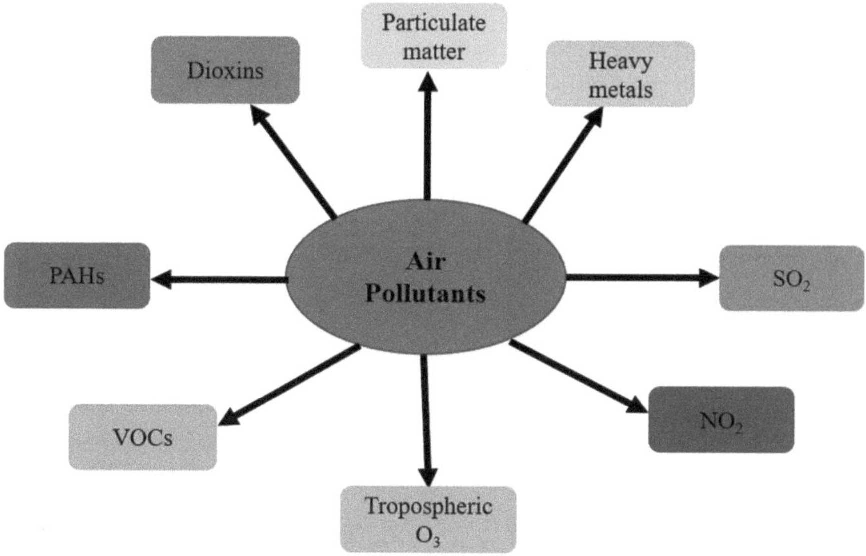


**Fig. 13.1** ***Different kinds of air pollutants present in the atmosphere.***

sources of CO emission are vehicular pollution which contributes up to 95% while the rest is contributed through the combustion of fossil fuels, biomass burning, and oxidation of hydrocarbons like methane and Isoprene in a limited supply of oxygen (US EPA, 2016b, 2016c; Seinfeld and Pandis, 2006). Nitrogen dioxide ($NO_2$) is one of the main constituents of vehicular air pollution and also acts as a precursor in tropospheric $O_3$ formation. The gas is reddish-brown and highly reactive in the ambient atmosphere. Its emission is either direct from primary sources like automobile engines, biomass burning, and coal-based power plants, or indirect through the chemical reactions occurring between secondary pollutants. The increase in the concentration of polycyclic aromatic hydrocarbons (PAHs) in ambient air raises the concern due to their mutagenicity and carcinogenicity. PAHs are a large group of organic compounds composed of two or more fused benzene rings arranged in a different configuration. They are generally emitted into the atmosphere due to incomplete combustion of fossil fuels (from automobiles, coking plants, asphalt production, and manufacturing facilities) (Kim et al., 2013). Particulate matter (PM) is a complex of solid and liquid matter suspended in the air. In an urban area, it is mainly composed of organic compounds, elemental carbon, and metals. Particulate matters are commonly formed mainly due to anthropogenic sources like the burning of gasoline and diesel, industrial processes, coal-based power plants, and road construction. These are commonly classified into three categories based on their aerodynamic particle diameter: ultrafine, fine, coarse, and super-coarse ranging in size from ≤0.1 μm to >10 μm. Particles that have a size of ≤10 μm are known as $PM_{10}$ while particles with a size ≤2.5 μm are known as $PM_{2.5}$. $PM_{2.5}$ due to its very small size can cross the respiratory and circulatory system (Ghorani-Azam et al., 2016; US EPA, 2016a; Laumbach, and Kipen, 2012). The emission of VOCs is one of the major contributors to air pollution. The most commonly found VOCs are alcohols, ketones, halogenated compounds, aromatic compounds, aldehydes, and ethers (Kamal et al., 2016).and their common sources of emission are both natural sources like emission from forest trees and wildfire and anthropogenic sources like residential combustion, paints or solvent usage, vehicular and industrial evaporation (Bari and Kindzierski, 2018). Pollutants like Polychlorinated dibenzodioxins and dibenzofuran which are commonly known as 'dioxins' are toxic chemical pollutants. And their sources of emission are the combustion of chlorinated products in industrial processes, like metallurgy, waste incineration, etc., and also by accidental forest events like a forest fire. Lead (Pb) is toxic heavy air-borne metal, whose source of emission varies from area to area. Major sources of lead emission in the air are ore and metals processing and aircraft operating on aviation fuel loaded with lead. Other sources are waste incinerators and lead-acid battery manufacturers. Due to the continuous efforts of EPA's regulatory actions including the removal of lead from the motor vehicle gasoline, lead concentration in the air decreased by 98 percent between 1980 and 2014 (US EPA, 2016b).

## 13.3 Air pollutants as a mutagen

Air pollutants like CO, NOx, $SO_2$, PM 2.5, PM 10, $O_3$, PAHs, and VOCs, affect plant species both directly as well as indirectly by interacting with other environmental stresses like water stress, nutrient scarcity, temperature stress, and pathogen attack. The plant leaves which are the main target of any kind of environmental stress are frequently used for biomonitoring purposes because of their respiration and exposure pathway for air pollution. Thus, the alteration in the physical appearance, chemical composition as well as genetic material of leaf can be observed in correlation with air pollution. There are various anomalies like an increase in micronucleus frequency, formation of chromatid bridges, fragments, an increased number of chromosomal aberrations were observed in plants that are exposed to the high concentration of air pollutants (Tuteja et al., 2009).

## 13.4 Mechanism of ROS generation in plants

Air pollutants affect the plant by entering inside the plant via leaf and induce the generation of reactive oxygen species (ROS), which results in the induction of oxidative processes that catalyzes the generation of ROS such as superoxide ($O_2^{\cdot -}$), hydroxyl radical ($^{\cdot}OH$), singlet oxygen ($^1O_2$) and hydrogen peroxide ($H_2O_2$) (Xie et al., 2019). Similarly, when pollutant like $O_3$ enters in the plant leaf, gets dissolved in the aqueous medium of the cell and reacts with cellular components. The common ROS generated after the breakdown of $O_3$ in pure water are superoxide ($O_2^{\cdot -}$), Peroxy ($HO^{\cdot}_2$), and hydroxyl radicals ($^{\cdot}OH$). $SO_2$ phytotoxicity mainly results in the production of intracellular $O_2^{\cdot -}$ and its detoxification mainly occurs by oxidative conversion of sulfite ion ($SO_3^{2-}$) and hydrogen sulfite ion ($HSO_3^-$) into non-harmful sulfate ($SO_4^{2-}$). While the oxides of nitrogen (NOx) also cause oxidative stress, where NO can react rapidly with $O_2^{\cdot -}$ to form $ONOO^-$ which may transform to $^{\cdot}OH$, the most reactive and toxic ROS. During the process of nitrate assimilation, more $^{\cdot}OH$ can be generated and cause free radical-induced injuries (Liu et al., 2016). Other pollutants like lead and $PM_{2.5}$ can produce free radicals like $OH^{\cdot}$ by Fenton's reaction (Lodovici and Bigagli, 2011)

$$O_3 + OH^{\cdot} \rightarrow O_2^{\cdot -}$$

$$O_2^{-} + H^+ \rightarrow HO_2$$

$$HO_2 + HO_2^{\cdot} \rightarrow H_2O_2 + O_2$$

$$O_2^{-} + HO_2^{\cdot} + H_2O \rightarrow H_2O_2 + OH^{-} + O_2$$

$$M^{2+} + H_2O_2 + H^+ \rightarrow M^{3+} + OH^{\cdot} + H_2O\left(M = Pb, Fe, etc\right)$$

## 13.5 Mutagenic impacts of air pollutants

The mutagenic effects of air pollutants can be studied under two main heads; at the molecular level and at the chromosomal level. Fig. 13.2 summarizes the different levels and ways through which air pollutants can bring about mutagenic changes in the plants.

### 13.5.1 Impacts at the molecular level

As discussed above, most of the air pollutants result in oxidative stress and act by enhancing the production of ROS (Galveias et al., 2021; Mukherjee

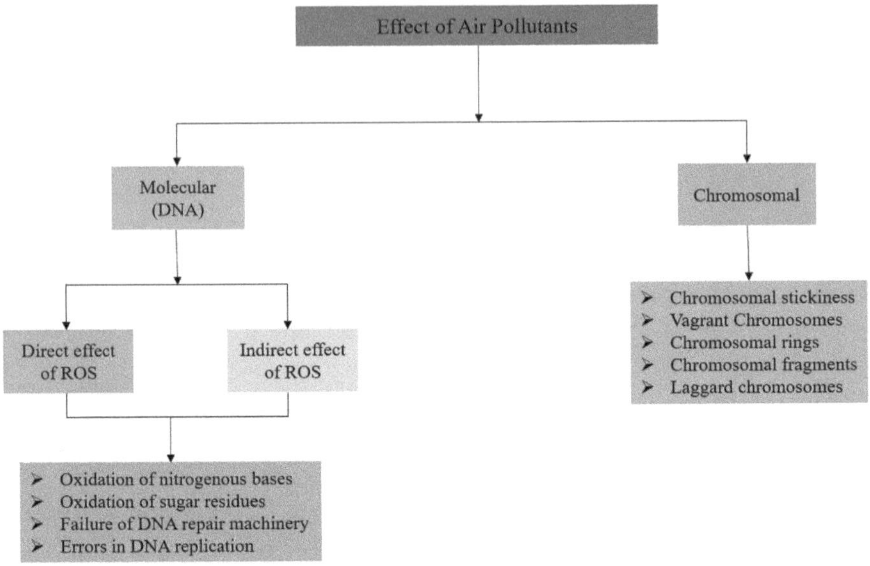


**Fig. 13.2** ***Effect of air pollutants on molecular and chromosomal aspects.***

et al., 2020). The ROS generated can readily oxidize most of the macromolecules like lipids, nucleic acids, and proteins (Juan et al., 2021). Apart from this, ROS also targets the cellular DNA with electron-rich bases, which undergoes oxidation to generate a range of modifications that are genotoxic (Foyer et al., 1994). The sessile nature of plants makes them more prone to environmental genotoxic agents like radiations and air pollutants. In addition, they do not possess a reticent germline and produce meiotic cells later in their development stage, so mutations that arise in somatic cells may be exhibited in gametes (Del Río et al., 2006). In plants, ROS is continuously generated as a byproduct of several metabolic reactions occurring in cellular organelles like chloroplasts, mitochondria, and peroxisomes (Sharma and Dubey, 2005). Active rate of ROS generation also occurs by oxidases and peroxidases in response to various environmental stresses such as pathogen attack and other abiotic stimuli, and they play various important roles in regulating the various development processes (Mishra et al., 2011; Tanou et al., 2009). The intrinsic ROS along with the extrinsically produced ROS induces an oxidative attack on DNA, resulting in altered bases like the apurinic and apyrimidinic sites and wrecked sugar residues that undergo fragmentation and ultimately leads to strand breakage (Maynard et al., 2009) (Fig. 13.3).

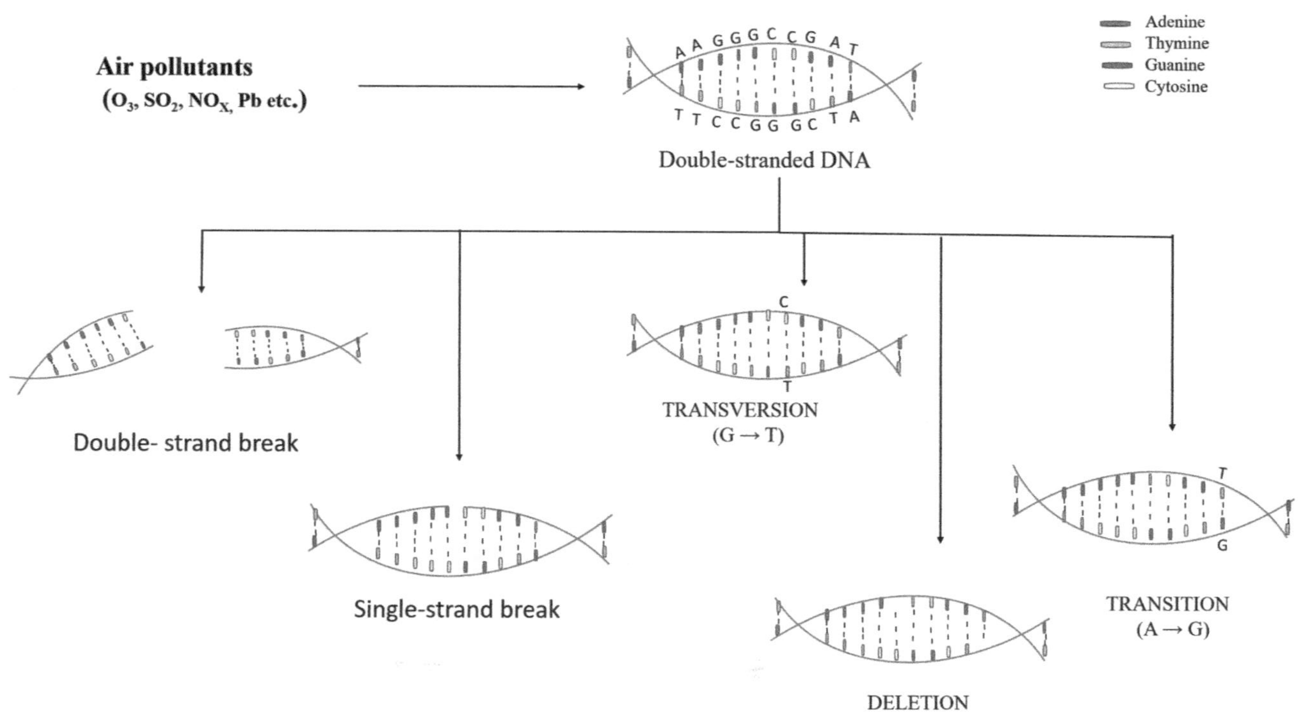


**Fig. 13.3** ***Air pollutants affect the DNA of plants and animals and induce various kind of DNA damages like double & single-stranded DNA breaks, transversion & transition mutations and deletion of base pairs.***

The oxidation of DNA bases generally entails the addition of hydroxyl (OH) group to the double bonds, while damage to sugar residues mainly results from hydrogen separation from deoxyribose (Chatterjee and Walker, 2017). Hydroxyl radical initiates damage to pyrimidine bases by attacking the double bond at C5-C6 position, via diffusion-controlled rates to generate 5-hydroxyl-6-yl and 6-hydroxy-5-yl radicals that react to form various stable DNA lesions (Bokhari and Sharma, 2019). The major products of pyrimidine oxidation are ring-saturated derivatives, particularly cytosine glycol, thymine glycol, and 5,6-dihydro-thymine (Cadet and Wagner, 2013; Gros et al., 2002). Cytosine products may get converted to uracil derivatives; thus, cytosine glycol gets converted to uracil glycol by deamination and 5-hydroxyuracil by deamination and dehydration (Tremblay and Wagner, 2008; Asada, 1987). Along with saturation or oxidation, the pyrimidine ring can also go through fragmentation to produce 5-hydroxy-hydantoin, methyltartronylurea and N-formamido-urea or urea (Apel and Hirt, 2004). $OH^{\cdot}$ radicals also react with purines by incorporating at C4-, C5-, and mkC8- positions (Apel and Hirt, 2004). Rozelle et al., (2021) have reported that the oxidative damage to DNA results in the production of 7,8-dihydro-8-oxoguanine (oxoG) and 7,8-dihydro-8-oxoadenine (oxoA) as two major lesions. Under oxidative conditions, oxoA, but not oxoG, readily reacts with an opposite base to produce interstrand cross-links (ICLs), highlighting a latent alkylating nature of oxoA (Rozelle et al., 2021). Nucleobase analog studies suggest C2-oxoA is covalently linked to N2-guanine and N3-adenine for the oxoA-G and oxoA-A ICLs, respectively (Rozelle et al., 2021). Among all the purine oxidation products, probably the best-studied one is 7-hydro-8-oxoguanine (8-oxoG), which results from the oxidation of the C8-OH adduct radical (Hackbarth et al., 2010; Hatz et al., 2007 and Krieger-Liszkay 2005). Another major guanine-derived abrasion is the imidazole open ring derivative 2,6-diamino-4-hydroxy-5-formamidopyrimidine (FaPyGua). Oxidized forms of adenines, which comprise 8-oxoadenine and 4,6-diamino-5-formamidopyramidine (FaPyAde) are less investigated byproducts of oxidative DNA damage (Kasai, 1997).

In addition to direct oxidation, DNA bases may also indirectly alter through reaction with active products generated by ROS attack on different macromolecules (Juan et al., 2021). For instance, the reactive products of lipid peroxidation, malondialdehyde reacts with the G residues of DNA to form the pyrimidopurinone adduct commonly called as M1G (Sachdev et al., 2021). Other lipid oxidation products acrolein and crotonaldehyde lead to the production of etheno modifications of DNA bases, like etheno-A and etheno-C (Nilsson and Liu, 2020; Halliwell and Gutteridge, 1984). Apart from the nitrogen bases, DNA sugars are also highly sensitive to ROS-mediated oxidative attack, resulting in single-strand breaks (SSB) (Fig. 13.3). This occurs frequently via the removal of the hydrogen atom from the C4' position of deoxyribose that further reacts to generate DNA strand breakage (Halliwell, 1977). DNA damage is also induced by endogenous ROS, the major type of DNA fragmentation observed is of single-strand type. However, double-strand DNA breaks (DSB) can also be generated by ionizing radiation mediated oxidation, which occurs through the production of radical clusters that affect sites that are in close proximity on both strands (Meriga et al., 2004). Table 13.1 summarizes the different ways through which air pollutants affect the DNA of the plants.

**Table 13.1** Air pollutants induced damage at the DNA level in various plants.

| Air pollutant | Dose | Species/cultivar/ variety | Type of DNA damage | References |
|---|---|---|---|---|
| Ozone ($O_3$) | 120 ppb | *Triticum aestivum* L. cv. Giza 168 | Transposition or deletion | Abdelhaliem & A. Al-Huqail, 2016 |
| Ozone ($O_3$) | 10 & 20 ppb | *Oryza sativa* L. | DNA strand breaks and altered bases | Sarkar et al., 2015 |
| Ozone ($O_3$) | 10 ppb | *Linum usitatissimum* L. cv. Padmini | Single strand breaks, mispairing of bases | Tripathi et al., 2011 |
| Ozone ($O_3$) | 1.5 times of ambient $O_3$ | *Populous tremuloides* Michx. | DNA lesions | Tai et al., 2010 |
| Lead ($Pb^{2+}$) | 40–240 mg $l^{-1}$ | *Lycopersicum esculentum* L. Falcon type | Pyrimidine dimers, 6-4 photoproducts | Aydin et al., 2015 |
| Lead ($Pb^{2+}$) | 10 μM | *Vicia faba L.*CV. "aguadulce" | DNA strand breaks | Pourrut et al., 2011 |
| Lead ($Pb^{2+}$) | 10–100 μM | *Triticum aestivum* L. cv. "Maruca" | DNA damage | Truta et al., 2011 |
| Mixed air pollutants<br>Mixed air pollutants | Ambient<br>Ambient | *Pseudovernia furfuracea* L.<br>*Euonymus japonicus* | Adducts, mutations, rearrangements, structural changes, large deletions<br>DNA-single strand break | Vardar et al., 2014; Canasaran-Duman et al., 2015<br>Li et al., 2016 |

The oxidized bases generated in DNA may have severe consequences on the affected cells. The failure of the DNA repair machinery leads to unrepaired bases which are encountered by the DNA replication fork leading to the blockage of the DNA polymerase (Cortez, 2019; Foyer et al., 1997). At times, the lesion can be side-stepped by the special type of DNA polymerases, which is potentially mutagenic because of its mispairing ability (Foyer et al., 1997). Oxidative lesions have various kinds of genotoxic effects that depend on the level of structural alteration endured by the original base. Extreme products of pyrimidine oxidation, like urea, confer strong blocks for DNA replication (Imlay, 2003). Some of the oxidized pyrimidines, that maintains an intact ring, such as dihydrothymine, generally do not block the synthesis of DNA and gets freely bypassed by DNA polymerases, whereas, thymine glycol actively blocks the main replicative DNA polymerases in vitro, but is usually bypassed by specialized DNA polymerases (Yoon et al., 2010; Takata et al., 2006). In purines, the chief oxidation product, 8-oxoG is a strong mutagenic abrasion that can pair with cytosine or adenine in vitro conditions (Valavanidis et al., 2009). In covenant with in vitro observation, the major type of mutations induced in plasmids in vivo modified with 8-oxodG is the main target for GC-TA transversion in both the prokaryotic and eukaryotic systems (Valavanidis et al., 2009). On the other hand, the FaPyGua, blocks DNA replication and is fetal but not mutagenic (Grąziewicz et al., 2000; Foyer et al., 1997). Fig. 13.3 depicts the different ways DNA can be disrupted under air pollution stress.

### 13.5.2 Impacts at the chromosomal level

As discussed above how the air pollutants affect the DNA, if these damaged DNA remains unrepaired, they may further get processed into chromosome structural changes such as chromosomal aberrations and sister chromatid exchanges (SCEs) (Fig. 13.4). Among the common DNA damages like SCEs and double-strand breaks (DSBs), Most of the SCEs were repaired by different repair mechanisms (Schuber et al., 2004). So, the DSBs remains are of major concern because of their critical role in the generation of structural changes in chromosomes by incorrect reciprocal recombination repair (Plante et al., 2019). Usually, two DSBs interact in cis or trans to form a chromosomal aberration. The commonly reported chromosomal aberrations in plants are isochromatid breaks, reciprocal translocation, duplication, deletion, open chromatid breaks, and intercalary deletions (Endo et al., 2008) (Fig. 13.4) while in the case of mammals open

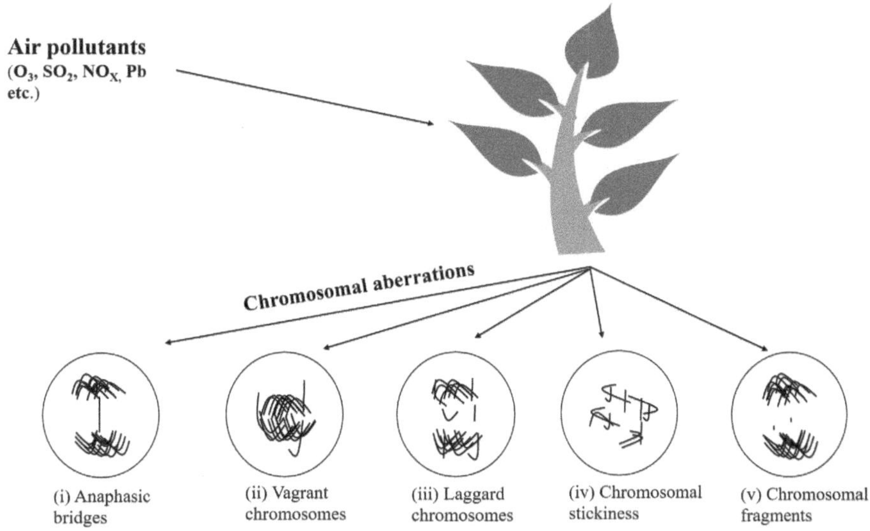


**Fig. 13.4** ***Air pollutants induced various kinds of chromosomal aberrations: (i) bridges at anaphase, (ii) vagrant chromosomes, (iii) laggard chromosomes, (iv) clumping or stickiness of chromosome, and (v) fragmented chromosomes.***

chromatid and isochromatid breaks are much more frequent than deletion and duplications and intercalary deletions (Schuber et al., 2004). Different kinds of chromosomal aberrations caused by air pollutants are enlisted in Table 13.2.

## 13.6 Indicators of air pollutants

The extent and magnitude of air pollution damage to plants and animals can be evaluated by studying the response of specific cellular parameters, which can act as indicators of air pollution. The indicators can be categorized as morphological, physiological, biochemical, cytological, or molecular indicators. Air pollutants primarily impair the translocation mechanisms or favors the production of genotoxic compounds rather than producing general oxidative stress on leaf tissue (Al Khateeb, 2018). Thus, observing the chromosomal aberrations in meristems appears to be a more sensitive indicator for air pollution impact on plants than the analysis of pigments and antioxidants in leaves. So, in the coming sections of this chapter, we will be discussing how the cytological and molecular changes can act as indicators for air pollutants.

**Table 13.2** Air pollutant-induced chromosomal aberrations in various plants.

| Air pollutant | Dose | Species/cultivar/variety | Type of chromosomal aberration | References |
|---|---|---|---|---|
| $O_3$ | 60 PPb | *Tradescantia pallida* (clone 4430) | increased micronuclei frequency | De Souza lima et al., 2009 |
| $O_3$ | 25–50 ppb | *Picea abies* (L.) Karst | Vagrant chromosomes, fragments, & chromosome stickiness | Wonish et al., 2003 |
| $O_3$ | 20–100 ppb | *Picea abies* (L.) Karst | Chromosome stickiness, breaks, & fragments | Muller et al., 1996 |
| $O_3$ | 50–100 ppb | *Tradescantia* (clone 4430) | increased micronuclei frequency | Rodrigues et al., 1996 |
| $O_3$ | 100 ppb | *Capsicum annuum* | Chromosomal bridges, laggard and micronuclei | Agrawal & Chaudhary, 1992 |
| $O_3$ | 200 ppb | Vicia faba L. | Chromosomal bridges, fragments, lagging, micronuclei & sticky chromosomes | Janakiraman & Harney, 1976 |
| $SO_2$ | 0.2 ppm | *Solanum nigrum* | Less chiasma frequency, bridges & laggards | Agrawal et al., 1995 |
| $SO_2$ | 5–150 ppm | *Vicia faba* | Chromosome stickiness, lagging, sticky bridges, fragments, multipolar anaphase & telophase | Amer et al., 1989 |
| $SO_2$ | 50 ppm | Tradescantia (clones 02 & 4430) | Chromatid aberration | Sparrow & Schairer, 1974 |
| $SO_2$ | 0.075 ppm | *Tradescantia paludosa* Anders. Sax clone-3 | Chromatid & Isochromatid breaks | Ma et al., 1974 |
| $Pb^{2+}$ | 1–100 μM | *Allium cepa* var. *agrogarum* | C-mitosis & chromosome stickiness | Jiang et al., 2014 |
| $Pb^{2+}$ | 10–100 μM | *Triticum aestivum* L. cv. *Maruca* | Micronucleus, tetrapolar anaphase with laggards and bridges | Truta et al., 2011 |
| $Pb^{2+}$ | 0.5–1 Mm | *Allium cepa* L. | Stickiness, bridges, chromosomal rupture, laggard chromosomes & micronuclei | Carruyo et al., 2008 |
| PAHs | ambient | *Allium cepa* L. | Chromosomal bridges, breaks, and micronucleus | Ramos de Rainho et al., 2013 |
| Mixed air pollution | ambient | *Betula pendula* (birch seed progeny) | Chromosome lagging, bridges, chromosome agglutination & asymmetric mitosis | Kalaev & Karpova, 2003 |

### 13.6.1 Molecular Indicators

Molecular indicators or markers refer to the evaluation of the response of the cellular DNA, the main benefit of estimating the genotoxicity at DNA level in plants is mostly associated with sensitivity and short response time. Molecular markers are advantageous over morphological, cytological, and biochemical markers and it has been extensively applied in life sciences (Fang and Roose, 1997). They provide us new tools for detecting genetic alterations in response to toxic chemicals at the level of DNA sequence and structure. Thus, it could act as one of the best ways to directly quantify the genotoxic effect of the pollutant on the living organism. Several studies described that environmental pollution induces different genetic damages over a range of plant species in various countries (Amato-Lourenco et al., 2017). Tuteja et al. (2009) described, how genotoxic stress in plants plays a critical role in affecting productivity and genomic stability. The frequently observed injuries in DNA bases are alkylation, oxidation, deamination, apurinic or apyrimidinic sites, and strand breaks, mainly single-strand breaks (Kaur et al., 2021; Tuteja et al., 2009). These alterations activate the general transduction signals for the cell cycle, arrest and regulation of genes repair expression, ultimately resulting in programmed cell death (Balestrazzi et al., 2013). According to Mittler (2002), the toxicity of air pollutants may arise due to the generation of ROS that might cause a wide range of damages to nucleic acids, lipids, proteins, which eventually leads to apoptosis.

Ozone induced toxicity in plants manifested, after its entry into the plant via stomata induces the generation of ROS, which triggers oxidative burst inside the plant cells (Sharma et al., 1996). This $O_3$ induced ROS acts as a signaling molecule in various signal transduction pathways, which are responsible for oxidative burst. ROS is a highly reactive molecule and its unrestricted production leads to lipid peroxidation, which was found to be correlated with DNA damage (Tripathi et al., 2011). DNA damage includes oxidative DNA lesions and the accumulation of high levels of DNA damage inhibits the key nuclear functions like replication and transcription which may lead to cell death and accumulation of deleterious mutations. These DNA lesions induced by ROS include the formation of oxidative adducts, strand breakage, deletion, and insertion of base pairs (Tai et al., 2010). Oxidative DNA damage includes the formation of single-strand breaks and 8-oxoguanine which results in G-A transversion, causing mispairing of bases (Tripathi et al., 2011). A study done on two rice cultivars Shivani and Malviya dhan 36 shows contrasting results, where Shivani with more ROS production showed lesser DNA damage than

Malviya dhan 36 which had lower ROS production. The probable reason for this finding could be that despite ROS causes more strand breaks in Shivani but the DNA repair mechanism of this cultivar, was possibly more efficient which results in faster recovery and showing lesser DNA damage than Malviya dhan 36 (Sarkar et al., 2015). While $SO_2$ induced toxicity is derived from sulfite and bisulfite ($HSO_3^-$), which are generated after the dissolution of $SO_2$ in the aqueous medium (Madamanchi and Alscher, 1991) further detoxification of sulfite to sulfate in plants leads to an increase in ROS such as hydroxyl radical, hydrogen peroxide and superoxide radical (Asada, 1987; Madamanchi and Alscher, 1991). The abrupt increase in ROS can oxidatively damage the DNA, proteins, and lipids (Laloi et al., 2004; Mittler, 2002). Experiments done on animals suggests that, if ROS-induced DNA damage remains unrepaired by base excision repair, it may lead to the formation of single-strand breaks or induction of mutation in DNA. These observations confirm that $SO_2$ is a clastogenic and genotoxic agent in cells, and is considered as a systematic oxidative DNA damaging agent (Meng et al., 2004).

$NO_2$ is also known to induce damage in living cells by various means like oxidation of lipid membrane, formation of N-nitrosamine, and single-strand DNA breaks. Not only that, $NO_2$ after free radical reactions in dark generates more stable compounds like nitric oxide (NO), and metal nitrosyl complexes like nitroprusside which upon light exposure gets converted to nitrosating species like $NO^+$ or reactive nitrogen oxides (Cooney et al., 1994). This NO is responsible for the nitrosative deamination of DNA, which represents an important method of genomic alteration. The common NO-induced genetic alteration is a base substitution, this substitution mainly involves C (cytosine) → T (thymine) transition mutation which is approximately 99% of total base substitution. NO under oxidizing environment nitrosative diazotities the primary aryl amine ($ArNH_2$) (Eq. 13.1), which leads to DNA damage. As the hydrolysis of diazonium ion would cause the replacement of nitrogen substituent with oxygen (Eq. 13.2), this reason explains that diazotitation and hydrolysis of cytosine and other bases in vivo might lead to heritable genetic alterations, which is mentioned in equation from (Eqs. 13.3–13.6).

$$ArNH_2\,(\text{arylamine}) + NO\,(\text{Oxidation}) \rightarrow ArN_2^{\;+} + H_2O \quad (13.1)$$

$$ArN_2^{\;+} + H_2O \rightarrow ArOH + N_2 + H^+ \quad (13.2)$$

$$dC + NO\,(pH7.4) \rightarrow dU + N_2 + H^+ \quad (13.3)$$

$$m^5dc(\text{methyl} - 2' - \text{deoxycytidine}) + NO(pH7.4) \rightarrow dT + N_2 + H^+ \quad (13.4)$$

$$dG + NO(pH7.4) \rightarrow dX(2' - \text{deoxyanthosine}) + N_2 + H^+ \quad (13.5)$$

$$dA + NO(pH7.4) \rightarrow dI(\text{deoxy Ionosine}) + N_2 + H^+ \quad (13.6)$$

The above equation shows that NO induces a mutation in living organisms, that is linked with a variety of disorders which are suspected to result from base deamination (Rolda´n-Arjona and Ariza, 2009; Wink et al., 1991).

Lead is generally considered to be a co-mutagen or weak genotoxic agent (Rojas et al., 1999). Pb-induced genotoxicity occurs via two pathways one is the direct interaction between metal and DNA while another one is indirect via oxidative stress (Mitler, 2002; Hartmann and Speit, 1994). During direct interaction, the association between metal and DNA takes place via non-covalent bonding which was quantified by several physical techniques (Tajmir-Riahi et al., 1993; Schaaper et al., 1987). Several studies suggest that the direct interaction pathway does not show DNA damage at low metal concentration, this shows that the Pb induced genotoxicity is either weak or constricted to high concentration (Pourrut et al., 2011 and Hartwig, 1998). So, it is now clear that most of the Pb-induced genotoxicity is mediated by the indirect pathway through oxidative stress (Rucinska et al., 2004). Oxidative stress-induced by Pb causes single-strand breakage. In an in vitro system, it has been demonstrated that singlet oxygen is the major participant in the process of inducing DNA strand-breakage and formation of 8-hydroxydeoxyguanosine adduct in Pb treatment (Yang et al., 1999). In addition to this, hydroxyl free radical (·OH) is considered to be the utmost ROS which interacts with DNA and causes genetic damage. Hydroxyl free radical attacks on the sugar residues of DNA and causes DNA fragmentation, strand breaks, base loss, and strand breaks with a terminal sugar residue fragment (Pourrut et al., 2011; Gichner et al., 2008).

### 13.6.2 Chromosomal markers

There is abundant evidence that shows how air pollutants affect the process of cell division and chromosomal segregation in living organisms (Abbas et al., 2016). According to some researches, the percentage of mutation in meristematic cells increases by 2-6 times in plants growing in polluted areas in comparison to those growing in less polluted environments (Li et al., 2016; Chehregani et al., 2004). Similarly cytological studies also showed that in pollutant exposed plants the clumping and stickiness of chromosomes is a commonly observed phenomenon at the metaphase while laggard chromosomes were observed in the highest percentage at the metaphase and anaphase stages of cell division (Glasenčnik et al., 2004). A study done on *Tradescantia pallida* by Prajapati and Tripathi (2008) to assess the genotoxic effects of urban air pollutants showed the increased micronuclei frequency in meiotic microspores of plants growing in the vicinity of agrochemical industries than the areas with high traffic density. These micronuclei are the DNA containing extracellular bodies enclosed with the plasma membrane, which are formed from chromosome fragments as a consequence of chromosomal breakage or chromosome lagging during anaphase or aneuploidy (Prajapati and Tripathi, 2008; Natarajan et al., 1982).

Ozone induces several kinds of chromosomal aberrations in the plant, the main kind of aberration is chromosome stickiness, which arises due to improper folding of chromosomes into single chromatids, as the chromosomes attached through sub-chromatid bridges (Wonish et al., 2003; Muller et al., 1994). The chromosome stickiness shows reverse kind of toxic effects, which can also lead to death. Other less frequent types of defects are fragments, rings, and breaks, which may be genotoxic and can pass this defect to descending cell generations (Fiskesjö, 1993). Druskovic (1988) suggests that short-term $O_3$ exposure may lead to single chromosome damages, like ring formation, breaks and fragments while long-term exposure causes chromosome stickiness (Muller et al., 1996 and Fetner, 1958). *Capsicum annuum* under $O_3$ exposure also, shows an increase in the frequency of unoriented and sticky or clumped chromosomes in metaphase I, while during anaphase and telophase I, anaphase bridges and laggards were frequently seen. Abnormalities during cytokinesis or karyokinesis also lead to the formation of monads, arrested diads, triads, and tetrads. (Agrawal and Chaudhary, 1992). Similar results were also reported by Ramos de Rainho et al., 2013 in *Allium cepa* exposed to pollutants like polycyclic aromatic hydrocarbons (PAHs). In presence of a high concentration of PAHs increased number of chromosomal aberrations like bridges, breaks and micronucleus were reported. These bridges probably might be formed due to breakage and reunion of chromosomes and chromatids. Another potential mutagen like sulphur dioxide is also responsible for inducing abnormalities in plants. In *Vicia faba*, $SO_2$ induced abnormalities were reported at different meiotic stages, but the level of abnormalities was found highest at stage I and II of anaphase. Chromosome stickiness was also commonly observed at different meiotic stages, but at metaphase, I stage it was the dominating abnormality that resulted in the formation of sticky bridges in the anaphase stages. The percentage of bridges increases in pollen mother cells (PMCs) as the concentration or duration of $SO_2$, fumigation decreases (Amer et al., 1989). Agrawal et al. (1995), in their study, performed on the different cytotypes (diploid, tetraploid, and hexaploid) of *Solanum nigrum* observed that the plant with a diploid set of chromosomes is more sensitive toward $SO_2$ than the hexaploid one. The diploid species of S. *nigrum* shows chromosomal aberration even at a very low concentration (0.2ppm) of $SO_2$, while species with higher ploidy need $SO_2$ exposure of high concentration. The abnormalities commonly observed in *S. nigrum* under $SO_2$ exposure are decreased frequency of chiasma, increased number of rod bivalents, lagging chromosome, bridges, and unequal separation. Laggards and bridges were mostly observed during the anaphase and telophase I & II stages of meiosis. Heavy metals are also known to cause chromosomal abnormalities. Heavy metals like $Pb^{2+}$induces chromosomal aberrations in plants like *Allium cepa*. It induces chromosomal aberration phenomenon which was described as chromosome agglutinations showing the sticky appearance and was reported as an indicator of high toxicity. Other observed anomalies were the formation of chromosomic bridges which can lead to, chromosomal rupture and laggard chromosomes. These anomalies arise due to chromosome stickiness, stickiness holds the two chromosomes together; during the separation process, it causes chromosomal rupture. It is also mentioned that chromosomic bridges, laggard chromosomes, and chromosomal rupture are considered as an inducer of aneuploidy and polyploidy. At a longer exposure (48 h), micronuclei are observed, possibly due to the development of the isolated chromosomes that result from an unequal distribution of genetic material. The clastogenicity induced by the metal can cause cellular death due to its blocking effect on enzymes involved in the process of repairing the genome; the cell cannot recover from the produced damage and thus increasing the number of aberrations as a new cycle begins. Among total chromosomal aberrations observed in *A. cepa,* there is a higher preponderance of colchicine mitosis (C-mitosis) and chromosome stickiness which is approximately more than 90% of total aberrations (Oladele et al., 2013). In *A. cepa* C-mitosis in the mitotic root tip cells leads to inactivation of the spindle fibers followed by the random scattering of condensed chromosomes, which were reported in the Pb treated root tip cells. The number of cells having C-mitosis and chromosomal stickiness increased with increasing Pb concentration and prolonging treatment duration. Besides chromosome stickiness, anaphase bridges, vagrants, fragments, and lagging chromosomes were also reported during low Pb concentration and treatment duration (Oladele et al., 2013). Among these aberrations, lagging chromosome can lead to micronuclei formation (Kumar and Rai, 2007). In an investigation, it was found that the microtubule arrangement was closely associated with chromosomal aberrations during the process of mitosis under Pb stress. In some Pb exposed cells, mitotic spindle microtubule lost their organization and gets randomly oriented and scattered in cells due to which sister chromatids could not move toward the cell poles, resulting in the formation of c-mitosis. It was also discovered that chromosomes at metaphase were normally distributed at the center of the cell under Pb stress. Although the spindle fibers could be normally formed in some cells under Pb stress, Pb still could affect them in varying degrees. These abnormal spindle fibers, during anaphase, were broken into short fragments. The serious toxic effect is that the spindle fibers get stick to each other, lumps were formed within cells, which results in polyploidy. Mitotic spindle MTs damage in telophase cells inhibits the formation of the phragmoplast, which affected the formation of daughter cells (Jiang et al., 2014; Jiang and Liu, 2000).

## 13.7 Conclusions

Air pollutants, along with producing the general oxidative stress in plants, also promote the production of genotoxic compounds, which may bring about changes in the genetic constitution of plants. Evaluating the DNA damage and chromosomal aberrations in plant meristems seems to be a more sensitive indicator for air pollution impact on plants rather than the analysis of pigments and antioxidants in leaves. Thus, these cytological and molecular indicators proved to be useful for the early detection of diseases, especially during development stages, or when the symptoms are not consistent, obscure or nonexistent. These indicators are advantageous over morphological, physiological, and biochemical indicators as these indicators prove to be useful, in disease diagnosis earlier during its development or when, symptoms are not consistent, uncertain, or nonexistent. As the response of these molecular and chromosomal parameters provides more authentic information, their use would be more beneficial for better detection of stress-induced damages in plants. However, since the studies related to this aspect are very rare, so one of the objectives of future research should focus more on the development of these molecular and chromosomal indicators, which may be used as an important tool in future biomonitoring programs.

## Acknowledgments

Authors are thankful to the Head, Department of Botany and Coordinators, Center of Advance Study (CAS) in Botany, Institute of Science, Banaras Hindu University, for providing library and internet facilities.

## References

Abbas, I., Verdin, A., Escande, F., Saint-Georges, F., Cazier, F., Mulliez, P., …, Garçon, G., 2016. In vitro short-term exposure to air pollution PM2. 5-0.3 induced cell cycle alterations and genetic instability in a human lung cell coculture model. Environ. Res. 147, 146–158.

Abdelhaliem, E., Al-Huqail, A.A., 2016. Detection of protein and DNA damage induced by elevated carbon dioxide and ozone in Triticum aestivum L. using biomarker and comet assay. Genet. Mol. Res. 15 (2), 1–19.

Agrawal, S.B., Singh, R.N., Sahi, A.N., 1995. Cytogenetic effect of sulphur dioxide on cytotypes of Solanum nigrum complex. Cytobios (United Kingdom) 83, 41–47.

Agrawal, S.B., Chaudhary, B.R, 1992. Cytogenetic effects of ozone in Capsicum annuum. CHROMATIN 2 (1), 99–104.

Agrawal, S.B., Chaudhary, B.R, 1989. The use of chromosomal aberrations in chilli plant as a monitoring system of sulphur dioxide pollution, *Man and His Ecosystem: Proceedings of the 8th World Clean Air Congress* 1989, The Hague 11-15 Sept., 1989, 21. Elsevier Publishing Company, pp. 143.

Al Khateeb, W., 2018. Plants genotoxicity as pollution bioindicator in Jordan using comet assay. Physiol. Mol. Biol. Plants 24 (1), 89–97.

Amato-Lourenco, L.F., Lobo, D.J.A., Guimarães, E.T., Moreira, T.C.L., Carvalho-Oliveira, R., Saiki, M., Mauad, T., 2017. Biomonitoring of genotoxic effects and elemental accumulation derived from air pollution in community urban gardens. Sci. Total Environ. 575, 1438–1444.

Amer, S.M., Mikhael, E., El-Ashry, Z.M., 1989. Cytogenetic effect of sulphur dioxide on Vicia faba plant I. Cytologia (Tokyo) 54 (2), 211–221.

Apel, K., Hirt, H., 2004. Reactive oxygen species: metabolism, oxidative stress, and signal transduction. Annu. Rev. Plant Biol. 55, 373–399.

Asada, K., 1987. Production and scavenging of active oxygen in photosynthesis. Photoinhibition, 227–287.

Aydin, S.S., KIliç, Z., Dönmez, Ç., Altunkaynak, E., Sümer, A., 2015. Assesment of genotoxicity induced by lead pollution in tomato Lycopersicum esculentum by molecular and population markers. Biyolojik Çeşitlilik ve Koruma 8 (1), 83–89.

Balestrazzi, A., Macovei, A., Donà, M., Carbonera, D., Confalonieri, M., 2013. Genotoxic stress, DNA repair, and crop productivityCrop Improvement Under Adverse Conditions. Springer, New York, NY, pp. 153–169.

Bari, M.A., Kindzierski, W.B., 2018. Ambient volatile organic compounds (VOCs) in Calgary, Alberta: sources and screening health risk assessment. Sci. Total Environ. 631, 627–640.

Bignal, K.L., Ashmore, M.R., Headley, A.D., Stewart, K., Weigert, K., 2007. Ecological impacts of air pollution from road transport on local vegetation. Appl. Geochem. 22 (6), 1265–1271.

Bokhari, B., Sharma, S., 2019. Stress marks on the genome: use or lose? Int. J. Mol. Sci. 20 (2), 364.

Cadet, J., Wagner, J.R., 2013. DNA base damage by reactive oxygen species, oxidizing agents, and UV radiation. Cold Spring Harb. Perspect. Biol. 5 (2), a012559.

Cansaran-Duman, D., Altunkaynak, E., Aslan, A., Büyük, İ., Aras, S., 2015. Application of molecular markers to detect DNA damage caused by environmental pollutants in lichen species. Genet. Mol. Res. 14 (2), 4637–4650.

Carruyo, I., Fernández, Y., Marcano, L., Montiel, X., Torrealba, Z., 2008. Correlation of toxicity with lead content in root tip cells (Allium cepa L.). Biol. Trace Elem. Res. 125 (3), 276–285.

Chatterjee, N., Walker, G.C., 2017. Mechanisms of DNA damage, repair, and mutagenesis. Environ. Mol. Mutagen. 58 (5), 235–263.

Chehregani, A.H., Majd, A., Moin, M., Gholami, M., Shariatzadeh, M., Mohsenzade, F., 2004. Effect of air pollution on some cytogenetic characteristics, structure, viability and proteins of Zinnia elegans pollen grains. Pak. J. Biol. Sci. 7 (1), 118–122.

Cooney, R.V., Harwood, P.J., Custer, L.J., Franke, A.A, 1994. Light-mediated conversion of nitrogen dioxide to nitric oxide by carotenoids. Environ. Health Perspect. 102 (5), 460–462.

Cortez, D., 2019. Replication-coupled DNA repair. Mol. Cell 74 (5), 866–876.

de Souza Lima, E., de Souza, S.R., Domingos, M., 2009. Sensitivity of Tradescantia pallida (Rose) Hunt. 'Purpurea' Boom to genotoxicity induced by ozone. Mut. Res. Genet. Toxicol. Environ. Mutagen. 675 (1-2), 41–45.

del Río, L.A., Sandalio, L.M., Corpas, F.J., Palma, J.M., Barroso, J.B., 2006. Reactive oxygen species and reactive nitrogen species in peroxisomes. Production, scavenging, and role in cell signaling. Plant Physiol. 141 (2), 330–335.

Druskovic, B., 1988. Cytogenetic bioindication 1-Utilization of cytogenetic analysis for detection of the effects of genotoxic pollutants on forest trees (material and method). Bioloski vestnik (Yugoslavia) 36, 1–17.

Endo, T.R., Nasuda, S., Jones, N., Dou, Q., Akahori, A., Wakimoto, M., …, Tsujimoto, H., 2008. Dissection of rye B chromosomes, and nondisjunction properties of the dissected segments in a common wheat background. Genes Genet. Syst. 83 (1), 23–30.

Fang, D.Q., Roose, M.L., 1997. Identification of closely related citrus cultivars with inter-simple sequence repeat markers. Theor. Appl. Genet. 95 (3), 408–417.

Fetner, R.H., 1958. Chromosome breakage in Vicia faba by ozone. Nature 181 (4607), 504–505.

Fiskesjö, G., 1993. The Allium test—a potential standard for the assessment of environmental toxicityEnvironmental Toxicology and Risk Assessment: Second Volume. ASTM International.

Foyer, C.H., Lopez-Delgado, H., Dat, J.F., Scott, I.M., 1997. Hydrogen peroxide-and glutathione-associated mechanisms of acclimatory stress tolerance and signalling. Physiol. Plant. 100 (2), 241–254.

Foyer, C.H., Harbinson, J., Mullineaux, P., 1994. Oxygen metabolism and the regulation of photosynthetic electron transportCauses of Photooxidative Stress and Amelioration of Defense Systems in Plants, 1–42.

Galveias, A., Arriegas, R., Mendes, S., Ribeiro, H., Abreu, I., Costa, A.R., Antunes, C.M., 2021. Air pollutants $NO_2$-and $O_3$-induced Dactylis glomerata L. pollen oxidative defences and enhanced its allergenic potential. Aerobiologia 37 (1), 127–137.

Ghorani-Azam, A., Riahi-Zanjani, B., Balali-Mood, M., 2016. Effects of air pollution on human health and practical measures for prevention in Iran. J. Res. Med. Sci 21 (65), 1–12.

Gichner, T., Žnidar, I., Száková, J., 2008. Evaluation of DNA damage and mutagenicity induced by lead in tobacco plants. Mut. Res. Genet. Toxicol. Environ. Mutagen. 652 (2), 186–190.

Glasenčnik, E., Ribarič-Lasnik, C., Savinek, K., Zaluberšek, M., Mueller, M., Batič, F., 2004. Impact of air pollution on genetic material of shallot (Allium cepa L. var. ascalonicum) exposed at differently polluted sites in Slovenia. J. Atmos. Chem. 49 (1), 363–376.

Gómez-Arroyo, S., Cortés-Eslava, J., Loza-Gómez, P., Arenas-Huertero, F., de la Mora, M.G., Bermea, O.M, 2018. In situ biomonitoring of air quality in rural and urban environments of Mexico Valley through genotoxicity evaluated in wild plants. Atmos. Pollut. Res. 9 (1), 119–125.

Grąziewicz, M.A., Zastawny, T.H., Oliński, R., Speina, E., Siedlecki, J., Tudek, B., 2000. Fapyadenine is a moderately efficient chain terminator for prokaryotic DNA polymerases. Free Radical Biol. Med. 28 (1), 75–83.

Gros, L., Saparbaev, M.K., Laval, J., 2002. Enzymology of the repair of free radicals-induced DNA damage. Oncogene 21 (58), 8905–8925.

Hackbarth, S., Schlothauer, J., Preuß, A., Röder, B., 2010. New insights to primary photodynamic effects–Singlet oxygen kinetics in living cells. J. Photochem. Photobiol. B 98 (3), 173–179.

Halliwell, B., Gutteridge, J.M, 1984. [4]Role of iron in oxygen radical reactions. Methods Enzymol. 105, 47–56.

Halliwell, B., 1977. Superoxide and hydroxylation reactions Superoxide and Superoxide Dismutases, 335–349.

Hartmann, A., Speit, G., 1994. Comparative investigations of the genotoxic effects of metals in the single cell gel (SCG) assay and the sister chromatid exchange (SCE) test. Environ. Mol. Mutagen. 23 (4), 299–305.

Hartwig, A., 1998. Carcinogenicity of metal compounds: possible role of DNA repair inhibition. Toxicol. Lett. 102, 235–239.

Hatz, S., Lambert, J.D., Ogilby, P.R., 2007. Measuring the lifetime of singlet oxygen in a single cell: addressing the issue of cell viability. Photochem. Photobiol. Sci. 6 (10), 1106–1116.

Hazari, M.S., Stratford, K.M., Krantz, Q.T., King, C., Krug, J., Farraj, A.K., Gilmour, M.I, 2018. Comparative cardiopulmonary effects of particulate matter-and ozone-enhanced smog atmospheres in mice. Environ. Sci. Technol. 52 (5), 3071–3080.

Imlay, J.A., 2003. Pathways of oxidative damage. Annu. Rev. Microbiol. 57 (1), 395–418.

Janakiraman, R., Harney, P.M, 1976. Effects of ozone on meiotic chromosomes ofVicia faba. Can. J. Genet. Cytol. 18 (4), 727–730.

Jiang, Z., Qin, R., Zhang, H., Zou, J., Shi, Q., Wang, J., Jiang, W., Liu, D., 2014. Determination of Pb genotoxic effects in Allium cepa root cells by fluorescent probe, microtubular immunofluorescence and comet assay. Plant Soil 383 (1-2), 357–372.

Jiang, W., Liu, D., 2000. Effects of Pb 2+ on Root Growth, Cell Division, and Nucleolus of Zea mays L. Bull. Environ. Contam. Toxicol. 65 (6), 786–793.

Juan, C.A., Pérez de la Lastra, J.M., Plou, F.J., Pérez-Lebeña, E., 2021. The chemistry of reactive oxygen species (ROS) revisited: outlining their role in biological macromolecules (DNA, lipids and proteins) and induced pathologies. Int. J. Mol. Sci. 22 (9), 4642.

Kalaev, V.N., Karpova, S.S, 2003. The influence of air pollution on cytogenetic characteristics of birch seed progeny. For. Genet. 10 (1), 11–18.

Kamal, M.S., Razzak, S.A., Hossain, M.M., 2016. Catalytic oxidation of volatile organic compounds (VOCs)—a review. Atmos. Environ. 140, 117–134.

Kasai, H., 1997. Analysis of a form of oxidative DNA damage, 8-hydroxy-2′-deoxyguanosine, as a marker of cellular oxidative stress during carcinogenesis. Mut. Res. Rev. Mut. Res. 387 (3), 147–163.

Kaur, G., Sharma, P., Rathee, S., Singh, H.P., Batish, D.R., Kohli, R.K., 2021. Salicylic acid pre-treatment modulates Pb 2+-induced DNA damage vis-à-vis oxidative stress in Allium cepa roots. Environ. Sci. Pollut. Res. 28 (37), 51989–52000.

Kim, K.H., Jahan, S.A., Kabir, E., Brown, R.J., 2013. A review of airborne polycyclic aromatic hydrocarbons (PAHs) and their human health effects. Environ. Int. 60, 71–80.

Krieger-Liszkay, A., 2005. Singlet oxygen production in photosynthesis. J. Exp. Bot. 56 (411), 337–346.

Kumar, G., Rai, P., 2007. Comparative genotoxic potential of mercury and cadmium in soybean. Turkish J. Biol. 31 (1), 13–18.
Laloi, C., Apel, K., Danon, A., 2004. Reactive oxygen signalling: the latest news. Curr. Opin. Plant Biol. 7 (3), 323–328.
Laumbach, R.J., Kipen, H.M, 2012. Respiratory health effects of air pollution: update on biomass smoke and traffic pollution. J. Allergy Clin. Immunol. 129 (1), 3–11.
Li, T., Zhang, M., Gu, K., Herman, U., Crittenden, J., Lu, Z., 2016. DNA Damage in Euonymus japonicus leaf cells caused by roadside pollution in Beijing. Int. J. Environ. Res. Public Health 13 (7), 742.
Liu, N., Cao, C., Sun, Z., Lin, Z., Deng, R., 2016. Pollutant-induced cell death and reactive oxygen species accumulation in the aerial roots of Chinese banyan (Ficus microcarpa). Sci. Rep. 6 (1), 1–8.
Lodovici, M., Bigagli, E., 2011. Oxidative stress and air pollution exposure. J. Toxicol. 2011, 1–9.
Loomis, D., Grosse, Y., Lauby-Secretan, B., El Ghissassi, F., Bouvard, V., Benbrahim-Tallaa, L., Guha, N., Baan, R., Mattock, H., Straif, K., 2013. The carcinogenicity of outdoor air pollution. Lancet Oncol. 4 (13), 1262.
Ma, T.H., 1982. Vicia cytogenetic tests for environmental mutagens: a report of the US environmental protection agency gene-tox program. Mut. Res. Rev. Genet. Toxicol. 99 (3), 257–271.
Ma, T.H., Isbandi, D., Khan, S.H., Tseng, Y.S., 1974. Low level of SO2 enhanced chromatid aberrations in Tradescantia pollen tubes and seasonal variation of the aberration rates. Mutat. Res.-Environ. Mutagen. Relat. Subj. 21 (2), 93–100.
Madamanchi, N.R., Alscher, R.G., 1991. Metabolic bases for differences in sensitivity of two pea cultivars to sulfur dioxide. Plant Physiol. 97 (1), 88–93.
Manisalidis, I., Stavropoulou, E., Stavropoulos, A., Bezirtzoglou, E., 2020. Environmental and health impacts of air pollution: a review. Front. Public Health 8, 14.
Maynard, S., Schurman, S.H., Harboe, C., de Souza-Pinto, N.C., Bohr, V.A., 2009. Base excision repair of oxidative DNA damage and association with cancer and aging. Carcinogenesis 30 (1), 2–10.
Meng, Z., Qin, G., Zhang, B., Bai, J., 2004. DNA damaging effects of sulfur dioxide derivatives in cells from various organs of mice. Mutagenesis 19 (6), 465–468.
Meriga, B., Reddy, B.K., Rao, K.R., Reddy, L.A., Kishor, P.K, 2004. Aluminium-induced production of oxygen radicals, lipid peroxidation and DNA damage in seedlings of rice (Oryza sativa). J. Plant Physiol. 161 (1), 63–68.
Miao, W., Huang, X., Song, Y., 2017. An economic assessment of the health effects and crop yield losses caused by air pollution in mainland China. J. Environ. Sci. 56, 102–113.
Mishra, S., Jha, A.B., Dubey, R.S., 2011. Arsenite treatment induces oxidative stress, upregulates antioxidant system, and causes phytochelatin synthesis in rice seedlings. Protoplasma 248 (3), 565–577.
Mittler, R., 2002. Oxidative stress, antioxidants and stress tolerance. Trends Plant Sci. 7 (9), 405–410.
Mukherjee, A., Agrawal, S.B., Agrawal, M., 2020. Responses of tropical tree species to urban air pollutants: ROS/RNS formation and scavenging. Sci. Total Environ. 710, 136363.
Müller, M., Köhler, B., Tausz, M., Grill, D., Lütz, C., 1996. The assessment of ozone stress by recording chromosomal aberrations in root tips of spruce trees [Picea abies (L) Karst]. J. Plant Physiol. 148 (1-2), 160–165.
Müller, M., Grill, D., 1994. Induction of chromosomal damage by ozone in the root meristems of Norway spruce. Proc. R. Soc. Edinburgh Section B Biol. Sci. 102, 49–54.
Natarajan, A.T., Obe, G., Altmann, H. (Eds.), 1982. DNA Repair, Chromosome Alterations, and Chromatin Structure: Proceedings of an International Meeting Held at Noordwijkerhout. The Netherlands, 4. Elsevier Biomedical Press 23-25 April 1981.

Nilsson, R., Liu, N.A., 2020. Nuclear DNA damages generated by reactive oxygen molecules (ROS) under oxidative stress and their relevance to human cancers, including ionizing radiation-induced neoplasia part I: physical, chemical and molecular biology aspects. Radiat. Med. Protect. 1, 140–152.
Oladele, E.O., Odeigah, P.G.C., Taiwo, I.A, 2013. The genotoxic effect of lead and zinc on Bambara groundnut (Vigna subterranean). Afr. J. Environ. Sci. Technol. 7 (1), 9–13.
Plante, I., Slaba, T., Shavers, Z., Hada, M., 2019. A bi-exponential repair algorithm for radiation-induced double-strand breaks: application to simulation of chromosome aberrations. Genes 10 (11), 936.
Pourrut, B., Shahid, M., Dumat, C., Winterton, P., Pinelli, E., 2011. Lead uptake, toxicity, and detoxification in plantsReviews of Environmental Contamination and Toxicology213. Springer, New York, NY, pp. 113–136.
Prajapati, S.K., Tripathi, B.D, 2008. Seasonal variation of leaf dust accumulation and pigment content in plant species exposed to urban particulates pollution. J. Environ. Qual. 37 (3), 865–870.
Ramos de Rainho, C., Machado Corrêa, S., Luiz Mazzei, J., Alessandra Fortes Aiub, C., Felzenszwalb, I., 2013. Genotoxicity of polycyclic aromatic hydrocarbons and nitro-derived in respirable airborne particulate matter collected from urban areas of Rio de Janeiro (Brazil). BioMed. Res. Int. 2013, 1–9.
Rodrigues, G.S., Madkour, S.A., Weinstein, L.H, 1996. Genotoxic activity of ozone in Tradescantia. Environ. Exp. Bot. 36 (1), 45–50.
Rojas, E., Herrera, L.A., Poirier, L.A., Ostrosky-Wegman, P., 1999. Are metals dietary carcinogens? Mut. Res. Genet. Toxicol. Environ. Mutagen. 443 (1-2), 157–181.
Roldan-Arjona, T., Ariza, R.R, 2009. Repair and tolerance of oxidative DNA damage in plants. Mut. Res. Rev. Mut. Res. 681 (2-3), 169–179.
Rozelle, A.L., Cheun, Y., Vilas, C.K., Koag, M.C., Lee, S., 2021. DNA interstrand cross-links induced by the major oxidative adenine lesion 7, 8-dihydro-8-oxoadenine. Nat. Commun. 12 (1), 1–16.
Rucińska, R., Sobkowiak, R., Gwóźdź, E.A., 2004. Genotoxicity of lead in lupin root cells as evaluated by the comet assay. Cell. Mol. Biol. Lett. 9 (3), 519–528.
Sachdev, S., Ansari, S.A., Ansari, M.I., Fujita, M., Hasanuzzaman, M., 2021. Abiotic stress and reactive oxygen species: generation, signaling, and defense mechanisms. Antioxidants 10 (2), 277.
Sarkar, A., Singh, A.A., Agrawal, S.B., Ahmad, A., Rai, S.P., 2015. Cultivar specific variations in antioxidative defense system, genome and proteome of two tropical rice cultivars against ambient and elevated ozone. Ecotoxicol. Environ. Saf. 115, 101–111.
Schaaper, R.M., Koplitz, R.M., Tkeshelashvili, L.K., Loeb, L.A, 1987. Metal-induced lethality and mutagenesis: possible role of apurinic intermediates. Mut. Res. Fundam. Mol. Mech. Mutagen. 177 (2), 179–188.
Schubert, I., Pecinka, A., Meister, A., Schubert, V., Klatte, M., Jovtchev, G., 2004. DNA damage processing and aberration formation in plants. Cytogenet. Genome Res. 104 (1-4), 104–108.
Seinfeld, J.H., Pandis, S.N., 2006. Atmospheric Chemistry and Physics: From Air Pollution to Climate Change, 2nd ed. John Wiley & Sons, Inc, New York, USA.
Sharma, P., Dubey, R.S, 2005. Drought induces oxidative stress and enhances the activities of antioxidant enzymes in growing rice seedlings. Plant Growth Regul. 46 (3), 209–221.
Sharma, Y.K., León, J., Raskin, I., Davis, K.R, 1996. Ozone-induced responses in Arabidopsis thaliana: the role of salicylic acid in the accumulation of defense-related transcripts and induced resistance. Proc. Natl. Acad. Sci. 93 (10), 5099–5104.
Sparrow, A.H., Schairer, L.A., 1974. Effects of chemical mutagens (EMS, DBE) and specific air pollutants ($O_3$, $SO_2$, $NO_2$, $N_2O$) on somatic mutation rates in Tradescantia. (No. BNL--19320). Brookhaven National Lab, pp. 1–40.

Tai, H.H., Percy, K.E., Karnosky, D.F, 2010. DNA damage in Populus tremuloides clones exposed to elevated O3. Environ. Pollut. 158 (4), 969–976.
Tajmir-Riahi, H.A., Naoui, M., Ahmad, R., 1993. The effects of Cu2+ and Pb2+ on the solution structure of calf thymus DNA: DNA condensation and denaturation studied by Fourier transform IR difference spectroscopy. Biopoly. Orig. Res. Biomol. 33 (12), 1819–1827.
Takata, K.I., Shimizu, T., Iwai, S., Wood, R.D, 2006. Human DNA polymerase N (POLN) is a low fidelity enzyme capable of error-free bypass of 5S-thymine glycol. J. Biol. Chem. 281 (33), 23445–23455.
Tanou, G., Molassiotis, A., Diamantidis, G., 2009. Induction of reactive oxygen species and necrotic death-like destruction in strawberry leaves by salinity. Environ. Exp. Bot. 65 (2-3), 270–281.
Tremblay, S., Wagner, J.R., 2008. Dehydration, deamination and enzymatic repair of cytosine glycols from oxidized poly (dG-dC) and poly (dI-dC). Nucleic Acids Res. 36 (1), 284–293.
Tripathi, R., Sarkar, A., Pandey Rai, S., Agrawal, S.B., 2011. Supplemental ultraviolet-B and ozone: impact on antioxidants, proteome and genome of linseed (Linum usitatissimum L. cv. Padmini). Plant Biol. 13 (1), 93–104.
Truta, E., Rosu, C.M., Bara, I.C., 2011. Lead-induced genotoxicity in wheat. J. Exp. Mol. Biol. 12 (1), 51–58.
Tuteja, N., Ahmad, P., Panda, B.B., Tuteja, R., 2009. Genotoxic stress in plants: shedding light on DNA damage, repair and DNA repair helicases. Mut. Res. Rev. Mut. Res. 681 (2-3), 134–149.
US EPA, OAR, 2014. Criteria air pollutants: (https://www.epa.gov/criteria-air-pollutants).
US EPA, OAR, 2016a. Sulfur dioxide Basics: (https://www.epa.gov/so2-pollution/sulfur-dioxide-basics).
US EPA, OAR, 2016b. Basic information about NO2: (https://www.epa.gov/no2-pollution/basic-information-aboutno2).
US EPA, OAR, 2016c. Sulfur dioxide basics: (https://www.epa.gov/so2-pollution/sulfur-dioxide-basics).
Valavanidis, A., Vlachogianni, T., Fiotakis, C., 2009. 8-hydroxy-2′-deoxyguanosine (8-OHdG): a critical biomarker of oxidative stress and carcinogenesis. J. Environ. Sci. Health C 27 (2), 120–139.
Vardar, Ç., Basaran, E., Cansaran-Duman, D., Aras, S., 2014. Air-quality biomonitoring: assessment of genotoxicity of air pollution in the Province of Kayseri (Central Anatolia) by use of the lichen Pseudevernia furfuracea (L.) Zopf and amplified fragment-length polymorphism Markers. Mut. Res. Genet. Toxicol. Environ. Mutagen. 759, 43–50.
Wagner, J.R., Madugundu, G.S., Cadet, J., 2021. Ozone-induced DNA damage: a Pandora's box of oxidatively modified DNA bases. Chem. Res. Toxicol. 34 (1), 80–90.
Wink, D.A., Kasprzak, K.S., Maragos, C.M., Elespuru, R.K., Misra, M., Dunams, T.M., Cebula, T.A., Koch, W.H., Andrews, A.W., Allen, J.S, 1991. DNA deaminating ability and genotoxicity of nitric oxide and its progenitors. Science 254 (5034), 1001–1003.
Wonisch, A., Tausz, M., Muller, M., Soja, G., Grill, D., 2003. Ozone-induced long-term effects on chromosomal aberration rates in root-tip meristems of spruce trees do not correspond to changes in tissue antioxidant status.(Engl.). PHYTON-HORN- 43 (1), 147–160.
Xie, X., He, Z., Chen, N., Tang, Z., Wang, Q., Cai, Y., 2019. The roles of environmental factors in regulation of oxidative stress in plant. BioMed. Res. Int. 2019, 1–11.
Yang, J.L., Wang, L.C., Chang, C.Y., Liu, T.Y, 1999. Singlet oxygen is the major species participating in the induction of DNA strand breakage and 8-hydroxydeoxyguanosine adduct by lead acetate. Environ. Mol. Mutagen. 33 (3), 194–201.
Yoon, J.H., Bhatia, G., Prakash, S., Prakash, L., 2010. Error-free replicative bypass of thymine glycol by the combined action of DNA polymerases κ and ζ in human cells. Proc. Natl. Acad. Sci. 107 (32), 14116–14121.

# CHAPTER 14

# Biomonitoring tools and bioprogramming: An overview

**Parvati Madheshiya, Gereraj Sen Gupta, Ansuman Sahoo, Supriya Tiwari**
Department of Botany, Institute of Science, Banaras Hindu University, Varanasi, Uttar Pradesh, India

## 14.1 Introduction

Upsurge in the industrial sector together with the advancement in modern-day agricultural practices have tremendously increased pollution levels in the biosphere (Naidu et al., 2021). Studies by Pure Earth and Green Cross Switzerland (2016) and Cribb (2021) have revealed that mining, farming, construction, mineral processing, and energy production sectors are the major contributor of hazardous chemicals that are induced by anthropogenic activities, while other studies indicated that the rise in the usage of chemicals since last centuries have greatly contributed to the release of several toxic compounds to the environment (Costa et al., 2006). The above measurements provide information on the potential of chemicals to cause adverse effects to the environment, crops, and thus the organisms. Traditionally, environmental pollution assessment was based on soil analysis and other estimates including the estimation of air, water, sediments, and, in certain cases, living organisms were considered. However, recent studies are using a new approach called, "biomonitoring," wherein living organisms, including plants and animals belonging to different taxa, are used to assess environmental contamination. Biomonitoring or biological monitoring is generally defined as "the systematic use of living organisms or their responses to determine the condition or changes of the environment" (Rosenberg, 1998; Gerhardt, 1999; Oertel, and Salánki, 2003).

The process of biomonitoring is accomplished using different organisms that show a characteristic response to particular stress conditions, and are referred to as bioindicators. According to Markert et al. (1997, 1999, 2003), a bioindicator is "an organism (or part of an organism or a community of organisms) that contains information on the quality of the environment (or a part of the environment)." According to Rosenberg (1992), Hilty (2000),

*New Paradigms in Environmental Biomonitoring Using Plants*
DOI: https://doi.org/10.1016/B978-0-12-824351-0.00015-8

and Fureder (2003), an "ideal" bioindicator should have at least the following characteristics:

**a.** Taxonomic soundness (easy to be recognized by a non-specialist).
**b.** High sensitivity to an environmental stressor.
**c.** Well-known ecological characteristics.
**d.** Low mobility (local indication).
**e.** Numerical abundance.
**f.** Suitability for laboratory experiments.
**g.** Wide or cosmopolitan distribution.
**h.** High ability for quantification and standardization.

The significance of bioindicators is immense and thus stays as an unavoidable segment of the environment. Apart from the significance of bioindicators, biomonitors and biomarkers are also important as far as contaminants are concerned. According to Hatje (2016), biomonitors are organisms that accumulate contaminants in their tissues and can be used to yield a relative measure of the total amount of contaminants in the environment integrated over a period of time whereas biomarkers are "any substance, structure, or process that can be measured in the body or its products and influence or predict the incidence of outcome or disease" (WHO 2001). To date, human biomonitoring programs have also been established in several countries where human biomonitoring makes use of biomarkers that are tools that allow estimation of exposure to a toxic agent, the extent of any toxic response, and prediction of likely to respond.

According to the World Health Organization (WHO), bioindicators reveal the presence or absence of contaminants by the occurrence of a particular type of symptoms or other measurable responses, biomonitors not only provide information about the presence of the contaminant, they also give additional information about the intensity of exposure and the concentration of the contaminant. Depending upon the ecological applications of bioindicators, there are three main categories in which they can be studied (Fig. 14.1).

Various bioindicators viz. ecological indicator, environmental indicator, and biodiversity indicator helps in the estimation of stresses at different sectors of the environment. The negative consequences of increasing environmental pollution and its impact on crops will be an important threat to global food security in the coming years. Therefore, several biomonitoring-based approaches have been adopted from time to time to evaluate the harmful effects of atmospheric pollutants on crop plants. In many developed and developing countries, different biomonitoring programs have

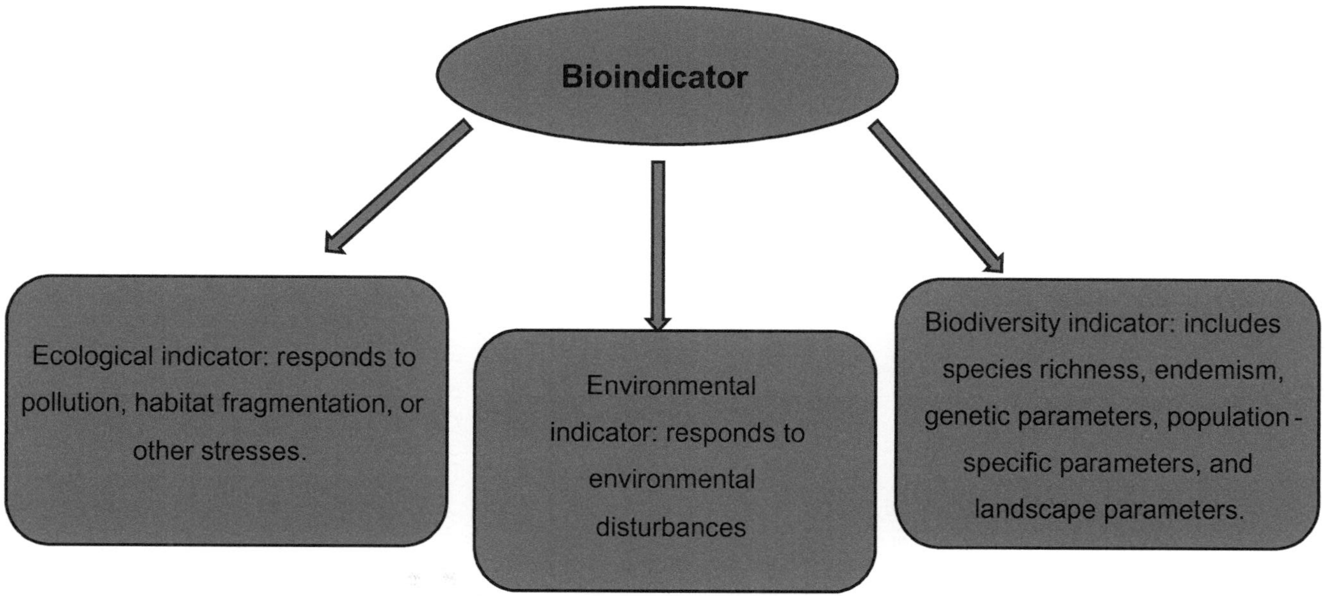


**Fig. 14.1** ***Categorization of bioindicators in different classes.***

been initiated, with the objective of identification of different pollutants and ways of their remedies. Till date, programs such as Global Invasive Species Programme (GISP), National Aquatic Resources Survey, Canadian Aquatic Biomonitoring Network (CABIN), National Ecosystem Survey (NES), Nationwide Aquatic Ecological Monitoring Programme (NAEMP), EuroBioNet Programme have been successfully established. The present chapter discusses the various ways of biomonitoring that have been done to date, the motive of the programs, and the fate of the same. As per the biomonitoring results, the regions of south and south-east Asia are specifically prone to higher magnitude of losses due to pollution stress and therefore, there is an urgent need to develop evaluation programs in these areas.

## 14.2 Traditional perspectives

In recent years, the concept of biomonitoring has gained attention and several studies have been conducted with respect to the studies related to the biomonitoring of air pollutants, trace elements, heavy metals, etc. The most commonly utilized organisms are those which are sensitive against certain pollutants, for instance, lichens against $SO_2$ pollution and *Nicotianatabacum* Bel-W3 against tropospheric $O_3$. Out of all the bioindicators, lichens and mosses are the most prevalent, as they are highly sensitive to atmospheric and edaphic pollution (Wolterbeek 2002). The primary concept behind the usage of lichens and mosses as bioindicators lies in the fact that they lack proper roots and limited absorption from substrate forces them to absorb nutrients and minerals from the atmosphere (Puckett and Burton 1981; Martin and Coughtrey 1982; Brown and Brown 1990; Ruhling and Steinnes 1998). Sloof (1995) also indicated passive intake of pollutants by mosses and lichens. Besides absorbing trace elements, these two plants groups have also been utilized in accumulating radioactive elements after Chernobyl nuclear disaster (Wolterbeek 2002). Mosses have also been used in the middle of the 20th century for biomonitoring of fluoride emitted by industries (Macintyre et al. 1952). In the late 20th century, they have also been utilized in biomonitoring programs for the accumulation of organic pollutants (Thomas and Herrmann, 1980).

## 14.3 Active and passive biomonitoring

The effects of tropospheric $O_3$ and heavy metals on crops and wild plants, for instance, are being addressed by the UNECE's ICP Vegetation using bioindicator plants mainly in rural and remote areas. In 1999, the local authorities and scientific institutions of eight nations and 12 cities of the European union together established an extensive biomonitoring program known as EuroBioNet (European Network of Air Quality by use of Bioindicator plants) (Tiwari et al. 2018). In this program, ozone, sulfurous compounds, heavy metals, polycyclic aromatic hydrocarbons (PAH), and mutagenic substances were monitored using the concept of biomonitoring. As per Klumpp et al. (2002), this program had the following objectives:

- establish the use of bioindicators for the assessment of urban air quality at the European level
- contribute to the standardization of methods
- analyses and evaluate the urban air quality in Europe
- compare pollution types among different cities and regions
- demonstrate the effects of air pollutants in a way easily comprehensible for laymen
- contribute to a sensitization of the urban population towards environmental issues
- stimulate initiatives in schools, companies, public authorities, and private households
- use the city's commitment to environmental protection in municipal marketing campaigns.

## 14.4 Concept of biomonitoring

Biomonitoring studies evaluate the response of plants in response to the changes in environmental conditions. The response of plants is then analyzed to assess the type and magnitude of the pollutant present. An $O_3$-sensitive plant, *Nicotianatabacum* L. Bel-W3, was grown across Europe to monitor the effects of tropospheric$O_3$. Four to six plants are exposed at each station for eight consecutive periods of two weeks each between the end of May and mid-September. At the end of the exposure, the percentage of $O_3$-induced lesions on three predetermined leaves is estimated in steps of 5% of leaf area with the help of a photo catalogue (Klumpp et al. 2002). The results of this monitoring program showed a direct correlation between $O_3$concentration indices (AOT20 and AOT40) and the visible foliar injury (Klump et al. 2006). Similar findings were also reported by Ribas and Penuelas (2003) during biomonitoring of tobacco plants in Spain. EuroBioNet Programme also used the ozone sensitive Poplar clones (*Populusnigra "Brandaris"*) that were continuously exposed to ambient ozone at the different biomonitoring stations between the end of May and mid-September. Visible injuries, the number of leaves, and shoot height were estimated and the data was used to establish the trends of $O_3$ concentration. Apart from this, PAH was monitored using Curly kale plants *Brassica oleracea cephala*), and sulphur concentration was analyzed through Italian ryegrass (*Lolium multiflorumitalicum*). These studies of the EuroBioNet Programme helped in estimating the local background levels for each element and each study area and on the basis of this 'European reference levels were determined by pooling all data. Another large-scale biomonitoring program was launched in the early 21st century that spans all across Europe and was known as ICP Forests Intensive Monitoring Network. In 1987, the international cooperative program on effects of air pollution on crops (ICP crops) was established as a subsidiary body of the Working Group on Effects of the UNECE Convention on long-range transboundary air pollution (LTRAP). In 1999, the ICP was renamed as the international cooperative program on the effects of air pollution on natural vegetation and crops (ICP vegetation). In 1985, a separate ICP was established to monitor the impacts of air pollution on forests (ICP Forests). Over time, ICP Vegetation and ICP Forests have generated a significant amount of information related to the trends and concentration of tropospheric $O_3$ across Europe. The main objective of these monitoring bodies, as per Harmens et al. (2015) are:

- Collate evidence of ozone impacts on vegetation, assess spatial patterns and temporal trends
- Develop dose–response relationships, establish critical levels for vegetation and provide European risk maps of $O_3$ impacts
- Publish thematic scientific reports and policy-relevant brochures on ozone impacts on vegetation by reviewing the literature.

The ozone monitoring program of ICP Vegetation included a multitude of plants of around 285 species and was monitored on almost 19 European nations. Out of the 285 plant species, 156 were found to be symptomatic and 74 plants displayed visible foliar injury (Michel and Seidling 2016). Like tobacco, white clover (*Trifoliumrepens* L.) plants were also utilized in another significant biomonitoring observation, known as ICP vegetation, in Europe from 1996 to 2006. The study reported $O_3$-induced foliar tissue damage even at low $O_3$ concentrations. From these findings, the $O_3$ flux-based monitoring studies were very much popularized along with concentration-based ones (Mills et al. 2011).

In 2000, ICP Vegetation took over the European moss survey, initiated in 1990. In this program, mosses were sampled every 5 years across Europe to study the spatial and temporal variations of heavy metal deposition to vegetation at a high spatial resolution (Harmens et al. 2010; 2015). The moss survey was later extended to include the determination of nitrogen (Harmens et al. 2011; 2015) and selected persistent organic pollutants (pops; Harmens et al. 2013), in naturally growing mosses.

Biomonitoring of heavy metals is another important aspect of monitoring studies and numerous reports exist that indicate the importance of plants and animals in these experiments. Calzoni et al. (2007) used *Rosa rugosa* plants to bio-monitor heavy metals (Pb) in a rural area of Italy. The elevated levels of lead in the leaves resulted in pollen abortiveness and their study suggested the usage of *Rosa rugosa* as a potential biomonitoring asset against heavy metals. Chandra et al. (2004) used all types of aquatic plants (free-floating species, submerged species, etc.) for monitoring chromium concentrations. Trees and shrubs were used by Ataabadi et al. (2010) to bio-monitor heavy metal exposure in an urban area of Iran and the results showed over-accumulation of iron and lead in the leaves. Ramadan, (2003) used various local plants for biomonitoring of four heavy metals (Zn, Pb, Cd, and Hg) in lake Manzala, Egypt. In a biomonitoring study done by Kumar et al. (1995), the results showed that *Brassica juncea* and *Brassica nigra* displayed high accumulation potential in the above-ground parts. Ebbs et al. (1997) used 30 plants species to study the accumulation of heavy metals, i.e., Cd, Cu, and Zn, and reported that *Brassica* sp. were hyper-accumulators in the shoot portion. Along with the above-mentioned examples, there exist plentiful more that reveals the importance of biomonitoring studies in maintaining pollution load below threshold levels.

## 14.5 Biomonitoring tools

Several plants (moss, fern, gymnosperm, and angiosperm) and animals including microorganisms (fish, protozoa, bacteria, daphnia, and algae) are used as biomarkers, bioindicator, biointegrator, and bioaccumulator (Rosenberg and Resh 1993; Badamasi, 2020) for environmental biotic and abiotic stress. Innate characteristics in plants help us to identify environmental changes from a smaller to a larger scale that why plants are used as bioindicators (Giordano et al., 2021) (Table 14.1). Several studies have been done for biomonitoring of environmental pollution and showed different bio-physicochemical and morphological characteristics (Moraes et al. 2003). Benthic Macroinvertebrates act as a good indicator for qualitative assessment for water quality (Rosenberg and Resh 1993). *Robinia pseudo acacia* L. leaves (Hasan et al., 2012) and Olive tree leaves (Turan et al., 2011)

**Table 14.1** Several selected plants and animals used for biomonitoring of different pollutants.

| Bioindicator | Pollutant | References |
|---|---|---|
| Alfalfa, Sesame, Moss, Spinach, Carrot, Sweet potato, and Apple tree | $SO_2$ | Shihong (2015) |
| Gladiolus, Garlic, Tulip, Tobacco, Plum, and Grape | Fluoride | Shihong (2015) |
| *Hypnumcupressiforme* | Heavy metals | Bargagli et al. (2002) |
| *Clostridium* species, *Salmonella tyhimurium* | Aquatic water logging | Saber et al. (2015b) |
| Plankton | P,N | Thakur et al. (2013) |
| Tobacco cultivar Bel W3, *Braccia oleracea, Phaseolus vulgaris* L., *Trifoliumalexandrium* L. | Tropospheric ozone | Calatayud et al. (2007); Agrawal et al. (2003); Chaudhary and Agrawal (2013) |
| Larch,Pinus, Alfalfa, Peach and Cabbage | Halogen ($Cl_2$, $Cl^-$) | Shihong (2015) |
| Tobacco, Oat, Wheat, Carrot, Corn, Citrus and Melon | $NO_2$ | Meng et al. (2015) |
| Petunia, Chickweed, Lettuce, Bean and Potato | PAN | Shihong (2015) |
| Ulva lactuca, Ostrea and *Balanusamphitrite,* Mussels, Nereis | Bioavailability of heavy metal in marine habitat | Phillips & Rainbow (1995) |
| Oyster (Crassostreagi gas), Crabs(Geoticadepressa) | Lead pollution | Uttah et al. (2008) |
| Grass shrimp | Aquatic heavy metal pollution | Key et al. (2006) |
| Cyanophyta phytoplankton | Eutrophication | Thakur et al. (2013) |
| Pleurocarpous moss | Atmospheric POPs | Wu et al. (2014) |
| *Hypnumcupressiforme* | Airborne priority pollutants | Capozzi et al. (2020) |
| Mosses and lichens | Radionuclide contamination | Nifontova and Alexashenko (1993) |
| Honey bee products and pollen | $^{137}Cs$ (radiocaesium) concentration | Bunzl and Kracke, (1981); Flood (1994) |

were used as a bioindicator for environmental pollution. Specific responses by plants can make them good bioindicators of pollution impacts, while tolerant species are ideally used as bioaccumulators, bioindicators can show a biochemical effect within the plant tissue that may eventually be expressed visibly (active biomonitors), bioaccumulators only indicate the level of pollutant in tissue (passive biomonitors) (Science Report – SC030175/SR2).

## 14.6 Parameters used as biomonitoring tools

### 14.6.1 Physiological parameters

The presence of different pollutants alters the physiological characteristics like photosynthetic gas exchange rate, stomatal conductance, and transpiration rate in plants and animals. Several air pollutants such as ozone, carbon dioxide, carbon monoxide, etc., flux inside the leaves are controlled by a small opening on leaves called stomata that have a significant effect on plant physiology. Surface $O_3$debilitates photosynthesis (Bell, 2017) and decreases stomatal conductivity in Arabidopsis (Vahisalu et al. 2010) and *Ruelliatuberosa* leaves as affected by coal smoke pollution (Partha 2014). The presence of trace metal also alters the physiological characteristics (Harmens et al. 2013) in *Taraxacum officinalis.* Ozone damages the sensitivity of stomatal cells to plant stress hormone signals (Conklin et al. 2004).

### 14.6.2 Visible leaf damage

Symptoms of visible damage can be seen in plants and animals. These symptoms include loss of chlorophyll (leaf chlorosis), tissue death (necrosis), or leaf loss (abscission) (Taylor et al. 1986), while symptoms in mammals have included dental lesions in voles (Boulton et al. 1994). When organisms are exposed to a higher concentration of toxin over a shorter period visible injuries will be acute and when exposure concentration is low but prolonged, symptoms will be chronic (science report – SC030175/SR2). Ozone is generally observed as the most phytotoxic atmospheric pollutant causing visible leaf injuries (Bell, 2017). Author (Shihong 2015) showed different visible leaves injuries in presence of different air pollutants like $SO_2$ $NO_2$ and chlorine showed interveinal necrotic blotches, fluoride showed red-brown distal necrosis, etc. on the other hand tropospheric ozone generated visible injury on the leaf which easily seen by naked eyes (Klumpp et al. 2006; Feng et al. 2014). Ozone stressed each plant species shows their specific leaves injury pattern (Manning et al. 2004; Feng et al. 2014). *Phaseolus vulgaris* L is one of the best examples showing bronze mottle on leaves (Sanz et al. 2020). *Nicotianatabacum* L. "Bel-W3" (hypersensitive bioindicator) has been frequently used in European countries and the American continent to estimate leaf injury index based on the percentage of the leaf area affected by necrosis (Assis et al. 2018). Hydrogen fluoride (HF) is more toxic than ozone, $SO_2$, or $NO_2$ (Weinstein et al. 1998) because small concentrations can damage the tip and margin of leaves (Weinstein and Davison 2004). Wan et al. 2014 estimated foliar injury symptoms 56.3% for the tree, 15.6% for shrubs, 21.9% for herb, and 6.3% for crop plants.

### 14.6.3 Estimate chemical element accumulation inside plant tissues

Estimating the accumulated pollutant inside the plant tissue is a common method for assessment of the effect of many harmful elements (He 1994) like for assessment of atmospheric nitrogen deposition, measuring the total tissue nitrogen in plant tissue for many years (Sutton et al. 2004). This method is relatively economical as other physical methods. *Vogesellaindigofera* bacterium is an important morphological marker of the quantity of heavy metals. This bacterium produces blue pigmentation in absence of heavy metal, which can be observed visually, on the other hand, hexavalent chromium blocked the production of blue pigment (Aslam et al. 2012).

### 14.6.4 Photosynthesis system and pigment deterioration

The leaf is an important part of plants and is the main organ for photosynthesis and transpiration (Shihong 2015). The presence of trace metal deteriorates the photosynthetic pigments (Ali, 1993) and showed a negative correlation with chlorophyll A in a purified strain of algae (*Chlorella ellipsoidea*), so this species is used as a suitable bioindicator in the monitoring of aquatic metal pollution (Zhou et al., 2008). Air pollutants reduced chlorophyll a, chlorophyll b, and carotenoids in *Ficusbenghalensis* L. (Prajapati and Tripathi, 2008) and 35%–60% reduction in photosynthetic pigments observed in *Shorearobusta*, *Albizialebbeck*, *Tectonagrandis*, *Lagerstroemia speciosa*, and *Delonixregia* exposed to industrial pollution (Nayek et al. 2011)

### 14.6.5 Production of antioxidant metabolites and change in antioxidant enzyme activities

During biotic and abiotic stress conditions plants can also indicate their stress via alteration in the production of antioxidant metabolites and change in antioxidant enzyme activities and minimized their oxidative stress (Heath 2008). Stomatal aperture is the first barrier acting against tropospheric ozone entering into mesophyll space and producing oxygen-free radicals that stimulate the enzymatic defense system that protects the cell (Conklin et al. 2004; Feng et al. 2010). A number of studies have measured antioxidant enzymes relative to exposure to pollutants in the field (De Luca-Abbott et al. 2005) and the laboratory (Brown et al. 2004).

The activities of superoxide dismutase (SOD), catalase (CAT), glutathione S-transferase (GST), glutathione (GSH) concentration, and malondialdehyde (MDA) production against metal pollution were reported in African catfish. Alteration in the enzymatic pool showed that African catfish (Farombi et al. 2007) and *Myriophyllumquitense* (Nimptsch et al. 2005) were used as good biomarkers of oxidative stress in biomonitoring of aquatic metal pollution. *Myriophyllumquitense* is also used for monitoring inorganic nitrogen loads combined with elevated lead and aluminum concentrations in the Auquia River basin (Nimptsch et al. 2005). The presence of air pollutants inside the plants that induced the formation of reactive oxygen species (ROS), which activated the enzymatic and non-enzymatic pool such as ozone stress increased the lipid peroxidation and phenolic content in radish (Tiwari and Agrawal, 2011) and UV-B stress increased SOD concentration in *Eclipta alba* (Rai and Agrawal, 2021). Low-cost enzymatic tests such as catalase and GST are used in laboratory studies and biomonitoring programs for fish (Romeo et al. 2000) and mussels (Romeo et al. 2003a, 2003b).

### 14.6.6 Tree ring analysis

The long-term pollution history in an area is analyzed by Tree ring. The analysis of tree ring patterns provides baseline data sets whereby alteration in growth response to different stresses can be detected (Shihong 2015). The potential of tree-ring analysis as a bioindication method in air pollution diagnosis is obvious. However, further research in this field is necessary to determine the credibility and future of the technique (Shihong 2015).

### 14.6.7 Community diversity variation

This method is relatively low cost compared to chemical tests of plant material. Variation in species diversity is a useful method for monitoring the health of an ecosystem. Any variation in community structure is a clear 'outcome' of pollutant impacts. The physical characteristics of species in polluted sites are the same as clean reference sites. The status of a community can be assessed. Variation in species composition is a direct measure of the biological response within an impacted community. The presence or absence of biomonitor species for air pollutants has been used mainly for atmospheric nitrogen deposition(Science report – SC030175/SR2).

In Addition, certain specific groups of an organism can directly be used for biomonitoring of environmental pollution

### 14.6.8 Using lichens to monitor air pollution and trace metal (Holt and Miller 2011)

The symbiotic relationship between bacteria and algae (lichen) can tolerate harsh environmental conditions. However, in the case of air pollutants mainly $SO_2$, $H_2S$ affect their growth and induce their death (Gombert et al. 2006). Niu 2000 and Shihong 2015 estimated lichen species, their number, and distribution in the area around the contaminated site and found as distance increased from the contaminated site number of the genus, species of lichen increased. Bryophytes and lichens are both good biomonitors of metal deposition (Science Report – SC030175/SR2). Lichen also accumulated the volatile heavy metals from atmospheric deposition more readily than in mosses (Evans and Hutchinson 1996) while metals accumulation from the soil (Al, Cr, Fe, Ti) more readily in mosses than in lichens (Loppi and Bonini 2000). Holt & Miller (2011) and Thakur et al. (2013) utilized lichens as biomonitors for quantitative analysis of metal concentrations inside individual lichens.

### 14.6.9 Using moss to monitor pollution especially air pollution and heavy metal

Mosses are sensitive indicators for different pollutants and accumulate more polluting than lichens. Moss and lichens are also useful for monitoring heavy metal and nitrogen pollution. Bargagli et al. (2002) estimated metal accumulation in two common species *Hypnumcupressiforme* (a moss) and *Parmeliacaperata* (a lichen) in a mining area in Italy. The mosses used in biomonitoring are mainly divided into two groups: pleurocarpous mosses (prostrate, matted, and highly branched) and acrocarpous mosses (erect, tufted, and sparingly branched) (Wu et al. 2014). During air pollution, moss will appear black or brown and species diversity will decline and will continue to decrease until it disappears in the long term. Moss-bag technique was used to monitor heavy metal, sulfur, and nitrogen in Wisconsin, USA, bags containing *Sphagnum russowii* (Makholm and Mladenoff 2005). Similar studies were done in Romania, Russia, and Bulgaria using *Sphagnum girgensohnii*; 36 elements were investigated (Culicov et al. 2005). In India, also using mosses as biomonitors to monitor trace element pollution (Chakrabortty et al. 2006). In India studies conducted in Mumbai, have identified *Pinnatellaalopccuroides, Pterobryopsisflexiceps*, and *Bryum spp.* as effective biomonitors for the region. The pollutant accumulation in moss occurs through different mechanisms such as layers of particles, attached to the upper surfaces of the cell, through the ion exchange method incorporated into the cell outer wall (Chakrabortty and Paratkar 2006).

### 14.6.10 Planktons

Plankton is highly sensitive to natural changes in diverse ecosystems that determine the pollution status of aquatic ecosystems. Scientists classified plankton into two categories phytoplankton (plants) and zooplankton (animals) which play crucial roles in the food chain for small and large biotas (Saber et al. 2015b). Cyanophyta plankton is the most significant indicator for rapid eutrophication of aquatic bodies via the creation of bloom formations (Thakur et al. 2013). Phytoplankton is a key source of aquatic ecosystem pollution. Their growth declined as absorbed or filtered the industrial pollutant (Malik and Bharti. 2012)

### 14.6.11 Earthworms

Earthworms serve as a powerful biological indicator in ecotoxicology risk assessment (Gao and Luo 2005), they enhance physical and chemical characteristics in ecosystems like activating microorganisms and changing bioavailability of pollutants. The density of earthworms indicates the health status of the ecosystem (Zaghloul et al. 2019).

### 14.6.12 Amphibians

Anurans (frogs and toads) are used as biological markers. Their skin and larvae gill membranes absorb toxic chemicals like pesticides that making them a model for evaluating the impact of ecological influences (Zhou et al. 2008).

## 14.7 Important biomonitoring programs

Several biomonitoring programs are occurring for monitoring of air pollutants, heavy metals, POPs, and radioactive nuclei across the world. The development of various biomonitoring programs led to the protection of aquatic and terrestrial ecosystems as well as human health. Several plants and animals including microorganisms called biomonitors or bioindicators used to detect the pollution at a site such as *Nicotiana tabaccum* L cultivar used to the monitoring of ozone phytotoxicity in EuroBioNet Programme in Europe (Klummp et al. 2006), GISP monitoring invasive species, etc. Several biomonitoring program on the aquatic ecosystem, forest, crop, and POPs describes below:

### 14.7.1 GISP

GISP was one of the monitoring programs on invasive species dealing with dynamic ecosystems started in 1997. GISP was part of DIVERSITAS, a program on biodiversity science. Its governing body was a consortium of the Scientific Committee on Problems of the Environment, CAB International, the World Conservation Union, and the United Nations Environment Programme (UNEP). It developed capabilities to employ rapid warning, better risk assessment methods of invasive, and strengthen international agreements. It recommends appropriate strategies to public education for a better understanding of ecology and the impact of invasive. GISP included every invasive species for management purposes as if it is potentially invasive, unless and until convincing evidence indicates that this is not so. This calls for urgent international action by a wide range of governmental, intergovernmental, private sector, and civil institutions (Mcneely, 2001).

### 14.7.2 National Aquatic Resources Survey

Major national biomonitoring programs in the USA and funded through the US Environmental Protection Agency (USEPA). The main objective of this is to examine the ecological status and trends in all US lakes, stream river wetlands, etc. (surface water bodies) under the Clean Water Act of 1972 (Buss et al. 2015). Assessment of whole water bodies is based on probability sampling and census of water bodies is logistically infeasible and previous attempts produced imprecise and inaccurate reports (Hughes et al. 2000)

### 14.7.3 CABIN

This aquatic biomonitoring program developed in Canada in 1999 (Reynoldson et al. 1999) relies on the Reference Condition Approach for assessment (Bailey et al. 2004). This program was first developed for research on the Great Lakes (Reynoldson et al. 1995) and in the Fraser River Basin, British Columbia (Reynoldson et al. 1997). CABIN provides assessment and reports on freshwater health via routine monitoring.

### 14.7.4 NES and NAEMP

These are Major national-scale macroinvertebrate biomonitoring programs for monitoring the ecological status of streams in South Korea. NES involved in monitoring species diversity in both terrestrial and aquatic ecosystems. One to two times surveyed of 3500 sites in every 5-year cycle in 1986. NES has reported 1875 freshwater ecosystems sites (Park et al. 2004). NAEMP was launched mainly for the assessment of aquatic ecosystems in 2008. Aiming specifically to assess five main river basins, where 640–960 sites are selected for surveying diatoms, macroinvertebrates, and fish two times per year. Based on a survey assessment report, Korean Saprobic Index sets multi taxa community organization and quality of water (Jun et al. 2012).

### 14.7.5 Aarhus protocol on POPs (UNECE, 1998)

Aarhus (Denmark) adopted this protocol on June 24, 1988. The major objective is to eliminate any discharge, emission, and losses of Pops. It focused mainly on 16 substances based on risk criteria. The substances include eleven pesticides, two industrial chemicals, and three by-products or contaminants.

### 14.7.6 Stockholm convention (UNEP, 2001, 2009)

During UNEP, 2001 the "dirty dozen" 12 POPs were listed as priority control and that banned. Further, 9 POPs were added to the new list at the Stockholm convention in 2009. Therefore, the major objectives of both conventions are to restrict the production and use of these dirty dozen Pops. Since the 1970s, National and international environmental monitoring programs have continued to measure POPs (Muir and Howard, 2006).

### 14.7.7 UNECE ICP vegetation program for POPs

As a part of ICP vegetation program review studies by Harmens et al. (2013) has mentioned that mosses are the most suitable biomonitor for POPs such as polycyclic aromatic hydrocarbons, polychlorobiphenyls, dioxins and furans, and polybrominated diphenyl ethers, etc. Mosses are collected at thousands of sites across Europe, and Asia (Badamasi, 2017). Norway, Poland, Slovenia, Spain, and Switzerland, conducted a pilot study on mosses as a biomonitor for pops in 2010 and mosses showed a different conception of POPs based on geographical distribution and atmospheric deposition patterns (Foan et al. 2014).

### 14.7.8 USFS (United States Forest Service) along with USEPA (United States Environmental Protection Agency)

This biomonitoring program included 24 states which were divided in two sub-regions: the northeast and the north-central region of the

USA from 1994 to 2009. Both subregions included 450 bio sites locations. This biomonitoring mainly focused on foliar injury due to tropospheric ozone in 46,000 ozone-sensitive plant species (*Prunusserotina, Apocymumandrosaemifoliumetc*). The concentration of ozone was examined on biosites as SUM06 (sum of all hourly average ozone concentration > 0.06 ppm) and N100 (number of hours of ozone > 100 ppm) Output of the program indicated that the indicator plant of all sites with higher ozone concentration showed more foliar damaged than species of lower concentration. Foliar injury index was determined for each site (tropospheric ozone its impact on crop plants (Tiwari and Agrawal. 2018). This program proved that foliar injury can be used as a potential tool for assessment of O3 concentration over a large area and $O_3$ injury was more prominent in the North-central US than in North-east United States (Smith2012).

### 14.7.9 EuroBioNet Programme, Europe

This program is European Network for the Assessment of Air Quality by the use of Bioindicator Plants in Europe in 1999 with financial assistance from the LIFE Environment Programme of the European Commission, which under the coordination of the University of Hohenheim has planned biomonitoring of air quality and increasing the environmental awareness of the urban population. The program provided a report on the spatial and temporal distribution of the effects of air pollutants both within local networks and at the European level. It includes research institutes and municipal environmental administrations from 12 cities in eight EU member states. In this program, tobacco cultivar Bel W3, poplar, ryegrass, curly kale are used for air pollutants (sulfurous compounds, trace metals, hydrocarbons, and mutagenic substances, and Polycyclic aromatic hydrocarbons) at different selected sites. Klumpp et al. (2006) determined the response of ozone-sensitive tobacco and tried to establish a relationship between ozone concentration and foliar damage on tobacco (Nicotianatabaccum L) cultivar at 100 biomonitoring sites in urban centers for three years. The output of this result showed a significant correlation between foliar damage level and various ozone exposure indices like AOT20 and AOT40 (Klumpp et al. 2006). At a rural site in Spain, a similar observation was recorded by Ribas and Penuelas. (2003). The small-scaled evaluation showed a linear relationship between ambient ozone and foliar damage as compared to a large area (Cuny et al. 2004). This programmed indicate very clearly that the gradient of O3 impact is increasing with plant injury levels from North and northwest Europe to Central and Southern regions

### 14.7.10 United States Forest Service (USFS): Forest Health Monitoring program (FHM)

The major aim is an assessment of forest health and air quality (mainly ozone-induced foliar injury on ozone-sensitive bioindicators) with public and scientific concerns (USFS 2000). It restores the health, diversity, and productivity of forest and grassland and makes a motto to care for the land and serve people ("The U.S. Forest Service - An Overview" (PDF). U.S. Forest Service). In the FHM program, 33 states were participating which included 918 ozone monitoring biosites and 2000 field seasons. FHM provided necessary baseline data that clearly showed the ozone air quality is relatively good across wide areas of the West (U.S. EPA 1996a). The bioindicator in the eastern site has provided data related to ozone field studies (Krupa and Manning, 1988) but in western bioindicator species have not provided data of ozone exposure, but received enough testing to justify inclusion in the FHM program, (Campbell et al. 2000). FHM intends to use consecutive 5-year periods (e.g., 1994–1998; 1999–2003) with variable $O_3$ levels, weather, wind flow, and precipitation patterns to examine regional statistics in $O_3$relative air quality over the long term.

### 14.7.11 UNECE-ICP vegetation for tropospheric ozone

It was large-scale biomonitoring across Europe by the International Co-operative Program on Effects of Air Pollution on Natural Vegetation and Crops (ICP Vegetation) since 1999 (Harmens et al. 2015). ICP vegetation also examines the impact of air pollution on trees that separate ICP was established in 1987 called ICP forest (Harmens et al. 2015). ICP Vegetation was mainly examined spatial patterns and temporal trends regarding ozone impacts on vegetation within the UNECE area and provided effective mitigation strategies. Hayes et al. (2007a) conducted a meta-analysis to identify the relationship between $O_3$ sensitivity and physiological and ecological characteristics of plants. ICP vegetation collected 644 databases of impacts of ambient on vegetation for 171 species (Mills et al. 2011) of the 644 records of visible injury, 39% were for crops (27 species), 38.1% were for (semi)–natural vegetation (95 species), and 22.9% were for shrubs (49 species; Mills et al. 2011). The author in ICP vegetation program used ethylene diurea (EDU the as an ozone-protectant at the experimental site during the years 1994, 1995, and 1996 (Ball et al. 1998). Studied two sets of several plant species like a bean (*Phaseolus vulgaris*), radish (*Raphanussativus*), white clover (*Trifoliumrepens*), red clover (*Trifolium pratense*), and tobacco (*Nicotianatabacum*), etc. were prepared and compared visible injury and/ or biomass. According to Hayes et al. (2007b), many-treated plants reduced visible injury and protected biomass reduction in ambient ozone conditions. ICP vegetation biomonitoring has been studied on white clover (sensitive to ozone) between 1996 and 2006 (Hayes et al. 2007b; Mills et al. 2011) in all regions of Europe. Ozone resistant varieties (NC-R) showed resistant character but some sensitive varieties (NC-S) showed small to large necrotic patches with leaves severely damaged (Harmens et al. 2015) at a site. Visible leaves damage data were collected from 45 biosites of 16 countries across Europe and Italy (Manes et al. 2003). ICP Vegetation 2008 has been used snap bean, which consists of the $O_3$-sensitive 101 genotype S156 and $O_3$-insensitive R123 that are used for large-scale biomonitoring across Europe (Harmens et al. 2015). However, the experience of ICP Vegetation examines the visible leaf injuries symptoms on leaves but with no clear correlation between $O_3$ exposure and visible leaf injury (Harmens et al. 2015). From a comparison of this discussion on snap beans with white clover, it seems that white clover is more efficient than snap beans for $O_3$ biomonitoring.

## 14.8 Conclusion

Studies conducted in different parts of the World provide a very transparent and clear-cut picture of the enormous negative impacts of environmental pollutants on crops and vegetation. The current levels of pollutants in nature trigger plant damage and are significant not only in developing but also in developed countries of the world, i.e., the problem has acquired a global outlook. Despite the implementation of current legislation, the pollution levels are predicted to follow an upward trend in the near future and thus it is likely that the effect of pollution on terrestrial as well as crop and vegetation will worsen. Due to the unplanned and uncontrolled emissions of pollutants in the environment, the developing countries especially those of south and east Asia are predicted to face a more severe consequence.

The biomonitoring programs have identified several sensitive species that can be used as bioindicators. Bioindicator plants are adequate means of communicating aspects of environmental pollution to the broad public, as especially visible injury symptoms are much more perceptible and comprehensible for laymen than data on ambient air pollutant concentrations can ever be. Bioindication makes air pollution problems visible and understandable to people directly and within their everyday life, whereas these problems are normally quite abstract. Although the concept of biomarkers has gained momentum in recent years, still this concept needs to be more focused upon. Biomonitoring studies provide relevant information that helps to frame policy issues related to the protection of animals, plants, and vegetation cover. Predictions by a few modelling studies indicated the global risk to the living beings and to the important staple crops which may experience the greater downfall in their productivity due to increasing levels of pollution in the atmosphere. Based on the recorded crop loss production and deterioration in the health of terrestrial and aquatic organisms, it is clear that the programs conducted during the last few decades are not sufficient, and more such programs and legislation are needed for getting a clear picture of environmental pollution on our planet.

## References

Agrawal, M., Deepak, S.S., 2003. Physiological and biochemical responses of two cultivars of wheat to elevated levels of CO2 and SO2, singly and in combination. Environ. Pollut. 121 (2), 189–197.

Ali, E.A., 1993. Damage to plants due to industrial pollution and their use as bioindicators in Egypt. Environ. Pollut. 81 (3), 251–255.

Aslam, M.,Verma, D.K., Dhakerya, R., Rais, S., Alam, M., Ansari, F.A., 2012. Bioindicator: a comparative study on uptake and accumulation of heavy metals in some plants leaves of M.G. Road, Agra City. Indian Res. J. Environ. Earth Sci. 4 (12), 1060–1070.

Assis, P.I., Moraes, R.M., Nakazato, R.K., 2018. Will the shift from crude oil to natural gas burning for power generation at an oil refinery increase ozone concentrations in the region of Cubatão (SE-Brazil)? Ecol. Indic. 85, 921–931.

Ataabadi, M., Hoodaji, M., Najafi, P., Adib, F., 2010. Evaluation of airborne heavy metal contamination by plants growing under industrial emissions. Environ. Eng. Manage. J. (EEMJ) 9 (7), 903–908.

Badamasi, H., 2017. Biomonitoring of air pollution using plants. MAYFEB J. Environ. Sci. 2, 27–39.

Badamasi, I., Odong, R., Masembe, C., 2020. Threats posed by xenoestrogenic chemicals to the aquatic ecosystem, fish reproduction and humans: a review. Afr. J. Aquat. Sci. 45 (3), 243–258.

Bailey, R.C., Norris, R.H., Reynoldson, T.B., 2004. Bioassessment of freshwater ecosystems-Bioassessment of Freshwater Ecosystems. Springer, Boston, MA, pp. 1–15.

Ball, G.R., Benton, J., Palmer-Brown, D., Fuhrer, J., Skärby, L., Gimeno, B.S., Mills, G., 1998. Identifying factors which modify the effects of ambient ozone on white clover (Trifoliumrepens) in Europe. Environ. Pollut. 103 (1), 7–16.

Bargagli, R., Monaci, F., Borghini, F., Bravi, F., Agnorelli, C., 2002. Mosses and lichens as biomonitors of trace metals. A comparison study on Hypnumcupressiforme and Parmeliacaperata in a former mining district in Italy. Environ. Pollut. 116 (2), 279–287.

Bell, J.N.B., 2017. Encyclopedia of Applied Plant Sciences. Plants Atmos. Pollut., 332–340. doi:10.1016/B978-0-12-394807-6.00017-4.

Boulton, I.C., Cooke, J.A., Johnson, M.S., 1994. Experimental fluoride accumulation and toxicity in the short-tailed field vole (Micvotusagrestis). J. Zool. 234 (3), 409–421.

Brown, D.H., Brown, R.M., 1990. Reproducibility of sampling for element analysis using bryophytesElement Concentration Cadasters in Ecosystems. 55, 62.

Brown, P.J., Long, S.M., Spurgeon, D.J., Svendsen, C., Hankard, P.K., 2004. Toxicological and biochemical responses of the earthworm Lumbricusrubellus to pyrene, a non-carcinogenic polycyclic aromatic hydrocarbon. Chemosphere 57 (11), 1675–1681.
Bunzl, K., Kracke, W., 1981. sup (239/240) Pu, 137 Cs, 90 Sr, and 40 K in different types of honey. Health Phys. 41 (3), 554–558.
Buss, D.F., Carlisle, D.M., Chon, T.S., Culp, J., Harding, J.S., Keizer-Vlek, H.E., Robinson, W.A., Strachan, S., Thirion, C., Hughes, R.M., 2015. Stream biomonitoring using macroinvertebrates around the globe: a comparison of large-scale programs. Environ. Monit. Assess. 187 (1), 1–21.
Calatayud, V., Sanz, M.J., Calvo, E., Cerveró, J., Ansel, W., Klumpp, A., 2007. Ozone biomonitoring with Bel-W3 tobacco plants in the city of Valencia (Spain). Water Air Soil Pollut. 183 (1), 283–291.
Calzoni, G.L., Antognoni, F., Pari, E., Fonti, P., Gnes, A., Speranza, A., 2007. Active biomonitoring of heavy metal pollution using *Rosa rugosa* plants. Environ. Pollut. 149 (2), 239–245.
Campbell, S.J., Smith, G., Temple, P., Pronos, J., Rochefort, R., Andersen, C., 2000. Monitoring for ozone injury in west coast (Oregon, Washington, California) forests. In: 1998. Department of Agriculture. Forest Service, Pacific Northwest Research Station, Portland.
Capozzi, F., Di Palma, A., Sorrentino, M.C., Adamo, P., Giordano, S., Spagnuolo, V., 2020. Morphological Traits Influence the Uptake Ability of Priority Pollutant Elements by Hypnumcupressiforme and Robiniapseudoacacia Leaves. Atmosphere 11 (2), 148.
Chakrabortty, S., Paratkar, G.T., 2006. Biomonitoring of trace element air pollution using mosses. Aerosol Air Qual. Res. 6 (3), 247–258.
Chakrabortty, S., Jha, S.K., Puranik, V.D., Paratkar, G.T., 2006. Use of mosses and lichens as biomonitors in the study of air pollution near Mumbai. Evansia 23, 1–8.
Chandra, P., Kulshreshtha, K., 2004. Chromium accumulation and toxicity in aquatic vascular plants. Bot. Rev. 70 (3), 313–327.
Chaudhary, N., Agrawal, S.B., 2013. Intraspecific responses of six Indian clover cultivars under ambient and elevated levels of ozone. Environ. Sci. Pollut. Res. 20 (8), 5318–5329.
Conklin, P.L., Barth, C., 2004. Ascorbic acid, a familiar small molecule intertwined in the response of plants to ozone, pathogens, and the onset of senescence. Plant Cell Environ. 27 (8), 959–970.
Costa, C., Teixeira, J.P., Silva, S., Roma-Torres, J., Coelho, P., Gaspar, J., Alves, M., Laffon, B., Rueff, J., Mayan, O., 2006. Cytogenetic and molecular biomonitoring of a Portuguese population exposed to pesticides. Mutagenesis 21 (5), 343–350.
Cribb, J., 2021. Earth Detox: How and Why We Must Clean Up Our Planet. Cambridge University Press, Cambridge.
Culicov, O.A., Mocanu, R., Frontasyeva, M.V., Yurukova, L., Steinnes, E., 2005. Active moss biomonitoring applied to an industrial site in Romania: relative accumulation of 36 elements in moss-bags. Environ. Monit. Assess. 108 (1), 229–240.
Cuny, D., Davranche, L., Van Haluwyn, C., Plaisance, H., Caron, B., Malrieu, V., 2004. Monitoring ozone by using tobacco, automated network and passive samplers in an industrial area in France. In: Klumpp, A., Ansel, W., Klumpp, G. (Eds.), Urban Air Pollution. Bioindication and Environmental Awareness. Cuvillier, Göttingen, pp. 97–108.
De Luca-Abbott, S.B., Richardson, B.J., McClellan, K.E., Zheng, G.J., Martin, M., Lam, P.K., 2005. Field validation of antioxidant enzyme biomarkers in mussels (Pernaviridis) and clams (Ruditapesphilippinarum) transplanted in Hong Kong coastal waters. Mar. Pollut. Bull. 51 (8-12), 694–707.
Ebbs, S.D., Kochian, L.V., 1997. Toxicity of zinc and copper to *Brassica* species: implications for phytoremediation. American Society of Agronomy. Crop Sci. Soc. Am. Soil Sci. Soc. Am. 26 (3), 776–781.

European Network for the Assessment of Air Quality by the Use of Bioindicator Plants - Final Technical Report (2004)

Evans, C.A., Hutchinson, T.C., 1996. Mercury accumulation in transplanted moss and lichens at high elevation sites in Quebec. Water Air Soil Pollut. 90 (3), 475–488.

Farombi, E.O., Adelowo, O.A., Ajimoko, Y.R., 2007. Biomarkers of oxidative stress and heavy metal levels as indicators of environmental pollution in African cat fish (Clariasgariepinus) from Nigeria Ogun River. Int. J. Environ. Res. Public Health 4 (2), 158–165.

Feng, Z., Wang, S., Szantoi, Z., Chen, S., Wang, X., 2010. Protection of plants from ambient ozone by applications of ethylenediurea (EDU): a meta-analytic review. Environ. Pollut. 158 (10), 3236–3242.

Feng, Z., Sun, J., Wan, W., Hu, E., Calatayud, V., 2014. Evidence of widespread ozone-induced visible injury on plants in Beijing, China. Environ. Pollut. 193, 296–301.

Flood, S., 1994. The Use of Pollen Pellets to Monitor Caesium in the Environment. School of Botany, Trinity College, Dublin.

Foan, L., Leblond, S., Thöni, L., Raynaud, C., Santamaria, J.M., Sebilo, M., Simon, V., 2014. Spatial distribution of PAH concentrations and stable isotope signatures (δ13C, δ15N) in mosses from three European areas—characterization by multivariate analysis. Environ. Pollut. 184, 113–122.

Füreder, L., Reynolds, J.D., 2003. Is austropotamobiuspallipes a good bioindicator? Bull. Franç. Pêcheet Piscic. 370-371, 157–163.

Gao, Y., Luo, Y., 2005. Earthworms as biological indicators of soil pollution and their potential for remediation of contaminated soils. EPA, NCEA, HERO 42, 140–148.

Gerhardt, A., 1999. Biomonitoring of polluted water - reviews on actual topicsEnvironmental Research Forum9. Trans Tech Publications - Scitech Publications, Uetikon-Zuerich, Switzerland, pp. 1–13.

Giordano, S., Spagnuolo, V., Capozzi, F., 2021. Biomonitoring of Air Pollution.

Gombert, S., Asta, J., Seaward, M.R.D., 2006. Lichens and tobacco plants as complementary biomonitors of air pollution in the Grenoble area (Isère, southeast France). Ecol. Indic. 6 (2), 429–443.

Harmens, H., Norris, D.A., Steinnes, E., Kubin, E., Piispanen, J., Alber, R., Aleksiayenak, Y., Blum, O., Coşkun, M., Dam, M., De Temmerman, L., 2010. Mosses as biomonitors of atmospheric heavy metal deposition: spatial patterns and temporal trends in Europe. Environ. Pollut. 158 (10), 3144–3156.

Harmens, H., Mills, G., Hayes, F., Norris, D., 2011. Air Pollution and Vegetation: ICP Vegetation Annual Report 2010/2011. NERC/Centre for Ecology & Hydrology.

Harmens, H., Norris, D., Mills, G., 2013. Heavy Metals and Nitrogen in Mosses: Spatial Patterns in 2010/2011 and Long-Term Temporal Trends in Europe. NERC/Centre for Ecology & Hydrology.

Harmens, H., Mills, G., Hayes, F., Norris, D.A., Sharps, K., 2015. Twenty eight years of ICP vegetation: an overview of its activities. Ann. Bot. 5, 31–43.

Harmens, H., Norris, D.A., Sharps, K., Mills, G., Alber, R., Aleksiayenak, Y., Blum, O., Cucu-Man, S.M., Dam, M., De Temmerman, L., Ene, A., 2015. Heavy metal and nitrogen concentrations in mosses are declining across Europe whilst some "hotspots" remain in 2010. Environ. Pollut. 200, 93–104.

Hasan, Z., Kamran, R.O., Mohammad, F., Ahmad, G., Hosein, F., 2012. Evaluation of different conditions on the mixing bitumen and carbon nano tubes. International Journal of Civil & Environmental Engineering IJCEE-IJENS 12 (06), 53–59.

Hatje, V., 2016. Biomonitors. In: Kennish, M.J. (Ed.), Encyclopedia of Estuaries. Encyclopedia of Earth Sciences Series. Springer, Dordrecht. https://doi.org/10.1007/978-94-017-8801-4_140.

Hayes, F., Jones, M.L.M., Mills, G., Ashmore, M., 2007. Meta-analysis of the relative sensitivity of semi-natural vegetation species to ozone. Environ. Pollut. 146 (3), 754–762.
Hayes, F., Mills, G., Harmens, H., Norris, D., 2007. Evidence of Widespread Ozone Damage to Vegetation in Europe (1990-2006). ICP Vegetation Programme Coordination Centre, CEH, Bangor, UK.
Heath, R.L., 2008. Modification of the biochemical pathways of plants induced by ozone: what are the varied routes to change? Environ. Pollut. 155 (3), 453–463.
He, Z.H., 1994. Environmental Monitoring. China AgriculturePress, Beijing, pp. 195–203.
Hilty, J., Merenlender, A., 2000. Faunal indicator taxa selection for monitoring ecosystem health. Biol. Conserv. 92, 185–197.
Holt, E.A., Miller, S.W., 2011. Bioindicators: using organisms to measure. Nature 3, 8–13.
Hughes, R.M., Paulsen, S.G., Stoddard, J.L., 2000. EMAP-Surface Waters: A multiassemblage, probability survey of ecological integrity in the USAAssessing the Ecological Integrity of Running Waters. Springer, Dordrecht, pp. 429–443.
Jun, Y.C., Won, D.H., Lee, S.H., Kong, D.S., Hwang, S.J., 2012. A multimetric benthic macroinvertebrate index for the assessment of stream biotic integrity in Korea. Int. J. Environ. Res. Public Health 9 (10), 3599–3628.
Key, P.B., Wirth, E.F., Fulton, M.H., 2006. A review of grass shrimp, Palaemonetes spp., as a bioindicator of anthropogenic impacts. Environ. Bioindic. 1 (2), 115–128.
Klumpp, A., Klumpp, G., Ansel, W. and Fomin, A., 2002. European network for the assessment of air quality by the use of bioindicator plants-the first year of Euro. Bio. Net.
Klumpp, A., Domingos, M., Klumpp, G., 2002. Foliar Nutrient Contents in Tree Species of the Atlantic Rain Forest as Influenced by Air Pollution from the Industrial Complex of Cubatão, SE-Brazil. Water Air Soil Pollut. 133, 315–333. https://doi.org/10.1023/A:1012914720934.
Klumpp, A., Ansel, W., Klumpp, G., Calatayud, V., Garrec, J.P., He, S., Penuelas, J., Ribas, A., Ro-Poulsen, H., Rasmussen, S., Sanz, M.J., 2006. Ozone pollution and ozone biomonitoring in European cities. Part I: ozone concentrations and cumulative exposure indices at urban and suburban sites. Atmos. Environ. 40 (40), 7963–7974.
Klumpp, A., Ansel, W., Klumpp, G., Vergne, P., Sifakis, N., Sanz, M.J., Rasmussen, S., Ro-Poulsen, H., Ribas, A., Penuelas, J., Kambezidis, H., 2006. Ozone pollution and ozone biomonitoring in European cities Part II. Ozone-induced plant injury and its relationship with descriptors of ozone pollution. Atmos. Environ. 40 (38), 7437–7448.
Krupa, S.V., Manning, W.J., 1988. Atmospheric ozone: formation and effects on vegetation. Environ. Pollut. 50 (1-2), 101–137.
Kumar, P.N., Dushenkov, V., Motto, H., Raskin, I., 1995. Phytoextraction: the use of plants to remove heavy metals from soils. Environ. Sci. Technol. 29 (5), 1232–1238.
Loppi, S., Bonini, I., 2000. Lichens and mosses as biomonitors of trace elements in areas with thermal springs and fumarole activity (Mt. Amiata, central Italy). Chemosphere 41 (9), 1333–1336.
Macintyre, W.H., Harden, L.J., Hester, W., 1952. Measurement of atmospheric fluorine. Ind. Eng. Chem. 44, 1365–1370.
Makholm, M.M., Mladenoff, D.J., 2005. Efficacy of a biomonitoring (moss bag) technique for determining element deposition trends on a mid-range (375 km) scale. Environ. Monit. Assess. 104 (1), 1–18.
Malik, D.S., Bharti, U., 2012. Status of plankton diversity and biological productivity of Sahastradhara stream at Uttarakhand, India. J. Appl. Nat. Sci. 4 (1), 96–103.
Manes, F., De Santis, F., Giannini, M.A., Vazzana, C., Capogna, F., Allegrini, I., 2003. Integrated ambient ozone evaluation by passive samplers and clover biomonitoring mini-stations. Sci. Total Environ. 308 (1-3), 133–141.
Manning, W.J., Godzik, B., 2004. Bioindicator plants for ambient ozone in Central and Eastern Europe. Environ. Pollut. 130 (1), 33–39.

Martin, M.H., Coughtrey, P.J., 1982. Biological Monitoring of Heavy Metal Pollution. In: Mellanby, K (Ed.), Land and Air, Pollution Monitoring Series. *Applied Sci. Publ., London*, pp. 93–96.

Markert, B., Wappelhorst, O., Weckert, V., Herpin, U., Siewers, U., Friese, K., 1999. The use of bioindicators for monitoring the heavy-metal status of the environment. J. Radioanal. Nucl. Chem. 240 (2), 425–429 Li L et al. /Procedia Environmental Sciences 2 (2010) 1510–1524 1517.

Markert, B., Oehlmann, J., Roth, M., 1997. General aspects of heavy metal monitoring by plants and animals. In: Subramanian, G, Iyengar, V (Eds.). Environmental Biomonitoring - Exposure Assessment and Specimen Banking. ACS Symposium Series, 654. American Chemical Society, Washington DC.

Markert, B, Breure, T, Zechmeister, H (Eds.), 2003. Bioindicators and Biomonitors - Principles, Concepts and Applications. Elsevier, Amsterdam.

Mcneely J.A., 2001. Global strategy for addressing the problem of invasive alien species. A result of the Global Invasive Species Programme. GISP.

Meng, X., Chen, L., Cai, J., Zou, B., Wu, C.F., Fu, Q., Zhang, Y., Liu, Y., Kan, H., 2015. A land use regression model for estimating the NO2 concentration in Shanghai, China. Environ. Res. 137, 308–315.

Michel, A. Seidling, W., 2016. Forest condition in Europe: 2016 technical report of ICP forests: Report under the UNECE convention on long-range transboundary air pollution (CLRTAP).

Mills, G., Hayes, F., Simpson, D., Emberson, L., Norris, D., Harmens, H., Bueker, P., 2011. Evidence of widespread effects of ozone on crops and (semi-) natural vegetation in Europe (1990–2006) in relation to AOT40-and flux-based risk maps. Global Change Biol. 17 (1), 592–613.

Moraes, R.M., Delitti, W.B.C., Moraes, J.A.P.V., 2003. Gas exchange, growth, and chemical parameters in a native Atlantic forest tree species in polluted areas of Cubatao, Brazil. Ecotoxicol. Environ. Saf. 54 (3), 339–345.

Muir, D.C., Howard, P.H., 2006. Are there other persistent organic pollutants? A challenge for environmental chemists. Environ. Sci. Technol. 40 (23), 7157–7166.

Naidu, R., Biswas, B., Willett, I.R., Cribb, J., Singh, B.K., Nathanail, C.P., Coulon, F., Semple, K.T., Jones, K.C., Barclay, A., Aitken, R.J., 2021. Chemical pollution: A growing peril and potential catastrophic risk to humanity. Environ. Int. 156, 106616.

Nayek, S., Satpati, S., Gupta, S., Saha, R.N., Datta, J.K., 2011. IndustrIAlpollutIon. J. Environ. Sci. Eng. 53 (1), 57–64.

Nifontova, M.G., Alexashenko, V.N., 1993. Content of 90 Sr and 134,137 Cs in fungi, lichens, and mosses in the vicinity of the Chernobyl nuclear power plant. Sov. J. Ecol. 23 (3), 152–156.

Nimptsch, J., Wunderlin, D.A., Dollan, A., Pflugmacher, S., 2005. Antioxidant and biotransformation enzymes in Myriophyllumquitense as biomarkers of heavy metal exposure and eutrophication in Suquía River basin (Córdoba, Argentina). Chemosphere 61 (2), 147–157.

Niu, C.Q., 2000. Introduction to contribution of animal and plant to air pollution monitoring. Jinzhong Normal Junior Coll. 17 (2), 27–28.

Oertel, N., Salánki, J., 2003. Biomonitoring and Bioindicators in Aquatic Ecosystems. In: Ambasht, RS, Ambasht, NK (Eds.), Modern Trends in Applied Aquatic Ecology. Kluwer Academic/Plenum Publishers, New York, pp. 219–246.

Park, Y.S., Chon, T.S., Kwak, I.S., Lek, S., 2004. Hierarchical community classification and assessment of aquatic ecosystems using artificial neural networks. Sci. Total Environ. 327 (1-3), 105–122.

Partha, P., 2014. Biomonitoring with special reference to visible damages in different plant species due to air pollution-a review. Int. Lett. Nat. Sci. 11 (1), 32–37.

Prajapati, S.K., Tripathi, B.D., 2008. Seasonal variation of leaf dust accumulation and pigment content in plant species exposed to urban particulates pollution. J. Environ. Qual. 37 (3), 865–870.

Puckett, K.J., Burton, M.A.S., 1981. The effect of trace elements on lower plantsEffect of Heavy Metal Pollution on Plants. Springer, Dordrecht, pp. 213–238.

Pure Earth; Green Cross Switzerland, 2016. World's Worst Pollution Problems—The Toxics Beneath Our Feet. Pure Earth and Green Cross, Switzerland NY, USA and Zurich, Switzerland. http://www.worstpolluted.org/2016-report.html.

Rai, K., Agrawal, S.B., 2021. An assessment of dose-dependent UV-B sensitivity in Eclipta alba: Biochemical traits, antioxidative properties, and wedelolactone yield. Environ. Sci. Pollut. Res. 28 (33), 45434–45449.

Rainbow, P.S., 1995. Biomonitoring of heavy metal availability in the marine environment. Mar. Pollut. Bull. 31 (4–12), 183–192.

Ramadan, A.A., 2003. Heavy metal pollution and biomonitoring plants in Lake Manzala, Egypt. Pakistan Journal of Biological Sciences (Pakistan).

Review and implementation study of biomonitoring for assessment of air quality outcomes-Science Report – SC030175/SR2 SCHO0907BNGY

Reynoldson, T.B., Bailey, R.C., Day, K.E., Norris, R.H., 1995. Biological guidelines for freshwater sediment based on BEnthic Assessment of SedimenT (the BEAST) using a multivariate approach for predicting biological state. Aust. J. Ecol. 20 (1), 198–219.

Reynoldson, T.B., Norris, R.H., Resh, V.H., Day, K.E., Rosenberg, D.M., 1997. The reference condition: a comparison of multimetric and multivariate approaches to assess water-quality impairment using benthic macroinvertebrates. J. North Am. Benthol. Soc. 16 (4), 833–852.

Reynoldson, T.B., 1999. Strategy for a Canadian Aquatic Biomonitoring Network. National Water Research Institute= Institut national de recherche sur les eaux.

Ribas, A., Peñuelas, J., 2003. Biomonitoring of tropospheric ozone phytotoxicity in rural Catalonia. Atmos. Environ. 37 (1), 63–71.

Romeo, M., Bennani, N., Gnassia-Barelli, M., Lafaurie, M., Girard, J.P., 2000. Cadmium and copper display different responses towards oxidative stress in the kidney of the sea bass Dicentrarchuslabrax. Aquat. Toxicol. 48 (2-3), 185–194.

Romeo, M., Mourgaud, Y., Geffard, A., Gnassia-Barelli, M., Amiard, J.C., Budzinski, H., 2003a. Multimarker approach in transplanted mussels for evaluating water quality in Charentes, France, coast areas exposed to different anthropogenic conditions. Environ. Toxicol. 18, 295–305.

Romeo, M., Hoarau, P., Garello, G., Gnassia-Barelli, M., Girard, J.P., 2003b. Mussel transplantation and biomarkers as useful tools for assessing water quality in the NW Mediterranean. Environ. Pollut. 122, 369–378.

Rosenberg, D.M., 1992. Freshwater biomonitoring and Chironomidae. Netherland J. Aquat. Ecol. 26 (2), 101–122.

Rosenberg, D.M., Resh, V.H, 1993. Freshwater Biomonitoring and Benthic Macroinvertebrates. Chapman and Hall, New York.

Rosenberg, D.M., 1998. A National Aquatic Ecosystem Health Program for Canada: We should go against the flow. Bull. Entomol. Soc. Can. 30 (4), 144–152.

Rühling, Å., Steinnes, E.eds., 1998. Atmospheric Heavy Metal Deposition in Europe 1995-1996. Nordic Council of Ministers.

Saber, M., Abouziena, H.F., Hoballah, E., Soad El-Ashry, Zaghloul, A.M., 2015. Phytoremediation of potential toxic elements in contaminated sewagedsoils by Sunflower (Helianthus annuus) and Corn (Zea mays L.) plants, 12th Inter Phytotechnology Conf. Manhattan, Kansas USA, 27-30 September (2015).

Sanz, M.J.; Calvo, E., 2020. Ozone injury in crops. Available online: http://ozoneinjury.org (Accessed January 17, 2020).

Sloof, J.E., 1995. Lichens as quantitative biomonitors for atmospheric trace-element deposition, using transplants. Atmos. Environ. 29 (1), 11–20.

Smith, G., 2012. Ambient ozone injury to forest plants in Northeast and North Central USA: 16 years of biomonitoring. Environ. Monit. Assess. 184 (7), 4049–4065.

Sutton, M.A., Pitcairn, C.E.R., Whitfield, C.P., Leith, I.D., Sheppard, L.J., van Dijk, N., Tang, S., Skiba, U., Smart, S., Mitchell, R., Wolsley, P., 2004. Bioindicator and biomonitoring methods for assessing the effects of atmospheric nitrogen on statutory nature conservation sites.Peterborough, JNCC, p. 230. (JNCC Report, 356).

Taylor, H.J., Ashmore, M.R., Bell, J.N.B., 1986. Air Pollution Injury to Vegetation (A Guidance Manual Commisioned by HM Industrial Air Pollution Inspectorate of the Health and Safety Executive). *Imperial College Centre for Environmental Technology (ICCET) Imperial College of Science and Technology, IEHO, Chadwick House, Rushworth Street, London.*

Thakur, R.K., Jindal, R., Singh, U.B., Ahluwalia, A.S., 2013. Plankton diversity and water quality assessment of three freshwater lakes of Mandi (Himachal Pradesh, India) with special reference to planktonic indicators. Environ. Monit. Assess. 185 (10), 8355–8373.

Thomas, W., Herrmann, R., 1980. Nachweis von Chlorpestiziden, PCB, PCA und Schwermetallen mittels epiphytischer Moose als Biofilter entlang eines Profils durch Mitteleuropa. Staub-Reinhaltung der Luft 40 (10), 440–444.

Tiwari, S., Agrawal, M., 2011. Assessment of the variability in response of radish and brinjal at biochemical and physiological levels under similar ozone exposure conditions. Environ. Monit. Assess. 175 (1), 443–454.

Tiwari, S., Agrawal, M., 2018. Tropospheric Ozone and Its Impacts on Crop Plants: A Threat to Future Global Food Security. Springer.

Turan, D., Kocahakimoglu, C., Kavcar, P., Gaygısız, H., Atatanir, L., Turgut, C., Sofuoglu, S.C., 2011. The use of olive tree (Oleaeuropaea L.) leaves as a bioindicator for environmental pollution in the Province of Aydın, Turkey. Environ. Sci. Pollut. Res. 18 (3), 355–364.

U.S. Environmental Protection Agency, 1996. Air Quality Criteria for Ozone and Related Photochemical Oxidants, Vol I of III, Section 4.0, Environmental Concentrations, Patterns, and Exposure Estimates. Office of Research and Development, Washington, DC EPA/600/P-93/004aF.20460.

USDA Forest Service, 2000. Forest Inventory and Analysis, National Core Field Guide, Volume I: Field Data Collection Procedures for Phase 2 & 3 Plots, Section 9 – Ozone Bioindicator Plants. Northeast Forest Experiment Station, New Town Square, PA and Pacific Northwest Research Station, Portland, OR.

Uttah, E.C., Uttah, C., Akpa, P.A., Ikpeme, E.M., Ogbeche, J., Usip, L., Asor, J., 2008. Biosurvey of plankton as indicators of water quality for recreational activities in Calabar River, Nigeria. J. Appl. Sci. Environ. Manage. 12 (2), 42–45. doi:10.4314/jasem.v12i2.55525.

Vahisalu, T., Puzõrjova, I., Brosché, M., Valk, E., Lepiku, M., Moldau, H., Pechter, P., Wang, Y.S., Lindgren, O., Salojärvi, J., Loog, M., 2010. Ozone-triggered rapid stomatal response involves the production of reactive oxygen species, and is controlled by SLAC1 and OST1. Plant J. 62 (3), 442–453.

Wan, W., Manning, W.J., Wang, X., Zhang, H., Sun, X., Zhang, Q., 2014. Ozone and ozone injury on plants in and around Beijing, China. Environ. Pollut. 191, 215–222.

Weinstein, L.H., Davison, A., 2004. Fluorides in the Environment: Effects on Plants and Animals. Cabi.

Weinstein, L.H., Davison, A.W., Arndt, U., 1998. Fluoride. In: Flagler, R.B. (Ed.), Recognition of Air Pollution Injury to Vegetation: A Pictorial Atlas. Air & Waste Management Association, Pittsburgh, PA.

WHO International Programme on Chemical Safety Biomarkers in Risk Assessment: Validity and Validation. 2001. Retrieved from http://www.inchem.org/documents/ehc/ehc/ehc222.htm

Wolterbeek, B., 2002. Biomonitoring of trace element air pollution: principles, possibilities and perspectives. Environ. Pollut. 120 (1), 11–21.
Wu, Q., Wang, X., Zhou, Q., 2014. Biomonitoring persistent organic pollutants in the atmosphere with mosses: performance and application. Environ. Int. 66, 28–37.
Zaghloul, A., Saber, M., El-Dewany, C., 2019. Chemical indicators for pollution detection in terrestrial and aquatic ecosystems. Bull. Natl. Res. Centre 43 (1), 1–7.
Zhou, Q., Zhang, J., Fu, J., Shi, J., Jiang, G., 2008. Biomonitoring: an appealing tool for assessment of metal pollution in the aquatic ecosystem. Analyticachimicaacta 606 (2), 135–150.

# Index

Page numbers followed by "*f*" and "*t*" indicate, figures and tables respectively.

## W